D0893315

# New Directions in Electrophoretic Methods

ACS SYMPOSIUM SERIES 335

# New Directions in Electrophoretic Methods

**James W. Jorgenson,** EDITOR
*University of North Carolina at Chapel Hill*

**Marshall Phillips,** EDITOR
*National Animal Disease Center*
*Agricultural Research Service*
*U.S. Department of Agriculture*

Developed from a symposium sponsored by
the Divisions of Agricultural and Food Chemistry
and Analytical Chemistry
at the 190th Meeting
of the American Chemical Society,
Chicago, Illinois,
September 8-13, 1985

American Chemical Society, Washington, DC 1987

**Library of Congress Cataloging-in-Publication Data**

New directions in electrophoretic methods.
(ACS symposium series, ISSN 0097-6156; 335)

Includes bibliographies and index.

1. Electrophoresis—Congresses.

I. Jorgenson, James, 1952- . II. Phillips, Marshall, 1932- . III. American Chemical Society. Division of Agricultural and Food Chemistry. IV. American Chemical Society. Division of Analytical Chemistry. V. American Chemical Society. Meeting (190th: 1985: Chicago, Ill.) VI. Series.

QD79.E44N49 1987 541.3′7 87-1777
ISBN 0-8412-1021-7



PRINTED IN THE UNITED STATES OF AMERICA

# ACS Symposium Series

## M. Joan Comstock, *Series Editor*

# Foreword

The ACS SYMPOSIUM SERIES was founded in 1974 to provide a medium for publishing symposia quickly in book form. The format of the Series parallels that of the continuing ADVANCES IN CHEMISTRY SERIES except that, in order to save time, the papers are not typeset but are reproduced as they are submitted by the authors in camera-ready form. Papers are reviewed under the supervision of the Editors with the assistance of the Series Advisory Board and are selected to maintain the integrity of the symposia; however, verbatim reproductions of previously published papers are not accepted. Both reviews and reports of research are acceptable, because symposia may embrace both types of presentation.

# Contents

INDEXES

# Preface

THE SEPARATION AND IDENTIFICATION of charged molecules based on migration in an electric field have had wide application to science and technology for more than 100 years. Electrophoretic methods play a key role in scientific advances in medicine, agriculture, and the chemical and biotechnology industries. The symposium from which this book was developed was designed to foster a critical awareness of recent developments in the area of electrophoretic separations. Our goal was to acquaint chemists with state-of-the-art electrophoretic technology and its applications.

Interest in electrophoresis falls into two camps: developers and users. Many scientists are concerned primarily with research into the basic theory and development of methods. An even larger interest involves scientists from a variety of fields who want to use the separation and identification tools provided by the developers. We hope the chapters in this book will be of value to both groups.

We wish to thank the following companies for their generous financial support of the symposium: Hewlett-Packard-Genenchem; The Upjohn Company; FMC Marine Colloids Division; Perkin-Elmer Corporation; Eldex Laboratories, Inc.; and Pharmacia, Inc. Additional financial support was provided by the Divisions of Agricultural and Food Chemistry and Analytical Chemistry of the American Chemical Society. We also greatly appreciate the assistance provided by members of the Electrophoresis Society of America. Finally, we wish to thank Janice Olson and Dorothy Olson for their assistance in preparing the symposium and this book.

JAMES W. JORGENSON
Department of Chemistry
University of North Carolina at Chapel Hill
Chapel Hill, NC 27514

MARSHALL PHILLIPS
National Animal Disease Center
Agricultural Research Service
U.S. Department of Agriculture
P.O. Box 70
Ames, IA 50010

November 20, 1986

# Chapter 1

# Overview of Electrophoresis

**James W. Jorgenson**

**Department of Chemistry, University of North Carolina at Chapel Hill, Chapel Hill, NC 27514**

Electrophoresis is the most powerful method available for separation and analysis of complex mixtures of charged biopolmers. This chapter provides an overview of modern electophoresis as a general introduction to the chapters which follow. The basic electrophoretic operating modes and formats for these modes are described. Means for detection of separated zones are reviewed. Finally, an approach to fully instrumental electrophoresis is discussed.

Electrophoresis is the premier method for separation and analyis of proteins and polynucleotides. Its importance to research in the life sciences is readily appreciated by looking through the pages of journals devoted to biochemistry and molecular biology. There one finds great numbers of photographs of electrophoretic separations. Now progress in biochemical research has spawned the rapidly growing biotechnology industry. The emergence of this industry promises remarkable new products, many of which are complex polypeptides and proteins. All aspects of the commercialization of biotechnology, from development of initial concepts in the laboratory to quality control of final product, will require analyses of potentially complex mixtures of labile macromolecules. Increasingly chemists face difficult analytical problems as the biotechnology industry grows. Methods such as HPLC are being studied and developed to improve their suitability for analysis of biopolymers. In this regard it is interesting to compare the present day capabilities of HPLC and electrophoresis. Both techniques are quite versatile and have many separation modes available. In HPLC, some of the most important include molecular exclusion, ion exchange, hydrophobic interaction (including reversed phase) and affinity. Electrophoretic modes include zone, gel sieving, isoelectric focusing, and isotachophoresis. But electrophoresis has a distinct advantage over HPLC for analysis of biopolymers: a vastly superior resolving power, especially in a two-dimensional format, where two separation mechanisms can be used in succession.

0097-6156/87/0335-0001$06.00/0

## Modes of Electrophoresis

The important separation modes in electrophoresis are moving boundary, isotachophoresis, zone (including zone with molecular sieving), and isoelectric focusing(1,2). Figure 1 shows schematically the course of separation of two substances in each of these four modes.

Moving Boundary Electrophoresis. In this mode, originally described by Tiselius, a long band of sample is placed between buffer solutions in a tube. Upon application of an electric field, sample components begin migrating in a direction and at a rate determined by each component's electrophoretic mobility (a solute's electrophoretic mobility is its velocity in an electric field of unit strength). Complete separation of sample components is never effected. Instead, only the fastest moving component (in each direction) is partially purified, while the remaining components overlap to differing degrees. The chromatographic analogue to this mode is frontal chromatography. Because sample components remain significantly overlapped, this technique is rarely used, and is generally reserved for measurement of electrophoretic mobilities.

Isotachophoresis. This name derives from the fact that in this technique all sample bands ultimately migrate at the same velocity(3). Sample is inserted between two electrolyte solutions, a leading and a terminating electrolyte. In a particular separation either cations or anions may be determined, but not both at once. If sample cations are being determined, the leading electrolyte contains a cation of higher mobility (such as hydrogen ion) than any of the sample cations of interest, while the terminating electrolyte contains a cation of lower mobility than any of the sample cations of interest. For this case of cation analysis, the leading electrolyte is on the side of the cathode, while the terminating electrolyte is toward the anode. When the electric field is applied, potential gradients evolve which ensure that all the ions eventually travel with the same velocity. In regions where cations of lower mobility are present, the electric field is stronger, moving these less mobile cations at the same velocity as the more mobile ions. This development of potential gradients is an intrinsic result of the nature of the system. If it were not the case, the leading electrolyte could pull completely ahead of any sample cations, leaving a region without ions in between. This is, of course, an impossibility as current would no longer be able to flow through this region. A characteristic of this mode is that eventually a steady-state condition is attained, where each individual cationic sample component is migrating as a "pure" band. Each pure band is sandwiched between the sample component of next highest and next lowest mobility. Thus the sample bands migrate in order of mobility, from most to least mobile. The potential gradient is constant within each band, but increases step-wise from each band to the next. The band with the highest mobility has the lowest electric field, the band with lowest mobility has the highest field. The product of mobility and electric field within each band is the same, so that all bands move with the same velocity. A

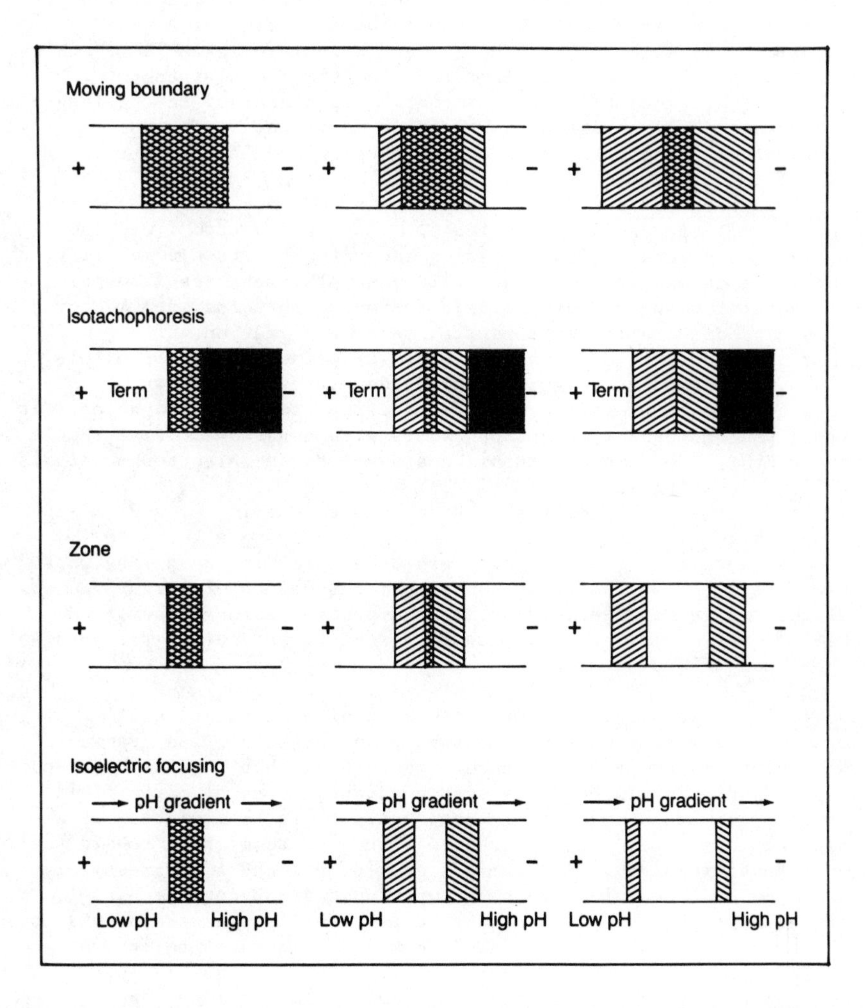


Figure 1. Schematic illustrating the separation of two substances by the four principal modes of electrophoresis. The cross-hatched bands represent the analytes. The black areas represent lead.

special characteristic of this technique is the isotachophoretic focusing which occurs at the boundary between bands, producing sharp boundaries. If a solute molecule diffuses forward into a preceding band, it is in a region of lower electric field strength, slows its rate of migration and rejoins its original band. If a solute diffuses backward into a following band, it is in a region of higher field strength, and speeds up and rejoins its original band. Analytical information is in an unconventional format. The identity of a band may be determined from the electric field strength within the band, while quantitative information is contained in the length of the band. The chromatographic analogue of this technique is displacement chromatography (See chapters 14 and 15).

Zone Electrophoresis. In this mode, sample is applied as a narrow zone (band), surrounded by buffer(1). As the electric field is applied, each zone begins migrating according to its own mobility. Ideally, each sample component will eventually separate from its neighbors, forming a "pure" zone. Achieving this end is a matter of trying to maximize the differential rate of migration while minimizing the rate of zone spreading (dispersion). In principle, each sample component migrates independently of the others, and no focusing effects are operative. In reality, zone electrophoresis is simply moving boundary electrophoresis with a narrow initial sample zone width. The chromatographic analogue of zone electrophoresis is elution chromatography.

If zone electrophoresis is done in a polymeric gel medium, it is possible to take advantage of molecular sieving effects to achieve separation(4). In this case, the pores in the gel network must be small enough so that the sample components of larger molecular size experience difficulty migrating through them, and their migration is retarded. In this way the rate of migration also depends on molecular size.

Isoelectric Focusing. This technique involves separation of amphoteric sample components in a pH gradient(2,5). The gradient must run from low pH at the anode end to high pH at the cathode end. When a sample is placed in this pH gradient and an electric field applied, the amphoteric sample components begin to migrate. If a component has a net negative charge it begins to migrate toward the anode (positive) end. As it migrates, it encounters progressively lower pH, and becomes increasingly protonated. Eventually it arrives at a pH where its net charge is zero (isoelectric point) and it ceases to migrate. Each sample component migrates to its own isoelectric point where it then stops. This technique is a true focusing technique. If a solute molecule from a focused band happens to diffuse away from the zone center, it immediately loses or gains protons, and thus acquires charge. In this charged state it migrates back toward the zone center. An important characteristic of this technique is that a steady state is eventually reached where the zones are stationary and sharply focused. They will remain so as long as the electric field is maintained.

The key to this method is formation of the pH gradient. This may be accomplished by a variety of methods, but the most effective is the use of a mixture of synthetic amphoteric molecules known as ampholytes. The ampholytes are a complex mixture of molecules whose isoelectric points span the pH range which is desired of the gradient. When an electric field is applied to the ampholytes they each migrate to their respective isoelectric point and in so doing form a relatively stable pH gradient. Ampholyte mixtures are available which allow generation of pH gradients from pH 2 to 11, and almost any portion thereof. Righetti and co-workers have shown that the stability of the pH gradient may be greatly enhanced by incorporation of the buffering ions (no longer ampholytes) into the molecular structure of a gel-forming medium. Once the pH gradient is established, the gel is polymerized to yield a permanently immobilized gradient. Isoelectric focusing is a powerful technique in that it permits resolution of substances differing in isoelectric point by as little as 0.001 pH unit. More information on isoelectric focusing is contained in chapters 3, 4, 7, 8, and 10.

## Formats for Electrophoresis

Electrophoresis may be done in a variety of experimental formats which can be divided into two major categories: "free solution", in which no stabilizers are used, or with anticonvection stabilizers such as paper or gels.

Electrophoresis in Free Solution. Much of the early work in electrophoresis was done without the presence of stabilizers. Indeed, Tiselius' original work on moving boundary electrophoresis was done in buffer-filled tubes without stabilizer. Electrophoresis in free solution has certain attractive features. First, electrophoretic mobilities can be measured without the introduction of complicating factors such as adsorption of analyte to the stabilizer, molecular sieving effects, or tortuousity of migration paths. Thus measurement of the true mobility of an analyte is more straightforward. Second, electrophoretic separation of large particles, including whole cells, is readily accomplished, whereas this is more difficult in the microporous network of a stabilizer. Unfortunately electrophoresis in free solution suffers from a serious problem with convection. The passage of electric current through the electrophoretic medium results in joule heating of the medium. This heat is dissipated only through the boundaries of the electrophoresis chamber, and a natural consequence is the evolution of temperature gradients within the electrophoresis medium. Because the density of the medium is a function of temperature, density gradients are established, leading to convective flows. These flows can easily disrupt a separation by mixing of zones. Of all the modes of electrophoresis, only analytical scale isotachophoresis is routinely done in free solution. This is due in part to the fact that the isotachophoretic process involves dynamic focusing at zone boundaries which helps to prevent mixing. Also, the technique is generally done in small bore capillary tubes, so that the magnitudes of temperature and density gradients remain rather small.

Electrophoresis With Anticonvective Stablizers. The original intent of the use of stabilizers in electrophoresis was to suppress the thermally driven convection currents in the electrophoretic medium. While this remains an important function of stabilizers, several additional benefits have been realized. Microporous gels permit separations based on molecular size through a sieving action(4). Gels create an "anchoring" point for buffering ions allowing immobilized (stable) pH gradients. And perhaps most important of all, stabilizers offer a convenient format for electrophoresis, where "strips" and "slabs" can be easily stained, destained, and otherwise manipulated in ways unthinkable for free solutions. Due to these significant advantages, virtually all electrophoresis is done with the use of a stabilizer.

Paper, cellulose acetate, and starch gels can all be used as effective stabilizine media, but gels made from polyacrylamide are the media of choice today (see chapters 2 and 4). By controlling the concentration of acrylamide and the relative proportion of the cross-linking agent bis-acrylamide, gels can be formed with well-defined molecular sieving properties(4). Polynucleotides may be run on these gels with spectacular results(6). Figure 2 shows the result of a separation of polydeoxyribonucleo-tides run on a polyacrylamide sieving gel. Polymer containing 250 bases is clearly separated from polymer containing 251 bases, and "unit" resolution of polymers with twice as many bases is easily realized in less concentrated gels (consider attempting such a separation in a chromatographic system!). Techniques for gel sieving in polyacrylamide are the heart of the powerful and efficient methods developed over the past decade for sequencing polynucleotides.

Gel sieving electrophoresis can also be applied to whole native proteins. However, the most effective application of gel sieving to proteins involves the use of denaturing ionic detergents such as sodium dodecyl sulfate (SDS). With remarkably few exceptions, proteins will "bind" a limiting amount of 1.4 grams of SDS per gram of protein. The SDS tends to break up protein aggregates (dimers, tetramers, etc.), denatures the protein, and generally imparts the same free solution mobility to all proteins, regardless of their identity. When these SDS-denatured proteins are subjected to electrophoresis on polyacrylamide sieving gels, separation is due principly to sieving effects, with lower molecular weight proteins migrating more rapidly through the gel. The rate of migration of a protein may be compared to the rate of migration of standard proteins, and fairly accurate estimates of protein molecular weight obtained. Polyacrylamide may also be cast in slabs or rods in which the concentration of acrylamide increases in a continuous manner over the length of the gel. This type of gel is known as a gradient gel. When SDS-denatured proteins are applied to the "low density" end of the gel and voltage applied, they migrate until they reach a point where the gel concentration is so high and the pores so small, that they can no longer migrate. Although they are not in fact focused at this point, the resulting zones are sharp. Again, by comparison with the migration of protein standards, the molecular weight of an unknown protein may be determined.

Polyacrylamide also provides an excellent anticonvective medium in which to do isoelectric focusing(5). Stable pH gradients can be achieved with ampholytes in such a stabilized medium with no disruption from convection. The lower portion of figure 3 shows the

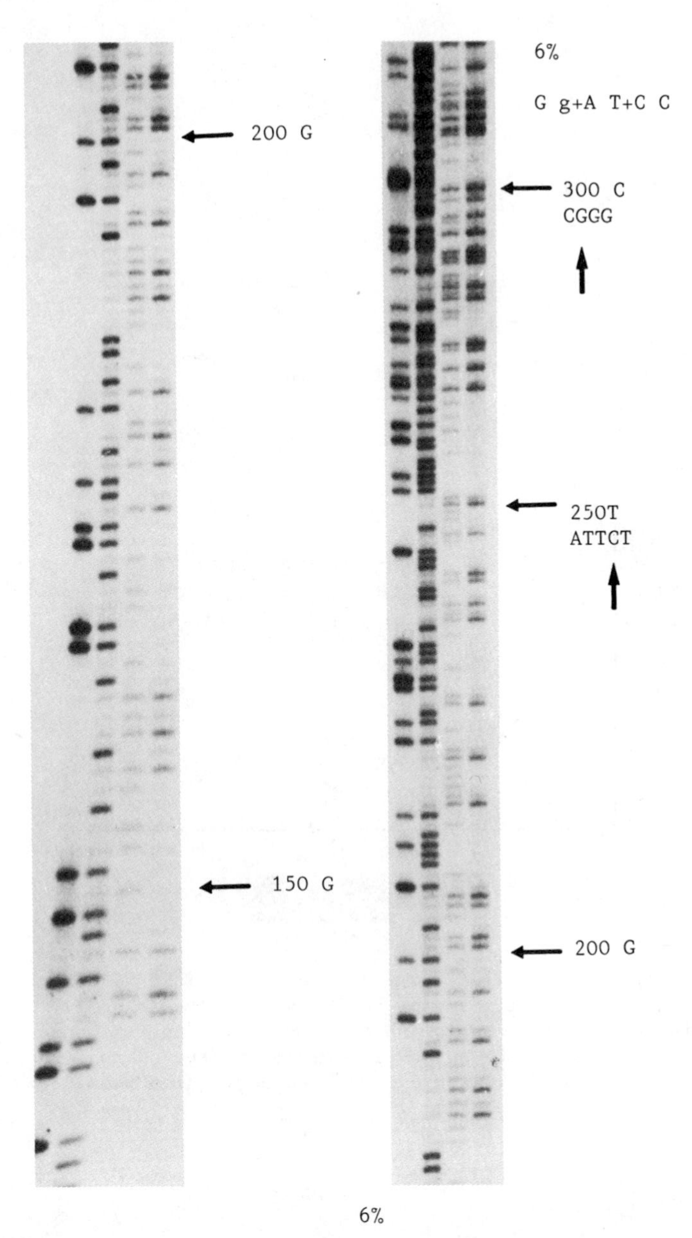


Figure 2. Section of a sieving gel on which polydeoxynucleotides have been separated. The polyacrylamide concentration is 6%. The numbers to the right of the arrows indicate the number of bases in the polymer at this point. The letters G, A, T, C indicate which base is present at that particular position in the sequence. Reproduced with permission from Ref. 6. Copyright 1984 Biochem. Biophys. Method.

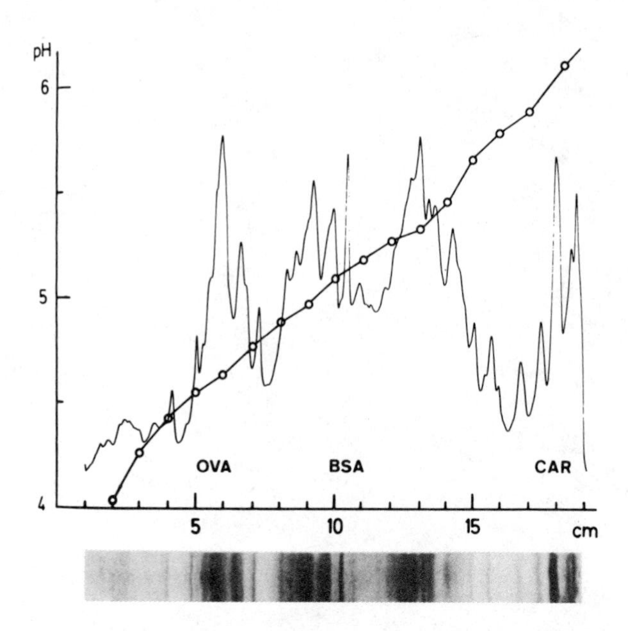


Figure 3. Lower: Isoelectric focusing gel stained with Ponceau S. The sample was a mixture of bovine serum albumin (BSA), ovalbumin (OVA) and carbonic anhydrase (CAR). Upper: Densitometric trace of gel. Line with points is a graph of the pH as a function of position in the gel. Reproduced with permission from Ref. 7. Copyright 1982 Electrophoresis.

result of isoelectric focusing of several proteins(7). The proteins were "fixed" and then stained with Ponceau S. The figure also shows the pattern which results from scanning a densitometer across this gel. The good resolution possible in isoelectric focusing is evident in this figure.

Perhaps the ultimate in protein separations is obtained when a two-dimensional electrophoretic approach is used. O'Farrell first demonstrated the truly spectacular resolution that may be achieved when proteins are first separated by isoelectric focusing, then transferred to the top of a gradient gel slab and separated by SDS gradient gel electrophoresis(8). Figure 4 shows the results of such a two dimensional separation of the proteins from the bacteria E. coli(9). The "spot capacity" (maximum number of protein spots which could possibly be resolved) in such systems is presently about 30,000. This is a phenomenal number when compared with even the highest resolution chromatographic methods. It must be pointed out that achieving such results requires considerable experimental skill, expertise, and time. Still this clearly represents the state-of-the-art in protein separations, and the existence of this technique has revolutionized the field of analysis of complex protein mixtures (see chapter 9).

For molecules of molecular weight in excess of one million, even the lowest concentration of polyacrylamide which forms a physically stable gel (approximately 3%) yields pores too small to permit migration. For such large molecules an alternative gel medium, derived from agar and known as agarose, is available. Gels formed from this material have larger pores than acrylamide, and permit separation of much larger molecules. Indeed, separations of very large DNA fragments may be obtained in agarose gels(10). In an interesting variation on electrophoresis in agarose gels, Schwartz and Cantor demonstrated electrophoretic separation of whole chromosomes from yeast(11). The molecular weight of these chromosomes spans the range of $3 \times 10^7$ to $1 \times 10^9$. Separation of such large DNAs by chromatographic methods is essentially impossible, as the slightest amount of flow (even the act of pipetting) creates shearing forces which cleave the molecules (see chapters 11 and 12).

Detection of Separated Zones

Techniques for detection and quantitation of zones are an important aspect of electrophoretic analysis. The most common approaches to detection involve the use of stains, autoradiography of radio-labelled analytes, or immunoreaction with specially prepared antisera. In the case of electrophoresis in free solution, on-line detection devices can be used.

Staining. A variety of dyes which bind to proteins have found use in electrophoresis(4). One of the most sensitive and frequently used dyes is Coomassie Blue. Usually the staining procedure involves a "fixative", such as trichloroacetic acid, which precipitates proteins and prevents their diffusion out of the gel. Gels are soaked in the staining solution, at which point the entire gel is stained uniformly and unbound dye must be removed in a

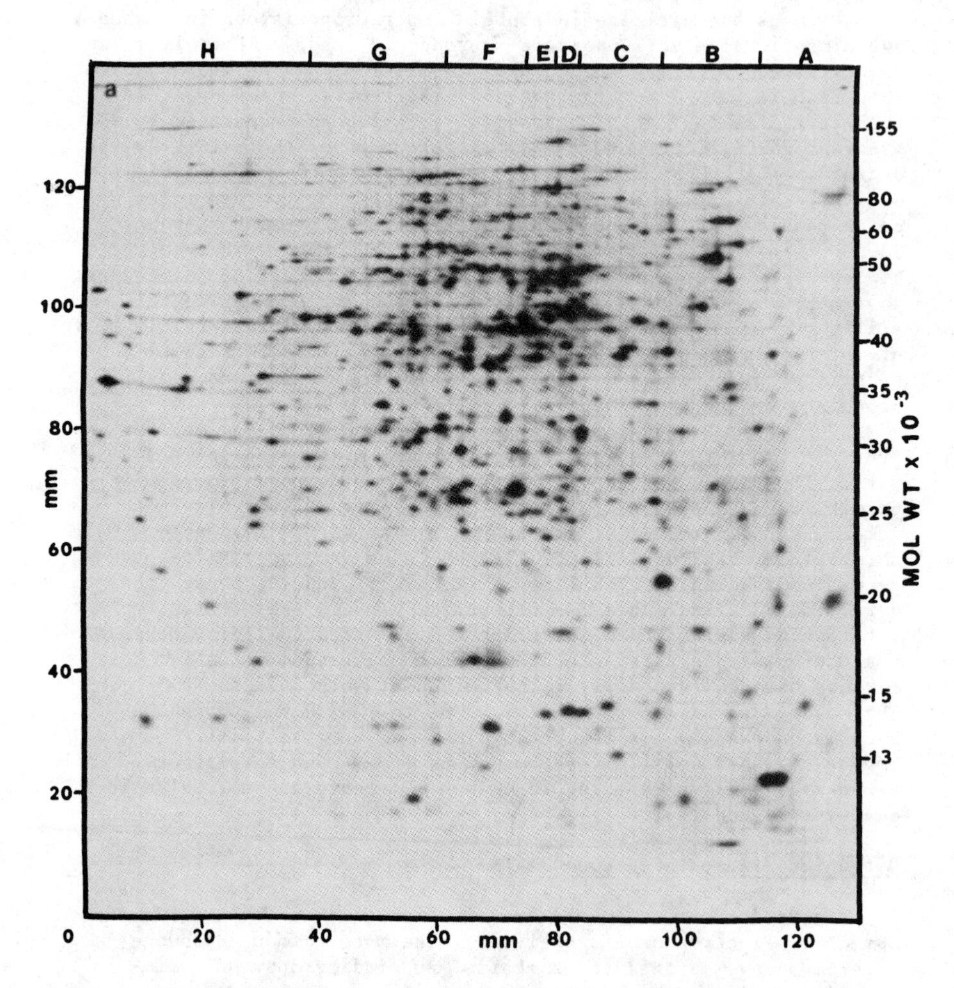


Figure 4. Two-dimensional electropherogram of proteins from E. coli. Isoelectric focusing in horizontal direction; SDS gradient gel electrophoresis in vertical direction. Detection by autoradiography; E. coli grown on $^{35}S$ sulfate in growth medium. Reproduced with permission from Ref. 9. Copyright 1984 Academic.

destaining rinse. The length of time for both the staining and destaining steps depends on the thickness of the gel, with thicker gels requiring as much as one day. Staining and destaining operations usually constitute a sizeable portion of the total time required for analysis by electrophoresis. The use of "ultra-thin" gel slabs (as thin as 20 μm) is thus of great interest. Detection limits for Coomassie Blue staining are roughly 100 nanograms of protein in a band.

A sensitive alternative staining procedure for proteins is the silver stain originally developed by Merril et al (see chapters 5 and 6. In one version of this technique, proteins are first fixed to the gel, and the gels then extensively rinsed. The gels are soaked in an acidic dichromate solution, and then soaked in a solution containing silver nitrate. The gel is then soaked in a basic solution containing formaldehyde, which reduces the silver ion to silver metal. Finally, the process is stopped by soaking in a dilute acetic acid solution. This procedure is roughly 100-fold more sensitive than Coomassie Blue staining. However, it is a rather involved procedure typically requiring one day to complete. Figure 5 shows the results which may be obtained by using a variation of the silver staining procedure developed by Sammons et al. The separation is two dimensional electrophoresis of human blood plasma proteins.

For staining DNA and RNA polymers the most popular reagent is ethidium bromide. This molecule is capable of intercalating in between the nucleotide bases. In this state the ethidium bromide fluoresces strongly under short wavelength UV illumination. No destaining is required, and a band containing as little as 50 nanograms of polynucleotide can be detected.

Quantification of stained bands can be accomplished with the use of a scanning densitometer. This is a relatively effective method of quantitation, but requires gels which are transparent and free of imperfections. Due to inherent non-linearity of the staining procedures, the linear response range rarely exceeds two orders of magnitude. Furthermore, because of the selectivity of stains, there is considerable variation in response factors from protein to protein. Extensive use of standards is crucial to achieving any degree of success in quantitation. In general the accuracy and precision obtainable are not on a par with what is routinely expected in HPLC.

Autoradiography. If the sample molecules contain radioactive atoms, or can be tagged with radioisotopes, then detection can be accomplished by autoradiography(4). In this procedure, the sample is first run on a slab gel. If the radioisotopes produce very high energy particles (e.g. $^{32}P$, $^{131}I$) the gel can simply be covered with a thin sheet of plastic and placed on top of x-ray film. Regions of the gel where radioactivity is localized will create corresponding exposed regions on the film. For isotopes producing less energetic beta particles ($^{14}C$, $^{3}H$), a variation on autoradiography known as fluorography can be done. In this case the gel is soaked in a scintillation fluid, dried, clamped on top of x-ray film between glass plates, and maintained at approximately -70°C. The film is exposed by the light produced as the scintillator molecules are

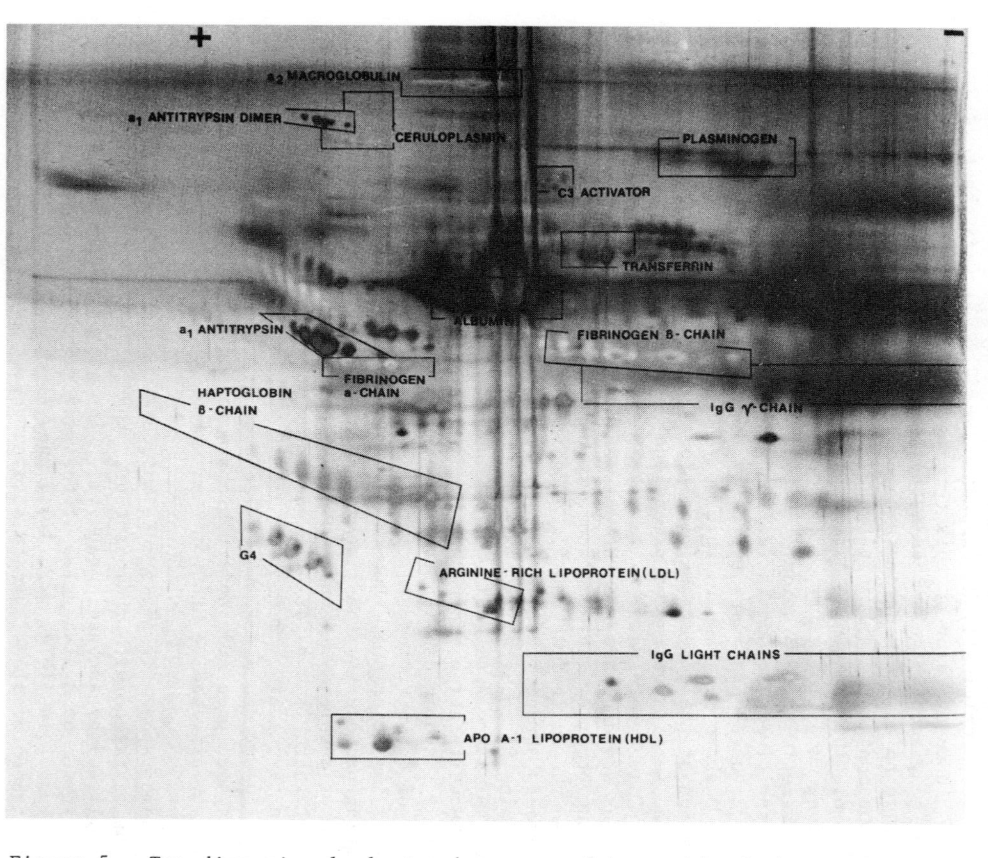


Figure 5. Two-dimensional electropherogram of human blood plasma proteins. Detection by silver staining procedure. Reproduced with permission from Ref. 12. Copyright 1981 Electrophoresis.

bombarded by beta particles. Where applicable, this method of detection can be quite sensitive, particularly if long periods of exposure of the film to the gel are possible. Exposure times of one day to a week are common. The developed x-ray film may be scanned with a densitometer to provide quantitation.

Detection With Immunoreagents. A number of electrophoretic techniques exist in which antibodies are used to detect antigenic substances of interest(1). The most impressive of these techniques is known as crossed immuno-electrophoresis. In this technique, a sample is separated electrophoretically in a narrow slab gel. Following this, antibody-containing gel is poured next to the electrophoresis gel. After this gel has solidified, the sample bands are electrophoretically migrated perpendicular to their original direction of migration, and into the antibody containing gel. Sample bands will continue to migrate through this gel until they have encountered sufficient antibody to result in precipitation of the antigen-antibody complex. Precipitated bands are then stained to permit their visualization. In principle, antibody to only one or a few of the sample components could be used, in order to effect very selective detection. In practice it is common to use polyspecific antisera which can react with virtually all of the sample components. Figure 6 shows the pattern obtained when human serum is first separated electrophoretically (horizontal direction) and then detected by electrophoresis into gel containing rabbit antiserum to whole human serum (verticle direction). Peak heights and areas are indicative of the quantity of protein in each band. The resolution of overlapping bands made possible by this unique detection mechanism is remarkable.

On-Line-Detection. This mode of detection is most useful when electrophoresis is done in free solution in tubes. Free solution electrophoresis virtually demands some form of on-line detection, as the previously described methods of staining, autoradiography and immunoreaction would hardly be feasible. Furthermore, gels may actually interfere with many forms of on-line detection, precluding use of this approach with gels. Many detection mechanisms have been used in on-line detection. These include optical (absorption, fluorescence, refractive index, scattering), electrical (electric field, conductivity) and thermal detection(3). Some of the optical detection principles were used in the early work on moving boundary electrophoresis. The modern techniques of capillary isotachophoresis and capillary zone electrophoresis rely on sensitive on-line detectors.

In isotachophoresis, electrical detection is most commonly used, although thermal and UV-absorption detection are also used(3). For capillary zone electrophoresis, UV-absorption and fluorescence detection have proven most useful so far. The principles behind the optical detection modes are fairly obvious. However, the electrical and thermal detectors deserve further explanation. As described earlier, in isotachophoresis, each zone is an individual "pure" band of sample ions. The zones travel in order of decreasing mobility. To compensate for each successive

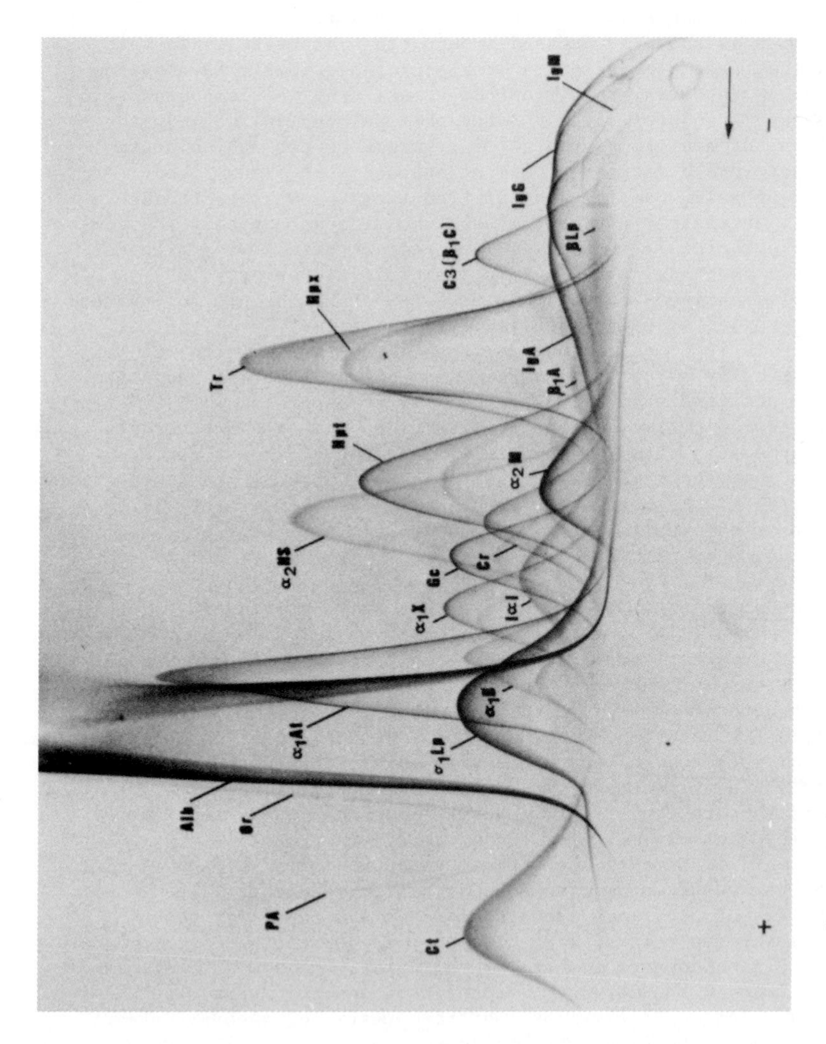


Figure 6. Crossed immuno-electrophoresis of a sample of human serum proteins. Reproduced with permission from Ref. 13. Copyright 1979 Clin. Chem.

zone's lower mobility, the electric field within each successive zone is higher. In this way all zones travel with the same velocity. If a pair of microelectrodes is placed side by side in the capillary channel, the electric field within a zone can be measured as a voltage between the electrode pair. As a new zone passes the electrodes the voltage will abruptly increase. The result is a plot as shown in figure 7. Each zone is a successive "step" in this plot. The zone's identity can be learned from the level of the step while the quantity of material is proportional to the length of the step. Also seen in figure 7 is a derivative plot, where the "peaks" serve to mark the boundaries between zones. As an alternative, the electrical conductivity of the migrating zones can also be monitored. As each zone passes, the conductivity will decrease, due to the decreasing mobility of ions in successive zones.

Thermal detection is made possible by the fact that the current is constant over the entire length of the capillary, but the electric field increases with each successive zone. The product of current and field, the power per unit length (heat generated), increases in successive zones. By placing a tiny thermocouple or thermistor on the capillary the temperature increase associated with each isotachophoretic zone can be measured. In general thermal detection has proven to be inferior to other detection techniques in terms of both sensitivity and spatial resolution (ability to distinguish narrow zones).

For capillary zone electrophoresis the electrical and thermal detection modes have insufficient sensitivity. This is because in capillary zone electrophoresis there is a relatively large background of supporting electrolyte (buffer) upon which a low concentration of sample ion is superimposed. Detecting the exceedingly small changes in electrical properties or temperature associated with sample zones is difficult. Thus UV absorption and fluorescence detection have been of greatest use in capillary zone electrophoresis.

## Instrumental Approaches To Electrophoresis

The previous figures have served to illustrate the extraordinary power and versatility of gel electrophoresis. It is sobering to realize the amount of work, expertise, and time involved in obtaining such results. Most simple analyses may require several hours to complete, while the more sophisticated techniques may require days. As a consequence, instrumental versions of electrophoresis have been developed. Capillary isotachophoresis is an excellent example of a fully instrumental form of electrophoresis. Although this technique works well in analyses of smaller ions, results with proteins have been less successful. Several clever and sophisticated instruments have been developed to do electrophoresis in free solution. Included are the rotating tube system developed by Hjerten(14), and "endless fluid belt" electrophoresis developed by Kolin(1). Catsimpoolas has constructed an on-line scanning detection system for following separations by isoelectric focusing in gel-filled tubes(1).

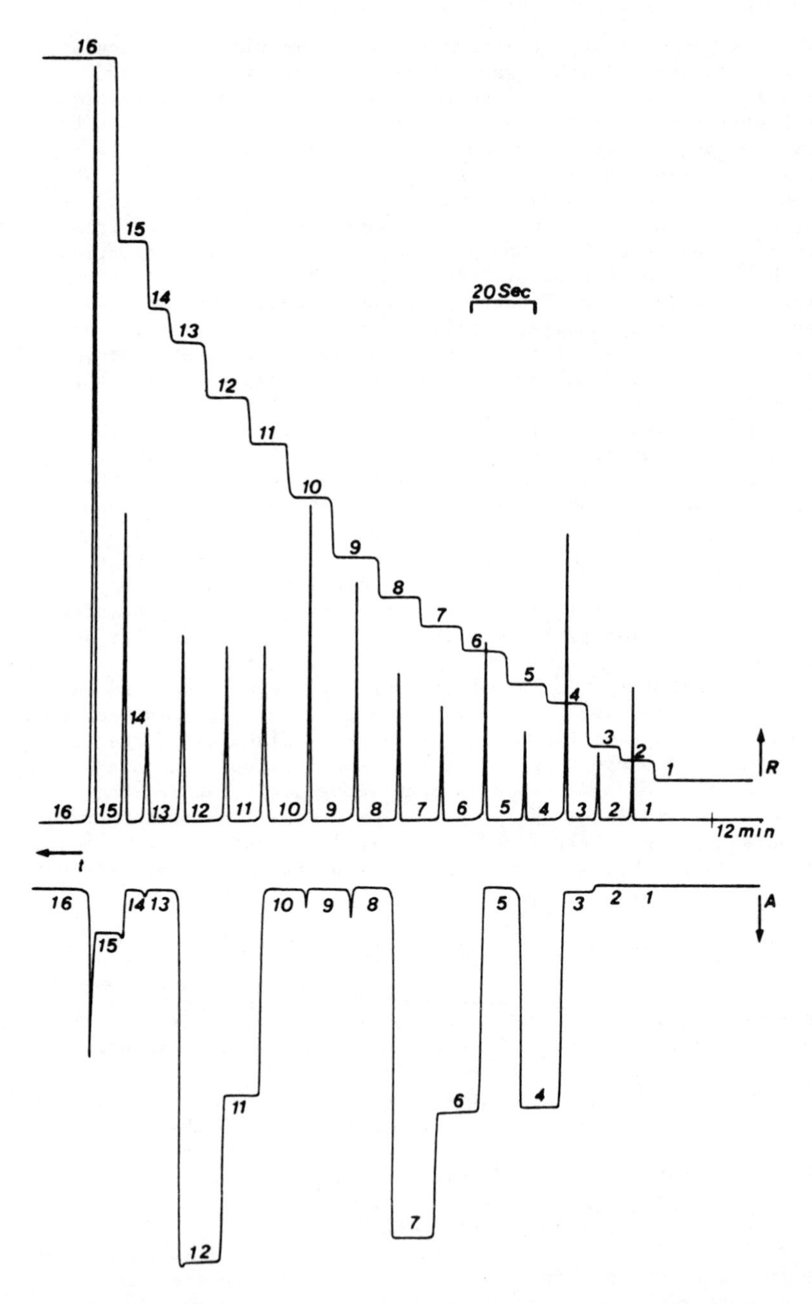


Figure 7. Isotachopherogram of small anions. Leading electrolyte, 1, chloride. Terminating electrolyte, 16, morpholinolethane-sulphate. Lower portion of figure, UV-absorption tracing obtained with on-line detector. Reproduced with permission from Ref. 1. Copyright 1979 Elsevier.

Unfortunately, these techniques have not come into widespread use, presumably due to their relative complexity.

During the 1980s several research groups, including my own, have been investigating the potential of instrumental zone electrophoresis in small bore capillary tubes (15-22) (see chapter 13). Using capillaries with inside diameters of only 25 to 75 µm, efficient dissipation of joule heat is possible, and negligible thermal and density gradients exist within the buffer solution. Using capillaries of one meter length and voltages to 30,000 volts, extremely high separation efficiencies and rapid analyses have been demonstrated for small molecules, such as fluorescent labeled amino acids and peptides. Efficiencies of 300,000 theoretical plates and analysis times of 20 minutes are easily obtained. Successful application of this technique to proteins has proven much more difficult(17,22). Proteins are especially troublesome in capillary zone electrophoresis for two main reasons. First, proteins tend to adsorb to the capillary wall, interfering with migration and seriously broadening zones. Second, proteins are relatively poor absorbers of UV light, making detection of zones at suitably low concentrations difficult. However, progress is being made on both of these fronts. Figure 8 is an example of an electropherogram of a mixture of proteins, which was obtained in my laboratory. The buffer system is an adaptation of a successful system described by Lauer and McManigill(22). Each peak in this electropherogram is from approximately one nanogram of protein introduced into the capillary. This result is encouraging as high resolving power and a rapid analysis time are achieved. Still, a great deal more work needs to be done in order to make such analyses routine. "Universal" systems, in which protein adsorption is eliminated for all proteins, would be highly desirable. And perhaps most important, further development of sensitive on-line detectors is crucial. To round out the capabilities of capillary zone electrophoresis, autosamplers and fraction collection devices are being developed.

At the other size extreme are electrophoretic "instruments" for separating large quantities (grams to Kilograms) of materials for preparative purposes. Here, due to the large size of the equipment, heat dissipation can become a major problem to be dealt with. More information on electrophoresis on a preparative scale can be found in chapter 16.

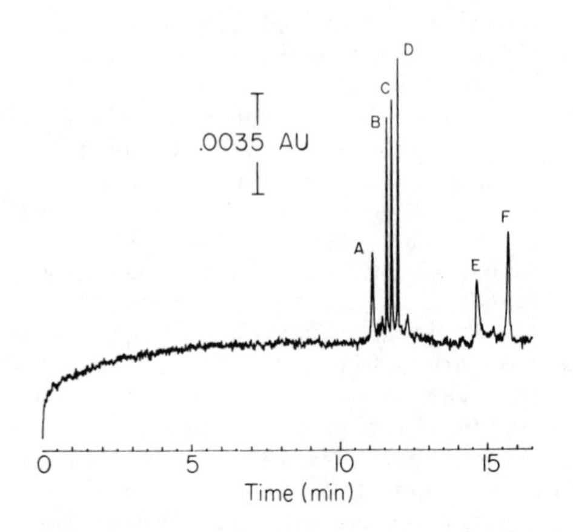


Figure 8. Electropherogram of six protein standards obtained by capillary zone electrophoresis. The standards are A: sperm whale myoglobin; B: horse myoglobin; C: human carbonic anhydrase; D: bovine carbonic anhydrase; E: ß-lactoglobulin B; F: ß-lactoglobulin A. Detection by UV absorption at 229 nm. Buffer: 10 mM, pH 8.24 Tricine; with 40 mM KCL.

## Acknowledgment

This work is largely adapted from a report originally appearing in Analytical Chemistry, 1986, Volume 58(7), p. 743A. I wish to acknowledge support for this work by the National Science Foundation under grant CHE-8213771, the Alfred P. Sloan Foundation, the Hewlett-Packard Corporation, and the Upjohn Company. I would also like to thank Yvonne Walbroehl, Jonathan Green, and Dorothy Olson for their assistance in preparing this manuscript, and Hank Lauer and Doug McManigill of Hewlett-Packard Laboratories for their helpful discussions on capillary zone electrophoresis.


## Literature Cited


1) "Electrophoresis: A Survey of Techniques and Applications; Part A: Techniques" ed. by Z. Deyl, Elsevier, Amsterdam, 1979.
2) Gaal, O., Medgyesi, G. A., Verczkey, K., "Electrophoresis in the Separation of Biological Macromolecules", Wiley-Interscience, Chichester, 1980.
3) Everaerts, F. M., Beckers, J. L., and Verheggen, Th. P. E. M., "Isotachophoresis: Theory, Instrumentation and Applications", Elsevier, Amsterdam, 1976.
4) Andrews, A. T., "Electrophoresis: Theory, Techniques, and Biochemical and Clinical Applications", Clarendon Press, Oxford, 1981, Chapters 4 and 5.
5) Allen, R. C., Saravis, C. A., Maurer, H. R., "Gel Electrophoresis and Isoelectric Focusing of Proteins: Selected Techniques", Walter de Gruyter, Berlin, 1984.
6) Ansorge, W., Barker, R., J. Biochem. Biophys. Method., 1984, 9, 33.
7) Frey, M. D., and Radola, B. J., Electrophoresis, 1982, 3, 216.
8) O'Farrell, P. H., J. Biol. Chem., 250, 4007 (1975).
9) Neidhardt, F. C., Phillips, T. A., in "Two-Dimensional Gel Electrophoresis of Proteins: Methods and Applications", ed. by J. E. Celis and R. Bravo, Academic Press, Orlando, 1984, Chapter 13.
10) Fangman, W. L., Nucl. Acids Res., 1978, 5, 653.
11) Schwartz, D. C., Cantor, C. R., Cell, 1984, 37, 67.
12) Sammons, D. W., Adams, L. D., Nishizawa, E. E., Electrophoresis, 1981, 2, 135.
13) Cline, L. J., and Crowle, A. J., Clin. Chem., 1979, 25, 1749.
14) Hjerten, S., Chromatogr. Rev., 1967, 9, 122.
15) Mikkers, F. E. P., Everaerts, F. M., Verheggen, Th. P. E. M., J. Chromatogr., 1979, 169, 11.
16) Jorgenson, J. W., Lukacs, K. D., Anal. Chem., 1981, 53, 1298.
17) Jorgenson, J. W., Lukacs, K. D., Science, 1983, 222, 266.
18) Tsuda, T., Nomura, K., Nakogawa, G., J. Chromatogr., 1983, 264, 385.
19) Hjerten, S., J. Chromatogr., 1983, 270, 1.
20) Gassmann, E., Kuo, J. E., Zare, R. N., Science, 1985, 230, 813.
21) Terabe, S., Otsuka, K., Ando, T., Anal. Chem., 1985, 57, 834.
22) Lauer, H. H., McManigill, D., Anal. Chem., 1986, 58, 166.


RECEIVED November 25, 1986

# Chapter 2

# Electrophoresis in Polyacrylamide Gels

**Michael J. Dunn**

**Jerry Lewis Muscle Research Centre, Royal Postgraduate Medical School, Ducane Road, London W12 0HS, United Kingdom**

Polyacrylamide gel has proved to be a versatile and popular matrix for the electrophoretic separation of protein mixtures. In this short review the techniques most commonly used for the analytical separation of proteins are described briefly. Recent developments in these techniques are documented in more detail in the subsequent articles in this symposium volume.

Proteins are charged at a pH other than their isoelectric point (pI) and thus will migrate in an electric field in a manner dependent on their charge density. If the sample is initially present as a narrow zone, proteins of different mobilities will travel as discrete zones and thus separate during electrophoresis. Such separations are best carried out in a support medium to counteract the effects of convection and diffusion during electrophoresis and to facilitate immobilization of the separated proteins. Polyacrylamide gel, the use of which dates back to 1959 (1-3), has proved to be a versatile and popular matrix for the electrophoretic separation of proteins. In the limited space available here I can do no more than introduce the techniques most commonly used for the analytical separation of proteins in polyacrylamide gels. Readers interested in more thorough treatments are referred to some of the excellent current texts (4-11).

## Properties of Polyacrylamide Gels

Polyacrylamide gels are formed by the copolymerization of acrylamide monomers with a crosslinking agent to form a three-dimensional (3-D) network. The most commonly used crosslinker is N,N'-methylene bis-acrylamide (Bis), but there are a variety of alternative reagents which can be used to impart special properties to the gels (see Table 3.1 in ref. 5). Of particular interest is a new crosslinker, AcrylAide (FMC Marine Colloids), an olefinic agarose derivative, which permits gels on polyester supports to be dried at 60°C without the use of vacuum. Gel polymerization is usually initiated by ammonium persulphate and N,N,N',N'-tetramethylethylenediamine (TEMED). Riboflavin can be used for uv-activated polymerization but this reaction is comparatively slow (12). Careful control of polymerization conditions is essential for reproducible results.

0097-6156/87/0335-0020$06.00/0

The pore size of polyacrylamide gels is dependent both on total monomer concentration (%T) and on the concentration of crosslinker (%C). Pore size can be progressively increased by reducing %T at a fixed %C, but very dilute gels are mechanically unstable and pore sizes >80 nm cannot be attained (13). The alternative approach is to progressively increase %C at fixed %T where the increase in pore size is thought to be due to the formation of a "bead-like" structure rather than a 3-D lattice (14). In this way, stable gels of high pore size (200-250 nm) can be obtained at 30%C Bis, but at higher concentrations the gels become hydrophobic and prone to collapse.

## Factors Involved in Protein Separations

Proteins can be fractionated by electrophoretic techniques on the basis of one or a combination of their three major properties: size, net charge and relative hydrophobicity. Electrophoresis under native conditions is ideal for soluble proteins, where biological properties can often be retained. In contrast, more vigorous and often denaturing conditions must be used for analysis of less soluble proteins. Electrophoretic separations can be carried out using either a continuous or discontinuous (Multiphasic) buffer system. The techniques are referred to as continuous zone electrophoresis (CZE) or discontinuous ("disc") electrophoresis (also known as multiphasic zone electrophoresis, MZE).

## Continuous Zone Electrophoresis (CZE)

CZE is a high resolution method provided that the sample is concentrated (>1 mg/ml) so that it can be loaded in a narrow zone. The buffer and pH for separation can be freely chosen, but buffer concentration should not exceed 0.01M to minimize Joule heating. In a homogeneous gel, separation occurs on the basis of both charge and size. In contrast, in a gradient polyacrylamide gel migration rates decrease until each protein species reaches its pore limit (15-17). This technique is termed "pore limit electrophoresis" and separates proteins on the basis of size.

## Discontinuous Electrophoresis

Discontinuous buffer systems are popular due to their ability to concentrate the sample into a narrow starting zone. This effect is due to the moving boundary formed between a rapidly migrating (leading) and a slowly migrating (trailing) species. Normally a lower gel phase contains the leading constituent while the trailing constituent is present in the upper buffer phase. If these constituents are chosen so that the sample proteins have intermediate mobilities, the latter will be concentrated in the moving boundary ("stack"). The most popular system is that of Ornstein and Davis (2,3) in which the leading and trailing ions are respectively chloride and glycinate. The moving boundary is formed in a low concentration, "stacking" gel and there is an increase in operational pH in the restrictive separating gel. This causes further dissociation of glycine, increasing its mobility so that it moves just behind the chloride ion. This effect, together with the decrease in gel pore size causes the proteins to be "unstacked" so that they are subsequently separated on the basis of size and charge.

The Ornstein-Davis system was developed for separation of serum proteins, and unfortunately there is no "universal" system suitable for all separations. Computational treatments have been developed to establish the constituents of moving boundary systems with the desired properties. The most powerful of these is that of Jovin (18) which has generated in excess of 4,000 buffer systems; the so-called "extensive buffer system output". A selection of 19 of these systems has been published to aid the investigator to identify the most suitable system for his particular separation problem (8, 19).

## Gel Concentration, Ferguson Plots and Molecular Weights of Native Proteins

Theoretically there is a gel concentration which is optimal for the resolution of any two proteins, but there is no single gel concentration which will give maximum separation of the components of a complex protein mixture. Linear or non-linear gradient polyacrylamide gels are ideal for the resolution of such samples, but if it is desired to separate a given pair of proteins it is essential to use the appropriate concentration of acrylamide in a homogeneous gel. This concentration can be selected using a series of gels of different concentrations and constructing a Ferguson plot (20) of $\log_{10}$ relative mobility ($R_f$) versus %T. In addition, such plots yield data concerning protein molecular size (slope = $K_R$, retardation coefficient) and net charge (intercept = $Y_o$) and a computer programme "PAGE-PACK" has been developed to facilitate this analysis (21). The Ferguson plot can also be used specifically for the determination of molecular weights of native proteins as there is a linear relationship between $K_R$ and molecular weight. However, this method is only valid if proteins used to construct the standard curve have the same shape, degree of hydration and partial specific volume. This methodology has recently been reviewed in detail (22).

## Molecular Weights of Denatured Proteins: Sodium Dodecyl Sulphate Polyacrylamide Gel Electrophoresis (SDS-PAGE)

Electrophoretic procedures generally separate proteins on the basis of *both* size and charge. A method which disrupted all noncovalent protein interactions and imposed a uniform charge on polypeptide chains would allow electrophoretic separation on the basis of polypeptide size *alone*. Such an ideal is largely fulfilled by the anionic detergent sodium dodecyl sulphate (SDS) which was first used for electrophoresis (SDS-PAGE) by Maizel (23). It was established (24-26) that, in SDS-PAGE, molecular weight was related to mobility and that molecular weights could be determined using a set of marker polypeptides of known molecular weight and plotting mobility verus $\log_{10}$ molecular weight. This relationship is linear for a given %T only over a limited range of molecular weight. Other linear and non-linear fitting functions have been described (27).

The majority of proteins bind 1.4 g SDS per 1 g protein (28). However, proteins containing non-protein moieties (e.g. glycoproteins, lipoproteins, nucleoproteins) can bind different amounts of SDS resulting in anomalous apparent molecular weight values. For glycoproteins, this problem can be alleviated by borate buffers due to binding of borate ions to cis-hydroxyl groups of sugars (27, 29). For optimal reaction with SDS, samples must be boiled for 3-5 min in the presence of reagents such as β-mercaptoethanol or dithiothreitol to disrupt disulphide bonds. DTT may be preferable in the presence of urea as β-mercaptoethanol can result in anomalous molecular weight values under these conditions (30).

Continuous buffer systems have been almost universally superseded by disctoninuous buffers for SDS-PAGE. The most popular system (31) is based on the Ornstein-Davis system (2,3) with the addition of SDS. It should be noted that the nature of stacking is modified in the presence of SDS and it is not necessary to have a discontinuity in pH as unstacking will occur by the change in gel concentration alone (32).

Gradient polyacrylamide gels for SDS-PAGE have considerable advantages for the resolution of complex protein mixtures. Linear and concave gradients are normally used, but the ability of sophisticated gradient forming devices to generate gradients of any desired shape can be exploited to optimise protein separations (33-35). It is also worth pointing out that SDS gels are analogous to DNA sequencing gels (34), so that a substantial increase in resolution can be obtained by increasing the length of the separating gel (36).

## Isoelectric Focusing (IEF) using Synthetic Carrier Ampholytes

Separation by IEF is based solely on protein charge so that polyacrylamide gels of low %T (3 to 5%) should be used to minimize any sieving effects. The technique is usually performed using slab gels run on a horizontal flat-bed apparatus with an efficient cooling platten. It is also advantageous to use thin or ultrathin gels (0.02 to 0.25 mm) cast on thin plastic sheets to improve heat dissipation (37, 38).

Although the origins of IEF can be traced back to the early years of this century (39), it is the theoretical work of Svensson (now Rilbe) (40, 41) which laid the foundation of IEF based on the formation of natural pH gradients. Svensson showed that a series of molecules with good conductivity in their isoelectric state and with different charged groups ("carrier ampholytes", CA) form a continuous pH gradient in an electric field. A protein in such a gradient will migrate according to its charge until it reaches the pH at which it has no net charge (i.e. its isoelectric point, pI) and attain a steady state of zero migration (i.e. be "focussed"). It was Vesterberg (42) who first described the synthesis of low molecular weight CA by coupling propanoic acid residues to polyethylene polyamines. Subsequently, other more acidic and basic CA have been synthesized (43) extending pH gradients from pH 2.5 to 11. These CA are marketed by LKB under the name Ampholine. Several alternative synthetic procedures for producing CA have been developed (reviewed in 5, 34) and some of these CA are available commercially under the names Servalyt (Serva) (44, 45) and Pharmalyte (Pharmacia) (46). As CA are synthesized by different procedures they will contain different species. Thus, blending of different CA preparations produces a mixture containing a greater diversity of charge varieties and so incorporates the advantages of each preparation. Using this strategy we have demonstrated enhanced resolution by IEF of human skin fibroblast proteins (35).

There are two characteristics of synthetic CA which result in problems. First, and perhaps most important, "cathodic drift" or the "plateau phenomenon" results in pH gradient instability. With time the pH gradient and the proteins within it migrates towards the cathode (anodic drift can occur under certain circumstances) causing decay of the pH gradient and loss of proteins. The effects of cathodic drift can be minimized but they cannot be totally overcome. Secondly, artefacts can arise in IEF due to interactions of certain proteins with CA (47).

### Rehydratable Gels for IEF

Recently, methods have been developed for producing rehydratable gels for IEF (48, 49) in which gels are cast, washed, dried and rehydrated prior to use in the desired solution. This approach has several advantages; (a) it is easy to prepare batches of gels thereby increasing reproducibility, (b) reagents such as unreacted catalysts, unpolymerized monomers, linear polymers and salts which interfere with IEF can be washed out, (c) additives which interfere with polymerization can be soaked into the gel prior to IEF, and (d) such gels are a practical solution to the use of immobilized pH gradients containing urea (50, 51). Various techniques are available (49), but we recommend casting gels (0.5 mm thick) in vertical cassettes on GelBond PAG using standard procedures. Gels are then removed from the cassettes, washed, dried under a fan at room temperature and stored at -20°C. We consider it better to use volume rather than weight to control rehydration (51), so that gels are reassembled in cassettes of the same dimensions as those in which they were cast and rehydrated in a controlled volume of the required solution.

### IEF using Simple Buffer Ampholytes

It is possible to generate pH gradients for IEF using simple, even non-amphoteric, buffers. A system consisting of 47 components, Poly/Sep 47 (Polysciences), has been described (52, 53) which generates broad pH 3 to 10 gradients. The advantages of this system are claimed to be pH gradient stability, absence of interactions with proteins and increased reproducibility due to the defined nature of the constituents. Although such buffer IEF systems are of potential interest, little practical use has so far been made of this technique. We have assessed its potential for use as the first-dimension separation procedure in two-dimensional electrophoresis (54) and our findings are discussed later.

### Immobilized pH Gradients for IEF

An exciting innovation in IEF has been the development of Immobiline reagents (LKB) for the generation of immobilized pH gradients. These are acrylamide derivatives of structure CH2=CH-CO-NH-R, where R is either a carboxyl or a tertiary amino group, forming a series of buffers with different pK values. A concentration gradient, stabilized with a density gradient (e.g. glycerol), of these reagents mixed with acrylamide and Bis is used to prepare gels so that the pH gradient is copolymerized and immobilized within the polyacrylamide matrix. As the species forming the pH gradient cannot migrate in the electric field the pH gradient is stable and cathodic drift is eliminated. In addition the pH gradients are not disturbed by high salt concentrations and separations can be carried out under conditions of controlled ionic strength and buffer power. Although a relatively new technique, it has already been the subject of several reviews (5, 47, 55, 56).

Very narrow pH gradients, spanning from 0.1 to 1 pH unit can be readily generated using Immobilines and it is claimed that a difference in pI of only 0.002 pH units can be detected. Such gradients are generated using only two Immobilines, one acting as a buffer and the other as a titrant. Recipes for 58 such 1 pH unit wide gradients are available (47, 57) and narrower pH gradients can be derived by linear interpolation. The generation of wider pH gradients is more complex as mixtures of several Immobiline species must be used. A complex 5-chamber gradient forming system was used initially (58), but computer-derived recipes for use with two-chamber gradient forming devices have been developed (59, 60) and a computer-controlled system has been described (33, 60) to pour high precision, reproducible gradients. In addition, it is possible to generate nonlinear gradients whose shape is matched to the distribution of protein pI values in the sample (61, 62).

Immobiline gels are cast in standard vertical cassettes on GelBond PAG supports. After polymerization the gel must be washed to remove unreacted catalysts and unpolymerized Immobilines. The gel swells during this process and must be dried back to its original weight. It is, therefore, much preferable to use a rehydratable gel system, where after washing the gel can be dried, stored and rehydrated for use (see above).

The Immobiline system can produce excellent separations but there are problems, particularly if wide pH gradients are used; (a) proteins enter the gels very slowly if a prerun is carried out, (b) lateral spreading of sample zones is common, and (c) very long focusing times are often required. These problems appear to be due to the inherently low conductivity of the Immobiline system. These problems can be overcome and better separations obtained by the addition of low concentrations (0.5 to 1.0% w/v) of synthetic CA of an appropriate range to the Immobiline gels (63, 64). Under these conditions the immobilized pH gradient dictates the separation, while the synthetic CA act to increase conductivity of the system. If higher concentrations of CA are used these can overpower the Immobiline gradient, but addition of 4% CA has been found to have advantages in the separation of membrane proteins by immobilized pH gradient IEF (65).

## Two-dimensional Polyacrylamide Gel Electrophoresis (2-D PAGE)

The resolving power of electrophoresis can be increased by combining two different techniques to produce a two-dimensional (2-D) method.

Ideally, the two procedures should separate proteins on the basis of independent parameters. This strategy results in the separated species being uniformly spread across the final 2-D maps rather than being aligned along a diagonal. Although the history of 2-D electrophoresis can be traced back 30 years (66), it was the adoption of the combination of IEF under denaturing conditions with SDS-PAGE (67-70) which has resulted in the almost univeral adoption during the last 10 years of 2-D PAGE for the analysis of complex protein mixtures. Discussion here will be limited to developments based on this procedure and a fuller account can be found elsewhere (34, 71-75).

The diverse nature of samples analysed by 2-D PAGE means that there is no one ideal method of sample preparation. Many samples (e.g. body fluids) require concentration, while others must be treated to remove interfering non-protein components (e.g. phenolic pigments from leaves). In all cases, precautions should be taken to avoid protein modifications which can result in artefacts. The original mixture of 9M urea and non-ionic detergent (NP-40, Triton X-100) (67) effectively solubilizes many samples, but certain proteins (e.g. histone, ribosome, and membrane proteins) are more resistent to disaggregation and solubilization. The zwitterionic detergent, 3-[(cholamidopropyl)-dimethylammonio]-1-propane sulphonate (CHAPS), improves solubilization of microsomal proteins (76) and we routinely use a urea/CHAPS mixture to solubilize samples for 2-D PAGE (54, 77). Procedures involving the use of SDS followed by the addition of NP-40 can be effective (67, 78, 79) but the possibility must be considered that some proteins might not remain in a soluble state when the SDS concentration is lowered by the addition of non-ionic detergent.

The first dimension of 2-D PAGE is usually carried out using weak (3 to 5%T) gels containing urea, non-ionic or zwitterionic detergents, and synthetic carrier ampholytes. Various commercial ampholytes give better resolution in different pH ranges (80). We have found that ampholyte mixtures can enhance resolution in 2-D PAGE (35, 81) and in addition can be used to engineer pH gradient shapes optimized for the resolution of particular protein mixtures. Traditionally, rod IEF gels are used which are simple to prepare and run, can be cast in batches (82, 83) and applied automatically to the second dimension (84). However, a major disadvantage of rod IEF gels is that they are subject to severe cathodic drift resulting in pH gradients which do not extend above pH 7 with consequent loss of basic proteins from 2-D maps. We obtained pH gradients extending to pH 10 by treating the tubes with methylcellulose to reduce electro-endosmosis and by the manipulation of anolyte pH (35, 85), but the cathodic proteins were poorly resolved. One approach to resolution at basic pH is non-equilibrium pH gradient electrophoresis (NEPHGE) (86) in which samples, loaded at the acidic ends of the gels, are electrophoresed for a short time. Reproducibility of this technique is difficult to control as proteins are separated in the presence of a rapidly forming pH gradient. Another approach which we have favoured is the use of horizontal, thin slab IEF gels run on a flat-bed apparatus as this technique can resolve proteins, in one step, over the whole pH 3 to 10 range. This technique is described in detail in (35, 54, 72, 77). Briefly, 0.5 mm IEF slab gels containing 8M urea and 0.5% (w/v) Triton X-100 or CHAPS and a mixture of synthetic carrier ampholytes are cast on GelBond PAG supports (FMC Marine Colloids). The gel is run covered with a plastic sheet in which sample holes have been punched, to minimize dessication and atmospheric effects. After IEF is complete, individual sample strips are simply cut from the gel on its plastic support so that the strips can be equilibrated and applied to the second dimension gels.

Recently, alternative procedures have been investigated in an attempt to overcome the problems associated with the use of synthetic carrier ampholytes. We have assessed buffer IEF using the 47 component mixture, PolySep 47 (52, 53), for the first dimension of 2-D PAGE (54). However, very unsatisfactory 2-D maps were obtained with this approach as during buffer IEF the proteins separated into four major bands with protein-free intermediate zones. Another, more promising alternative, is IEF using immobilized pH gradients. Unfortunately, full range (pH 3 to 10) gradients can only be generated using two non-buffering Immobilines which are not yet commercially available, but recently a formulation for a pH 4 to 10 gradient using the standard Immobiline reagents has been published (87). Righetti and his coworkers have published a series of papers on the use of immobilized pH gradients for 2-D PAGE analysis of serum proteins (61, 88, 89). Good separations of serum proteins were obtained but it is important to note that the IEF gels contained 8M urea but did not contain any detergent. IPG gels containing 8M urea have also been used by Görg and her colleagues for 2-D separations of soluble protein extracts from legume seeds (90-92).

Recently, we have used Immobiline gels (pH 4 to 7 and pH 4 to 10 ranges) for 2-D separations of total fibroblast proteins (93, 94). This protein mixture requires the presence in the IEF gel of non-ionic detergent to ensure solubility during focusing. A rehydratable gel system was used as this is the most practical approach to the incorporation of 8M urea into Immobiline gels. Unfortunately, rehydration of polyacrylamide gels in the presence of both urea and nonionic detergents is difficult (50). Thus, as a compromise 0.5% Triton X-100 was included in the gels when they were cast and in all subsequent washing solutions, whereas 8M urea was incorporated into the gels at the rehydration stage. A low concentration (0.5%) of synthetic carrier ampholytes was also included in the gels in view of the improvement this confers upon the separation parameters and patterns (63, 64). Although promising 2-D separations of fibroblast proteins were obtained, considerable vertical streaking was observed which appears to be due to problems of elution of proteins from the first-dimension Immobiline gels into the second-dimension SDS gels. This problem was not resolved by increasing the ionic strength of the equilibration buffer, whereas increasing the pH of equilibration to pH 8.6 produced improved 2-D maps (94). Inclusion of 8M urea in the equilibration buffer resulted in significantly reduced streaking (94, 95), but the 2-D separations were still inferior to those obtained using our standard method of 2-D PAGE.

Görg and her coworkers have experienced similar problems of streaking (90, 96). They attributed this to the presence of fixed charges on the Immobiline gel matrix, leading to increased electro-endosmosis in the region of contact between the first and second dimension gels, and resulting in disturbance of migration of proteins from the first to the second dimension. In a recently published paper (92) the problem of increased electroendosmosis was partially overcome by including 15% glycerol in the solution used for rehydration of the Immobiline gels and by using an equilibration buffer containing both 30% glycerol and 6M urea. Righetti's group have adopted a different strategy to reduce streaking (97) in which they "fix" the Immobiline gel strips in 12% acetic acid, 50% methanol before equilibration in Tris buffer, pH 8.8, and application to the second-dimension SDS gels.

Second dimension SDS gels for 2-D PAGE are almost invariably of the discontinuous Laemmli type (31). As in one-dimensional SDS-PAGE gradient gels often give a significant increase in resolution, particularly if the gradient shape is chosen to distribute the proteins uniformly across the 2-D maps (35, 54). The latter flexibility can be readily achieved using various gradient forming systems (33, 35, 98) to produce large batches of gels, thereby increasing reproducibility of 2-D maps particularly if the gels are also electrophoresed in batches (98, 99).

In the original O'Farrell method (67), 13 cm long IEF gels were used in combination with 16.4 cm long SDS gels. However, a substantial increase in resolution can be achieved by an increase in the length of the two dimensions. Young and his coworkers, in a technique termed "Giant Gels", used 32 cm IEF gels and 40.64 cm SDS gels (100, 101). With this approach it is claimed that 5,000 to 10,000 proteins can be detected in maps of single cell types (101-103). This now places the resolution capacity of 2-D PAGE in the range predicted by polysomal mRNA-DNA hybridization experiments which suggest that a single cell type may contain between 5,000 and 20,000 mRNA species. At the other extreme, microscale 2-D systems have been advocated (104, 105) but the main advantage of this approach is increased speed and reduced cost gained at the possible expense of resolution.

At the completion of 2-D PAGE, gels are fixed and the separated proteins visualized by any appropriate technique (e.g. silver staining, autoradiography). It then remains to extract qualitative and quantitative data from the resulting complex 2-D protein patterns. Simple visual inspection can provide only limited information and it is usually necessary to use sophisticated computer analysis systems (71, 106, 107).

## Acknowledgments

I am grateful to Mrs. C. Trand for typing the manuscript. Financial support of the Muscular Dystrophy Group of Great Britain and Northern Ireland is acknowledged.

## Literature Cited

1. Raymond, S.; Weintraub, L. Science 1959, 130, 711-713.
2. Ornstein, L. Ann. N.Y. Acad. Sci. 1964, 121, 321-349.
3. Davis, B.J. Ann. N.Y. Acad. Sci. 1964, 121, 404-427.
4. Hames, B.D.; Rickwood, D. Eds ; Gel Electrophoresis of Proteins; IRL Press: Oxford, 1981.

5. Righetti, P.G. "Isoelectric focusing: theory, methodology and applications"; Elsevier: Amsterdam, 1983.
6. Allen, R.C.; Saravis, C.A.; Maurer, H.R. Eds.; "Gel Electrophoresis and Isoelectric Focusing of Proteins: Selected Techniques"; de Gruyter: Berlin, 1984.
7. Neuhoff, V. Ed.: "Electrophoresis '84", Verlag Chemie: Weinheim, 1984
8. Chrambach, A. "The Practice of Quantitative Gel Electrophoresis"; VCH Verlagsgesellschaft: Weinheim, 1985.
9. Dunn, M.J. Ed.: "Gel Electrophoresis of Proteins"; Wright: Bristol, 1986.
10. Dunn, M.J. Ed.: "Electrophoresis '86"; VCH Verlagsgesellschaft: Weinheim, 1986.
11. Andrews, A.T. "Electrophoresis: Theory, Techniques and Biochemical and Clinical Applications"; Clarenden: Oxford, 1986.
12. Righetti, P.G.; Gelfi, C.; Bianchi-Bosisio. A. Electrophoresis 2, 291-295.
13. Righetti, P.G.; Brost, B.C.W.; Snyder, R.S. J. Biochem. Biophys. Meths. 1981, 4, 347-363.
14. Righetti, P.G. In "Electrophoresis '81"; Allen, R.C. and Arnaud, P., Eds.; de Gruyter: Berlin, 1981, pp3-16.
15. Slater, G.G. Fed. Proc. 1965, 24, 225.
16. Slater, G.G. Anal. Biochem. 1968, 24, 215-217.
17. Margolis, J.; Kenrick, K.G. Nature 1969, 221, 1056-1057.
18. Jovin, T.M. Biochemistry 1973, 12, 871-890.
19. Chrambach, A.; Jovin, T.M. Electrophoresis 1983, 4, 190-204.
20. Ferguson, K.A.; Wallace, A.L.C. Nature 1961, 190, 629-630.
21. Rodbard, D.; Chrambach, A. In "Electrophoresis and Isoelectric Focusing in Polyacrylamide Gels"; Allen, R.C.; Maurer, H.R. Eds.; de Gruyter: Berlin, 1974, pp28-61.
22. Rothe, G.M.; Maurer, W.D. In "Gel Electrophoresis of Proteins", Dunn, M.J. Ed.; Wright: Bristol, 1986, pp37-140.
23. Maizel, J.V. Science 1966, 151, 988-990.
24. Shapiro, A.L.; Scharff, M.O,; Maizel, J.V.; Uhr, J.W. Proc. Natl. Acad. Sci. USA 1966, 56, 216-221.
25. Shapiro, A.L.; Vinuela, E.; Maizel, J.V. Biochem. Biophys. Res. Commun. 1967, 28, 815-820.
26. Weber, K.; Osborn, M. J. Biol. Chem. 1969, 244, 4406-4412.
27. Poduslo, J.F.; Rodbard, D. Anal. Biochem. 1980, 101, 394-406.
28. Reynolds, J.A.; Tanford, C. J. Biol. Chem. 1970, 243, 5161-5165.
29. Poduslo, J.F. Anal. Biochem. 1981, 114, 131-139.
30. Shah, A.A. Electrophoresis 1984, 5, 180-182.
31. Laemmli, U.K. Nature 1970, 227, 680-685.
32. Wyckoff, M.; Rodbard, D.; Chrambach, A. Anal. Biochem. 1977, 78, 459-482.
33. Altland, K.; Altland, A. Electrophoresis 1984, 5, 143-147.
34. Dunn, M.J.; Burghes, A.H.M. Electrophoresis 1983, 4, 97-116.
35. Burghes, A.H.M.; Dunn, M.J.; Dubowitz, V. Electrophoresis 1982, 3, 354-363.
36. Anderson, D.; Peterson, C. In "Electrophoresis '81" Allen, R.C.; Arnaud, P. Eds.; de Gruyter: Berlin, 1981, pp41-48.
37. Görg, A.; Postel, W.; Westermeier, R. Anal. Biochem. 1978, 89, 60-70.
38. Radola, B.J. Electrophoresis 1980, 1, 43-56.

39. Ikeda, K.; Suzuki, S. U.S. Patent 1 015 891, 1912.
40. Svensson, H. Acta Chem. Scand. 1961, 15, 325-341.
41. Svensson, H. Acta Chem. Scand. 1962, 16, 456-466.
42. Vesterberg, O. Acta Chem. Scand. 1969, 23, 2653-2666.
43. Vesterberg, O. Acta Chem. Scand. 1973, 27, 2415-2420.
44. Pogacar, P.; Jarecki, R. In "Electrophoresis and Isoelectric Focusing in Polyacrylamide Gel"; Allen, R.C.; Maurer, H. Eds.; de Gruyter: Berlin, 1974; pp153-158.
45. Grubhofer, H.; Borja, C. In "Electrofocusing and Isotachophoresis" Radola, B.J.; Graesolin, D. Eds.; de Gruyter: Berlin, 1977, pp111-120.
46. Williams, K.W.; Söderberg, L. Intl. Lab. 1979, Jan/Feb, 45-53.
47. Righetti, P.G.; Gelfi, C.; Gianazza, E. In "Gel Electrophoresis of Proteins"; Dunn, M.J., Ed.; Wright:Bristol, 1986, pp141-202.
48. Frey, M.D.; Atta, M.B.; Radola, B.J. In "Electrophoresis '84"; Neuhoff, V. Ed.; Verlag Chemie: Weinheim, 1984, pp122-125.
49. Frey, M.D.; Kinzkofer, A.; Atta, M.B.; Radola, B.J. Electrophoresis 1986, 7, 28-40.
50. Gelfi, C.; Righetti, P.G. Electrophoresis 1984, 5, 257-262.
51. Altland, K.; Banzhoff, A.; Hackler, R.; Rossman, U. Electrophoresis 1984, 5, 379-381.
52. Cuono, C.B.; Chapo, G.A. Electrophoresis 1982, 3, 65-75.
53. Cuono, C.B.; Chapo, G.A.; Chrambach, A.; Hjelmeland, L.M. Electrophoresis 1983, 4, 404-407.
54. Burghes, A.H.M.; Patel, K.; Dunn, M.J. Electrophoresis 1985, 6, 453-461.
55. Righetti, P.G.; Gianazza, E.; Bjellqvist, B. J. Biochem. Biophys. Meths. 1983, 8, 89-108.
56. Righetti, P.G.; Gianazza, E.; Gelfi, C. In "Electrophoresis '84" Neuhoff, V. Ed.; Verlag Chemie: Weinheim, 1984, pp29-48.
57. LKB Application Note, no. 324, 1984.
58. Gianazza, E.; Dossi, G.; Celentano, F.; Righetti, P.G. J. Biochem. Biophys. Meths. 1983, 8, 109-137.
59. Gianazza, E.; Celentano, F.; Dossi, G.; Bjellqvist, B.; Righetti, P.G. Electrophoresis 1984, 5, 88-97.
60. Altland, K.; Altland, A. Clin. Chem. 1984, 30, 2098-2103.
61. Gianazza, E.; Giacon, P.; Sahlin, B.; Righetti, P.G. Electrophoresis 1985, 6, 53-56.
62. Altland, K.; Hackler, R.; Banzhoff, A.; Von Eckardstein, A. Electrophoresis 1985, 6, 140-142.
63. Fawcett, J.S.; Chrambach, A. Protides Biol. Fluids 1985, 33, 439-442.
64. Altland, K.; Rossman, U. Electrophoresis 1985, 6, 314-325.
65. Rimpilainen, M.A.; Righetti, P.G. Electrophoresis 1985, 6, 419-422.
66. Smithies, O.; Poulik, M.D. Nature 1956, 177, 1033-1034.
67. O'Farrell, P.H. J. Biol. Chem. 1975, 250, 4007-4021.
68. Klose, J. Humangenetik 1975, 26, 231-243.
69. Klose, J. In "New Approaches to the Evaluation of Abnormal Embryonic Development"; Neubert, D.; Merkes, H.J. Eds.; G. Thieme: Stuttgart, 1975, pp 375-387.
70. Scheele, G.A. J. Biol. Chem. 1975, 250, 5375-5385.
71. Dunn, M.J.; Burghes, A.H.M. Electrophoresis 1983, 4, 173-189.

72. Dunn, M.J.; Burghes, A.H.M. In "Gel Electrophoresis of Proteins" Dunn, M.J. Ed.; Wright: Bristol, 1986, pp203-261.
73. Celis, J.E.; Bravo, R. Eds; "Two-Dimensional Gel Electrophoresis of Proteins"; Academic: New York, 1984.
74. Clinical Chemistry Special Issue, Clin. Chem.1982, 28, 737-1092.
75. Clinical Chemistry Special Issue, Clin. Chem. 1984, 30, 1897-2108.
76. Perdew, G.H.; Schaup, H.W.; Selivonchick, D.P. Analyt. Biochem. 1983, 135, 453-455.
77. Dunn, M.J.; Burghes, A.H.M.; Witkowski, J.A.; Dubowitz, V. Protides Biol. Fluids 1985, 32, 973-976.
78. Ames, G.F.L.; Nikaido, K. Biochemistry 1976, 15, 616-623.
79. Garrels, J.I. J. Biol. Chem. 1979, 254, 7961-7977.
80. Burghes, A.H.M.; Dunn, M.J.; Dubowitz, V. Electrophoresis 1982, 3, 185-196.
81. Burghes, A.H.M.; Dunn, M.J.; Witkowski, J.A.; Dubowitz, V. In "Electrophoresis '82"; Stathakos, D. Ed.; de Gruyter: Berlin, 1983, pp371-380.
82. Garrels, J.I. Methods Enzymol. 1983, 100, 411-423.
83. Anderson, N.L.; Anderson, N.G. Anal. Biochem. 1978, 85, 331-340.
84. Ramasamy, R.; Spragg, S.P.; Jones, M.I.; Amess, R. Electrophoresis 1985, 6, 43-46.
85. Burghes, A.H.M.; Dunn, M.J.; Statham, H.E.; Dubowitz, V. In "Electrophoresis '81"; Allen, R.C.; Arnaud, P. Eds.; de Gruyter: Berlin, 1981, pp295-308.
86. O'Farrell, P.; Goodman, M.M.; O'Farrell, P.H. Cell 1977, 12, 1133-1142.
87. Gianazza, E.; Astrua-Testori, S.; Righetti, P.G. Electrophoresis 1985, 6, 113-117.
88. Gianazza, E.; Frigerio, A.; Tagliabue, A.; Righetti, P.G. Electrophoresis 1984, 5, 209-216.
89. Gianazza, E.; Astrua-Testori, S.; Giacon, P.; Righetti, P.G. Electrophoresis 1985, 6, 332-339.
90. Westermeier, R.; Postel, W.; Weser, J.; Görg, A. J. Biochem. Biophys. Methods 1983, 8, 321-330.
91. Görg, A.; Postel, W.; Weser, J. Protides Biol. Fluids 1985, 33, 467-470.
92. Görg, A.; Postel, W.; Günther, S.; Weser, J. Electrophoresis 1985, 6, 599-604.
93. Dunn, M.J.; Burghes, A.H.M.; Patel, K. Protides Biol. Fluids 1985, 33, 479-482.
94. Dunn, M.J.; Patel, K.; Burghes, A.H.M. In "Progrès Récents en Electrophorèse Bidimensionelle"; Galteau, M.M.; Siest, G. Eds; Presses Universitaires: Nancy, 1986, pp3-9.
95. Dunn, M.J.; Patel, K. Protides Biol. Fluids 1986, 34 (in press).
96. Görg, A.; Postel, W.; Weser, J.; Westermeier, R. In "Electrophoresis '83"; Hirai, H. Ed.; de Gruyter: Berlin, 1984, 525-532.
97. Righetti, P.G.; Gelfi, C.; Gianazza, E. Chimica Oggi 1986, March, 55-60.
98. Anderson, N.L.; Anderson, N.G.; Anal. Biochem. 1978, 85, 341-354.
99. Jones, M.I.; Massingham, W.E.; Spragg, S.P. Anal. Biochem. 1980, 106, 446-449.

100. Voris, B.P.; Young, D.A. Anal. Biochem. 1980, 104, 478-484.
101. Young, D.A. Clin. Chem. 1984, 30, 2104-2108.
102. Klose, J.; Zeindl, E. Clin. Chem. 1984, 30, 2014-2020.
103. Taylor, J.; Anderson, N.L.; Anderson, N.G. Electrophoresis 1983, 4, 338-346.
104. Rüchel, R. J. Chromatogr. 1977, 132, 451-468.
105. Poehling, H.H., Neuhoff, V. Electrophoresis 1980, 1, 90-101.
106. Spragg, S.P. In "Gel Electrophoresis of Proteins"; Dunn, M.J. Ed.; Wright: Bristol, 1986, pp363-394.
107. Garrels, J.I.; Farrar, J.T.; Burwell, C.B. In "Two-Dimensional Gel Electrophoresis of Proteins"; Celis, J.E.; Bravo, R. Eds.; Academic: New York, 1984, pp37-91.

RECEIVED October 30, 1986

# Chapter 3

# Immobilized pH Gradients: Recent Developments

**Pier Giorgio Righetti, Cecilia Gelfi, and Elisabetta Gianazza**

**University of Milano, Via Celoria 2, Milano 20133, Italy**

The present review deals with recent developments of isoelectric focusing in immobilized pH gradients in three main areas: a) "mixed bed" Ampholine-Immobiline gels; b) two dimensional maps and c) preparative runs. Topic (a) presents the advantages of using a combination of a primary, immobilized gradient whith a superimposed secondary, carrier ampholyte pH gradient. Zymogramming of membrane bound enzymes and pH measurements are described. Topic (b) deals with two-dimensional maps not only of cytoplasmic soluble proteins but also of membrane components, with particular emphasis on the use of non-ionic and zwitterionic detergents. Topic (c) deals with optimization of preparative runs and different elution systems (in hydroxyapatite, in ion-exchangers, in Sephadex beads or into a free-liquid phase).

The term 'immobilized pH gradients' (IPG) is a misnomer: to be able to graft a proton onto a polyacrylamide gel is quite a difficult proposition. Nevertheless, by immobilizing in the gel network buffering ions and titrants in given ratios, it is possible to ensure pH constancy in the surrounding liquid elements, i.e. to keep constant the local proton concentration (the pH tout court). Briefly, IPGs represent the latest evolutionary event in isoelectric focusing (IEF) (1), a fractionation technique based on the electrophoretic transport and condensation of amphoteric species (from simple amino acids to large proteins) at their isoelectric point (pI) along a pH gradient. Conventional IEF has been plagued, among other problems, by the instability of pH gradients with time (cathodic drift) (2): IPGs have solved this, and most of the drawbacks inherent to IEF in amphoteric buffers (carrier ampholytes) (CA). The other spectacular advance of IPGs is their capability of engineering any type of pH gradient (linear, non-linear, step-wise, from extremely shallow to very broad) along the separation axis,

0097-6156/87/0335-0033$06.00/0

thus rendering them potentially able to solve any fractionation problem involving charge differences, no matter how minute, among amphoteric macromolecules.

The outline of the present review: after a literature survey (largely on the work developed in our laboratory), we will deal with some modern developments of the IPG technique, in particular: a) the improved performance of 'mixed Ampholine-Immobiline' gels and their capability of allowing direct pH measurements; b) generation of two-dimensional (2-D) maps in the IPG-DALT version; c) newest aspects of the preparative IPG technique, with particular emphasis on elution and recovery of proteins from Immobiline matrices.

## A literature survey

We will start this review with a list of fundamental references in the ield of IPGs: they contain practically all the basic developments and the information the readers will seek in this new electrophoretic technique:

(a) ref. 3: the milestone article, laying the foundations of IPGs;
(b) refs. 4-8: description of extended pH intervals, spanning 2-6 pH computer programs and formulations); ref. 9: a new, acidic pH 3-4 interval; ref. 10: comparison among different focusing techniques in non-amphoteric buffers;
(c) refs. 11-15: strategies for optimizing preparative runs (manipulations of gel thickness, pH gradient width, buffering power, ionic strength; the discovery of the high loading capacity of 'soft gels'; retrieval of proteins from Immobiline matrices);
(d) refs. 16-18: co-polymerization kinetics of Immobilines into polyacrylamide gels; swelling kinetics of dried IPG gels and stability of Immobilines and pre-cast IPG gels;
(e) refs. 19-23: the first two-dimensional (2-D) maps and the use of urea and detergents in IPG systems;
(f) refs. 24-27: early reviews in the field of IPG and 2-D maps;
(g) ref. 28: the first (and we hope the last) artefact with the IPG technique;
(h) refs. 29-31: zymograms and substrate gradients in IPGs ;
(i) refs. 32 and 33: first examples of focusing in very acidic pH gradients (pH 3-4) of dansylated amino acids and computer modelling of separations in very acidic and alkaline milieus.

## 'Mixed Ampholine-Immobiline' gels

For membrane protein analysis. We have just stated that the old system of IEF in carrier ampholytes (CA, or Ampholine) should be abandoned in favor of the much advanced IPG technique. Yet, recently, we have resorted to a melange of the two, a primary, immobilized pH gradient with an overimposed, secondary, CA-pH gradient. We stumbled into that when trying to analyze in IPGs membrane proteins dissolved in 2% Nonidet P-40 (NP-40, a neutral detergent): very few bands focused, the remaining being smeared on the gel surface or precipitated at the application point (Figure 1A). We then resorted to an original idea we described years ago:

'CA - Immobiline' gels (34). Both sample and gel contained 2% NP-40 and 4% Ampholine (pH 4-8 range) superimposed to a pH 4-8 Immobiline gradient grafted in the matrix (35). As shown in Figure 1B, the results were astonishing: not only the total solubilization level was increased (from barely 40% in the absence to more than 60% in the presence of CAs) but the bands were all sharply focused and almost no precipitate was present at the application point. As the solubilization mixture was urea-free, it was possible to zymogram a membrane ATP-ase, which seems to consist of two dark-brown bands in Figure 1C (the fainter bands representing Pb-chelates with the CA chemicals, subsequently precipitated as PbS in the zymogram protocol). The reasons? There are at least two mechanisms involved: a) CAs form mixed micelles with NP-40, thus producing a zwitterionic detergent, which has a greater solubilizing power on membrane proteins; b) the excess free CAs increase the background conductivity of the IPG gel, allowing for quick migration of the proteins to the pI position.

For zymogramming. The above seems to be a more general mechanism for focusing not only membrane proteins, but in general sparingly soluble proteins from different sources. We have recently started a collaboration with Dr. P.K. Sinha (Freie Universitat, Berlin) on the IPG analysis followed by in situ zymograms of several hydrolases from different tissues (e.g. kidney, intestinal brush borders, lungs, epidydimus etc.). As shown in Figure 2, the results in IPGs alone were quite poor, in general a bad smear and lack of focusing for most of the enzymes investigated (dipeptidyl peptidases II and IV, y-glutamyl transferase, alkaline phosphatase), while in CAs alone all the enzymes focused, but in a compressed pattern at the acidic gel end. In a mixture of the two, IPGs soaked either in 1% or in 4% CAs, an incredible array of sharply focused bands was obtained. It should be noted that these microvillar hydrolases were integral parts of the membranes of the brush borders, from which they have been solubilized by clipping off their anchoring domains. While it appears that most membrane proteins have about the same overall hydrophobicity as ordinary water-soluble proteins, it cannot be excluded that their hydrophobic amino acids could be concentrated in certain regions of the molecule, or be more exposed on the surface, thus imparting to them a strong tendency to aggregate once in solution. The presence of CAs could be beneficial to their solubility both, in the transient and steady-state, as it is known that, especially when focused in the pI zone, their backbone exhibits more hydrophobic characteristics, as most of the ionizable groups tend to be in the uncharged state (36). It is a fact that, when focusing cytoplasmic enzymes in our system, good focusing patterns were obtained even in the absence of CAs, in agreement with (29). In addition, CAs could also quench possible ionic interactions with the matrix, if any. In most of our work dealing with IPGs no such interactions were observed, even when working in gels of higher charge density (13, 14), except in the case of a class of proteins, the high-mobility group, known to have a highly polarized charge structure (28). In the case of the present class of membrane bound enzymes such a possibility cannot be a priori excluded, especially if charges are clustered in some domains, rather than being more

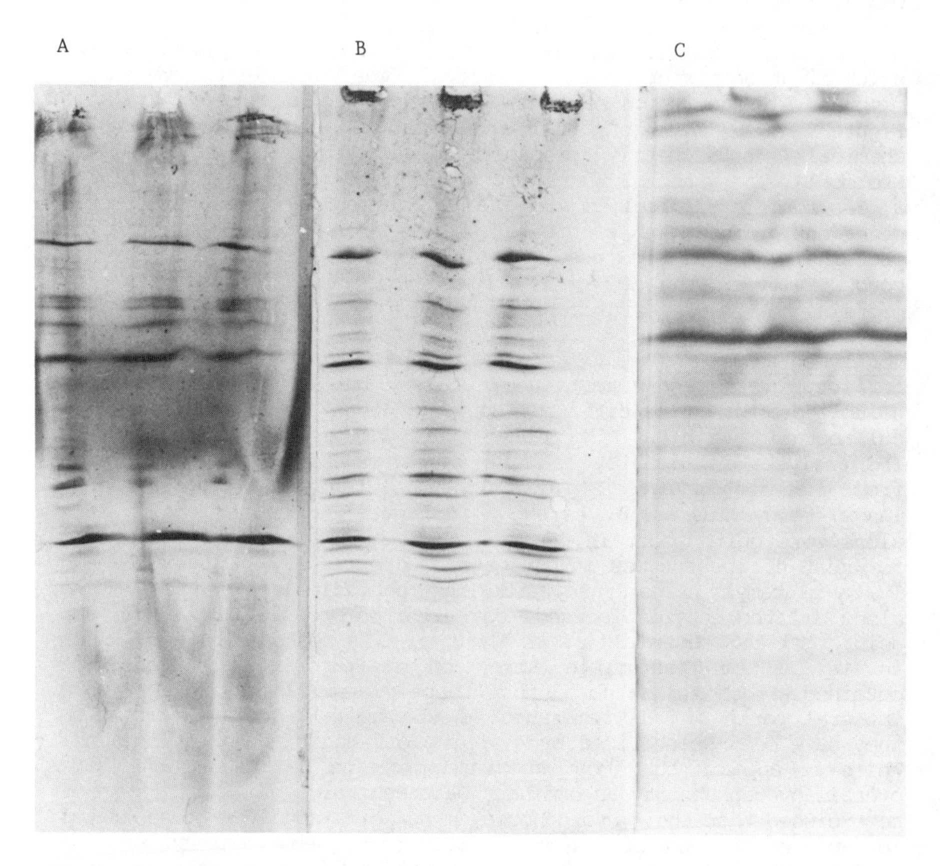


Figure 1. IEF of plasma membrane proteins of a capsulated strain of S. cremoris in a pH 4-8 IPG. (A) 3%T, 4%C polyacrylamide gel containing 8M urea and 2% NP-40. Sample: 75 µg protein in 30 µl of 2% NP-40 and 10 mM Tris-HCl, pH 8.0. (B) same as in (A) but with sample and gel containing 4% Ampholine pH 4-8 and no urea. (C) pattern of focused carrier ampholytes as in (B) visualized by complexing with Pb++ and precipitating in situ PbS. All gels were focused for 30,000 Vh at 10*C (Reproduced with permission from Ref. 35. Copyright 1985, Verlag Chemie).

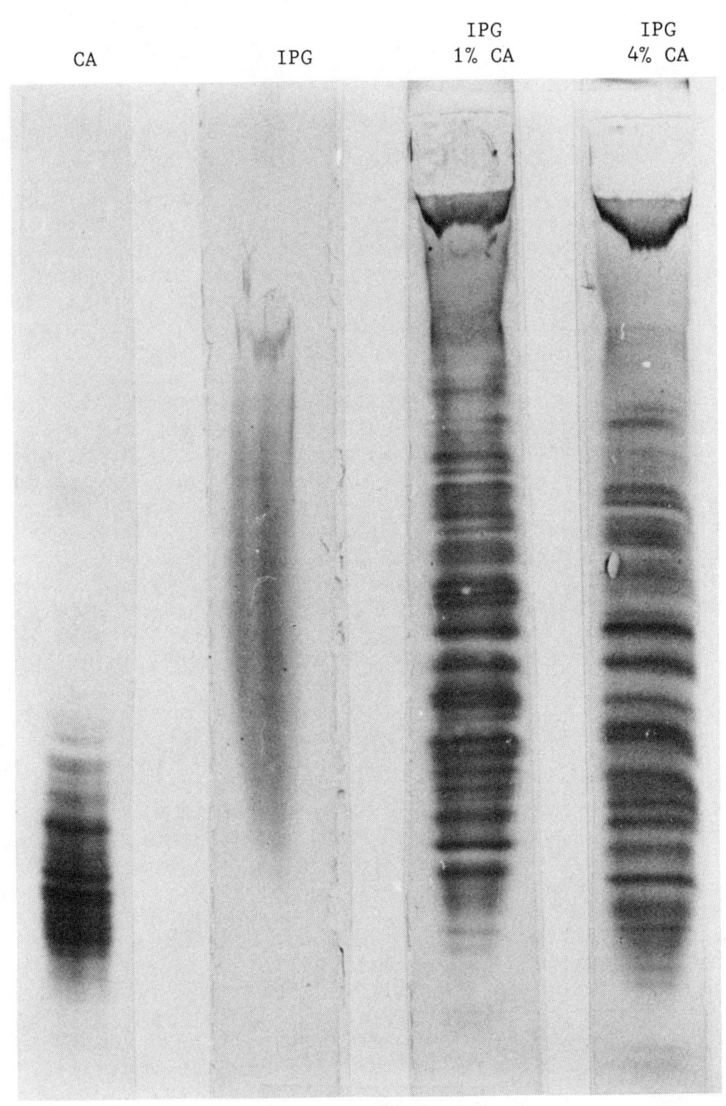


Figure 2. Comparison of bands of enzymatic activity using Gly-Pro-4-methoxy-2-naphtylamine as a substrate for dipeptidyl peptidase IV under different electrophoretic conditions. CA: conventional IEF with carrier ampholytes in the pH 4-6.5 range; IPG: immobilized pH gradient in the pH 4-6.5 range; mixed IPGs pH 4-6.5 with added 1% Ampholine (IPG-1% CA) and mixed IPGs pH 4-6.5 with added 4% Ampholine (IPG-4% CA). Cathode at the top. Conditions for IPGs: run at 2 W constant power for 12 hours at 10*C using 10 mM Glu and 10 mM Lys as anolyte and catholyte, respectively (Reproduced with permission from Ref. 30. Copyright 1986, from Elsevier).

randomly distributed. The presence of CAs would thus be beneficial, as they would effectively compete for the charged groups in the matrix, and split such ion pairs, if any. It is also quite possible that the different mechanisms here illustrated are simultaneously operative to some extent.

For pH measurements. It is impossible to measure pH values in immobilized pH gradients either by a surface electrode or by cutting gel slices and eluting in 10 mM KCl. The use of reversible gels, cross-linked with bisacrylylcistamine, improves the measurements in the acidic region, but gives false values in the alkaline region, due to the buffering power of added 2-mercaptoethanol or dithiothreitol. By using mixed-type gels, containing Immobilines and 1% carrier ampholytes, accurate and reliable pH measurements can be obtained (37). The discrepancy between the theoretical slope of the immobilized pH gradient and the actual pH values obtained by reading the pH of eluted Ampholine cofocusing in the same gel fragment is less than ±0.1 pH units over a 1 pH unit span. The effects of temperature and of CO2 adsorption on pH readings have been demonstrated and evaluated (Figure 3).

It should be appreciated that, in mixed-type gels, it is the primary, IPG matrix that dictates the width and shape of the pH gradient. Thus, if the secondary, CA-generated pH gradient is wider than the former, its width will be reduced to the span of the immobilized gradient, the excess of carrier ampholytes with higher and lower pIs collecting at the cathode and anode, respectively. We have in fact demonstrated that it is possible to convert a wide (2-3 pH unit) carrier ampholyte interval into a narrow (0.2-0.3 pH unit) span. In this case, however, it is imperative to use a highly heterogeneous mixture of CAs, obtained by combining several commercial sources, as no single buffering ampholyte cocktail will guarantee an even distribution of different pI species over such a narrow pH interval, but will be quite unevenly spread along the pH interval, with several gaps, giving quite erroneous pH readings (Figure 4). Over such narrow pH gradients, more accurate pH readings are obtained with the aid of a differential pH meter (38). We have described three ways of incorporating CAs in an IPG matrix: (a) the gel is washed, dried and reswollen in the desired concentration of CAs (35), according to our standard protocol; (b) the gel is washed, excess solvent evaporated to the original gel weight (before washing) and the carrier ampholytes driven electrophoretically into the gel from the electrodic strips (39) (in this case the wicks should be soaked simply in a CA solution, ca. 8 times more concentrated than the wanted final gel level, i.e. according to the ratio strip volume/gel volume); (c) the buffering ampholytes are directly mixed with the light and dense solutions needed to cast the IPG matrix and incorporated during the gelling process. In this case the gel is not washed and used as such (31) (warning: ca. 15% of the monomers are unreacted, so there will be plenty of hazardous acrylic double bonds in the matrix; adopt this procedure only in acidic ranges, where proteins will be quite insensitive to double bond addition and wear gloves in handling the gel!).

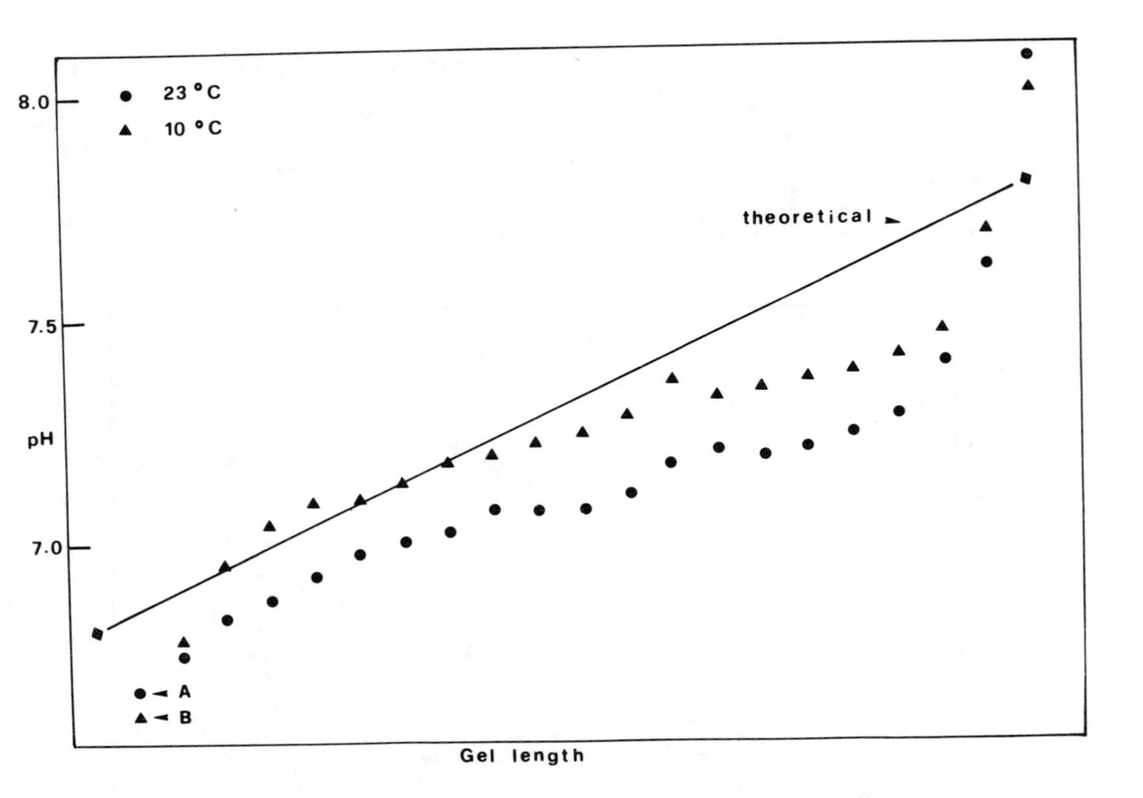


Figure 3. pH measurements in mixed Ampholine-Immobiline gels. A pH 6.8-7.8 IPG gel, containing 1% Ampholine pH 6-8, run 6 hrs at 2000 V, 10*C, was segmented into 22 slices and added with 300 µl of 10 mM KCl. Solid line: theoretical pH slope; line with dots (A): pH readings in slice eluates at 23*C; line with triangles(B): readings in slice eluates at 10*C, in air (Reproduced with permission from Ref. 37. Copyright 1986 Elsevier).

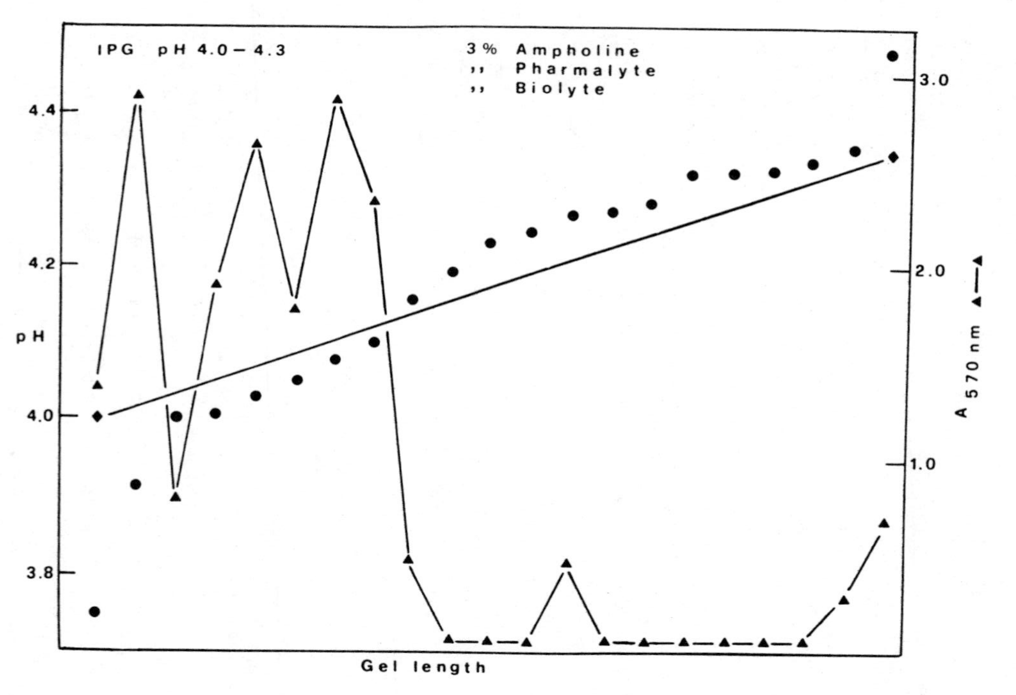


Figure 4. pH gradient determination in an acidic, pH 4.0 to 4.3 IPG gel, dried and re-swollen in a mixture of 3% each Pharmalyte, Biolyte, and Ampholine. The gels were run at 1000 V overnight at 10 *C, segmented in 21 strips (2.5 x 0.07 x 0.05 cm) and eluted with 300 µl of 10 mM kCl. Solid line: theoretical pH slope; dotted line: pH measurements at 10 *C in air; line with triangles: ninhydrin assay for primary amino groups of focused carrier ampholytes (1:10 dilution of each slice eluate), read at 570 nm (Reproduced with permission from Ref. 38. Copyright 1986 Elsevier).

Two-dimensional maps

It had first to be demonstrated that proteins would quantitatively elute from IPG strips and run in SDS gel electrophoresis according to their molecular mass and irrespective of their focusing position along the pH gradient. The absence of interference from ionic interaction or adsorption effects by the charged IPG matrix was in fact demonstrated by others (19, 40) for narrow and by us (21) for wide pH ranges.

The use of a slab for the 1st-D separation instead of the customary gel rods offers several advantages. The most relevant is the size stability given by the permanent backing from the Gel Bond film. This has a most positive influence on the overall reproducibility of the spot position. We have in fact undertaken a thorough investigation of this quality parameter. It has long been demonstrated that IPGs are stable up to several days of continuous electrophoretic run (3). We have then shown that, provided the slabs are washed free of catalysts and unreacted acrylamide monomers and dried, polymerized IPG slabs can be stored, without alteration, for at least 6 weeks (18). The reproducibility of banding position along the pI axis thus rests on the precision of blending the two limiting solutions and of pouring the gradients. The first parameter may be controlled by repeated checks on the pH of the solution after the addition of each buffer (in a sequence chosen to maximize discrepancies between expected and found, had a mistake occurred) (5) and by preparing large batches of gels, to be rehydrated in the solvent of choice when needed. The second point was specifically addressed in our investigation by comparing the results with the use of either a conventional two-vessel gradient mixer or of a set of step-motor burettes with computer-driven pumps (Desaga equipment). Molds of controlled thickness (kindly provided by Desaga) were used for IPG polymerization. The standard deviation on band position was found to be 20% smaller with the latter experimental approach in comparison with the former. Typical results were: with SDS electrophoresis alone (on a 7.5-17.5% polyacrylamide gradient) = ±0.6 mm (for a total gel length of 135 mm); with IPG on a 3 pH unit range = ±0.6 mm (for a total gel length of 180 mm); with 2-D separations (1st-D on a 4 pH unit IPG) = ±0.8 mm on the pI axis, ±0.6 mm on the Mr axis (see Figure 5) (41).

We had run at first only soluble proteins through the IPG-DALT system: markers (20) and serum components (21-23). Recently we have extended this technique to the analysis of particulate components. This requires the incorporation into the gel phase also of solubilization additives, namely urea and non-ionic detergents. This can be done either indirectly, i.e. by polymerizing an empty gel, drying and reswelling (17, 20) or directly, upon gel pouring. The latter approach, which allows for a substantial saving in time, and which results in a more complete sample migration from the application site, requires, as an alternative to the removal of ionic species by washing, their exchange for carrier ampholytes from the electrodic strips (39). The pH range of the latter should be identical, or narrower, than the span of the IPG; their concentration should be $>$ 0.3%.

Some unexpected problems were encountered, however, when the 1st-D IPG matrix was impregnated with non-ionic detergents. A direct

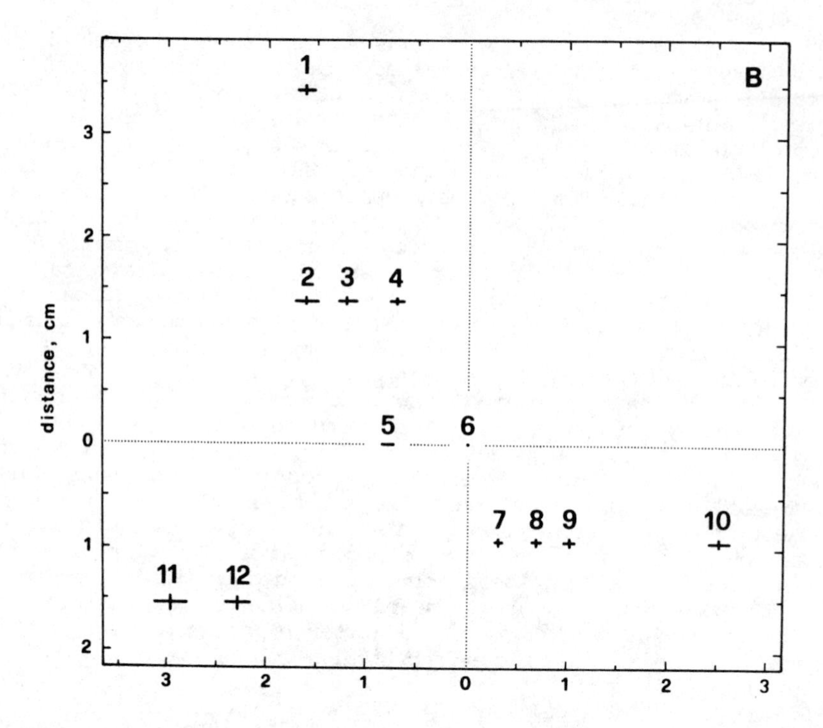


Figure 5. Reproducibility of spot position in the IPG-DALT technique. 1st-D: pH 4-8 IPG gel of marker proteins in 8M urea; 2nd-D: 7.5 to 17.5 %T linear porosity gradient in SDS. Each spot is represented by a symmetrical cross with the intersection corresponding to the band average positional value and the arms representing the standard deviation. In this computation the origin of the axes is set on spot No. 6. The standard deviation represents the average of six different gels (Reproduced with permission from Ref. 41. Copyright 1986, from Verlag Chemie).

transfer from one dimension to the other after the usual incubation in 3% SDS turned out to be impossible: heavy streaking, band splitting and distortion of the low Mr spots resulted, possibly due to the formation of mixed SDS-neutral detergent micelles, with depletion of the former from the electrophoresis medium and insufficient adsorption onto proteins (42). Among different protocols tested to overcome these difficulties, one was chosen requiring: fixation for 1 h in 12% acetic acid - 50% methanol, washing 2 x 15 min in 4 mM Tris base, equilibration for 15 min in 250 mM Tris/Gly, pH 8.8, 1% SDS, denaturation for 15 min in 25 mM Tris/Gly pH 8.8, 5% SDS - 2% 2-mercaptoethanol or 15 mM dithiothreitol (43). Washing may be done for 30 min in plain water when the pH range does not extend above 8 (42). Alkaline ranges also give some problems along the overall equilibration procedure (the basic end overswells and is loosened from the support) if they are prepared by the drying-reswelling procedure; no similar difficulty is to be faced with either IPGs polymerized to include all additives or with CA-IEF slabs also prepared by the drying-reswelling (35). We have run in this way two kinds of samples, while comparing the relative efficiency of different detergents among them, Nonidet P-40 (NP-40), CHAPS and sulfobetaine (SB 3-12) as solubilizers. With the first sample, plasma membranes from Streptococcus sp. MLS96, we obtained virtually complete solubilization with both NP-40 and CHAPS (plus 8 M urea) and about 95% yield with SB 3-12 (plus 4.5 M urea). However, when counting the resolved spots in the 2-D pattern, the three systems scored quite differently: NP-40 = 205 spots, CHAPS = 168 spots, SB 3-12 = 115 spots. With another sample, microvilli from beef kidney cortex, the solubilization was 97% with NP-40, 95% with CHAPS and 83% with SB 3-12. Hundreds of spots could be seen in the resulting 2-D patterns (Figure 6). No major differences were evident between the two former detergents, while virtually no focusing was obtained with the latter.

## Preparative aspects

Strategy for optimizing a preparative run. There are at least four environmental parameters which can be manipulated for maximizing protein loads in Immobiline matrices. By increasing the ionic strength of the gel from 1.25 to 7.5 mequiv. $L^{-1}$ a four-fold increment in load capacity is obtained; above this level, a plateau is abruptly reached around 10-12 mequiv.$L^{-1}$. By increasing the gel thickness from 1 to 5 mm a proportional five-fold increment in protein load ability is achieved; the system does not level off, however a 5 mm thickness seems to be optimal since thicker gels begin to develop thermal gradients in their transverse section, generating skewed zones. Finally, by progressively decreasing the width of the pH interval, there is a linear increase in protein load capability. Here too the system does not reach a plateau, however, due to the very long focusing times required by narrow pH gradients, aggravated by the high viscosity of protein zones at high loads, it is probably unwise to attempt to fractionate large protein amounts in pH ranges narrower than 0.5 pH units. The fourth parameter, unexpectedly, turned out to be the amount of matrix (%T) in the polyacrylamide gel. At 5%T, the maximum load ability is 40 mg

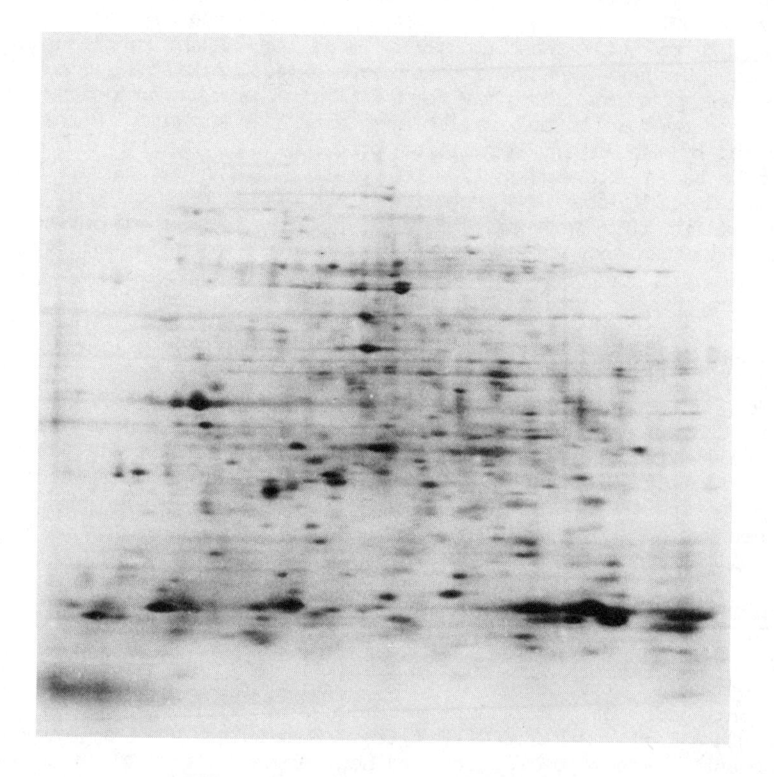

Figure 6. IPG-DALT of membranes from Streptococcus cremoris solubilized with urea and Nonidet P40. 1st dimension: IPG in the pH range 4-8, 2nd dimension: SDS-electrophoresis in a 10-17.5% polyacrylamide gradient. (Unpublished).

protein/ml gel, while, upon diluting the matrix, as much as 90 mg protein/ml gel can be applied to a 2.5%T gel. This has been interpreted as a competition for the available water between the two polymers, the polyacrylamide coils and the protein to be fractionated.

These highly diluted gels have two additional advantages: a) by diluting the matrix, while keeping constant the amount of Immobiline (the conventional ca. 10 mM buffering ion) the charge density on the polymer coil is in fact increased and this results in sharper protein zones and increased protein loading capacity; b) below 3%T, the visco elastic forces of the gel are weakened, allowing the osmotic forces in the protein zone to predominate and draw more water from surrounding gel regions: this results in a further increment in load ability within a given protein zone due to local gel swelling and concomitant increase in cross-sectional area (12, 13).

Elution systems. The recovery from IPG matrices would have to be electrophoretic for two reasons: (a) an IPG gel would behave as a weak ion-exchanger; (b) even extensively washed gels would still contain short, uncross-linked polyacrylamide-Immobiline chains which would be co-eluted with the protein if the latter were to be extracted directly from an excised and ground gel zone. There are at least four elution modes, which are given below.

a) In hydroxyapatite. Originally, Ek et al. (11) had described, for elution, a zone electrophoresis system based on embedding the excised Immobiline gel segment, containing the purified protein zone, into an agarose bed, followed by electrophoretic retrieval into a layer of hydroxyapatite (HA) beads (Figure 7). The protein was then recovered from HA crystals by elution with 0.2M phosphate buffer, pH 6.8. However, while calcium phosphate crystals are an excellent ion-exchange material for separation of nucleic acids (44), they have relative poor sorption capacity for proteins. This transfer technique had been adopted from Ziola and Scraba (45) and Guevara et al. (46), but it was then realized that, while it would work satisfactorily for small protein loads (in the mg/ml range) it would not perform properly on a larger scale (tens of mg/ml range), i.e. at loads compatible with Immobiline matrices. At these high loads, the HA grains would be quickly saturated and the protein zone would cross the entire layer of resin and be lost in the anodal agarose layer embedding it.

b) In DEAE- and CM- Sephadexes. For the above reasons, Casero et al. (14) described a new transfer system based on electrophoretic recovery into true ion-exchangers utilized for protein separations. The IPG gel strip containing the zone of interest is transferred to a horizontal tray and embedded in 1%, low-gelling (37*C) agarose. For acidic to neutral proteins (up to pI 7.7) the electrophoretic transfer is from the IPG strip into a layer of DEAE-Sephadex, buffered at pH 8.5 in 100 mM Tris-acetate (Figure 8A). Recovery (better than 90% in all cases studied) was achieved by titrating the resin at pH 9.5, in 200 mM Tris-Gly buffer, containing 200 mM salt. For basic proteins (pI>7.7) the electrophoretic retrieval is from the IPG strip into a zone of CM-Sephadex, buffered at pH 6.0, in 50 mM citrate (cathodic migration; Figure 8B). Recovery (again better

Figure 7. Recovery of protein zones from Immobiline gels into hydroxyapatite (HA) beads. The IPG gel strip, containing the protein of interest, is cut along the contours of the main band (still supported by the Gel Bond PAG). The 0.8% agarose layer, 5 mm thick, was made to contain 100 mM Tris-Gly buffer, pH 9.1. Just before gelling, the IPG gel strip is embedded in the liquid agarose layer. Upon gelling, 5 mm in front of the IPG gel strip, a 2 cm wide trough is cut in the agarose and filled with grains of HA-Ultrogel. The protein is recovered in the beads of hydroxyapatite contained in the central trough by applying a constant power of 30 W for 60 min (420 V initial voltage drop) at 10*C (Reproduced with permission from Ref. 11. Copyright 1983 Elsevier).

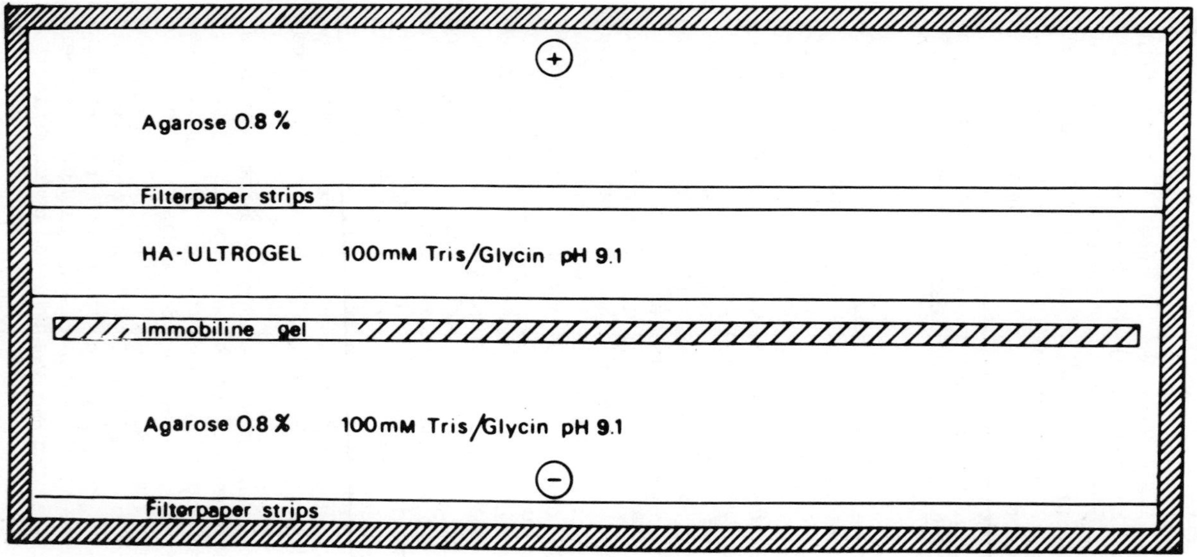

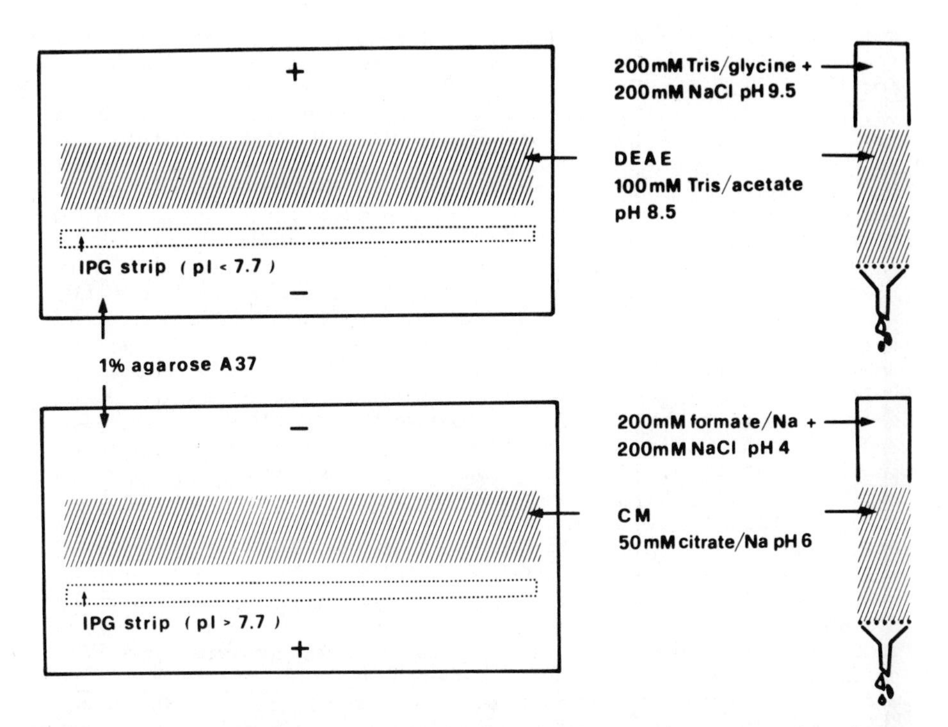


Figure 8. Recovery of protein zones from IPG gels into ion-exchangers. Left side: after IEF, the IPG strip is cut out along the protein contour (still supported by the Gel Bond foil) and embedded in a 5 mm thick layer of 1% agarose A-37 (in the 250 x 110 mm size tray for preparative IEF in granulated gels). In front of it, a 2 cm wide, 22 cm long (or matching the length of the IPG strip) trench is dug into the gelled agarose, and filled with the ion-exchanger (the distance between the IPG gel and the resin should be barely 3 to 5 mm). For proteins with pI's < 7.7, a DEAE-Sephadex in 100 mM Tris-acetate, pH 8.5, is used for the transfer (anodic migration); for species with pI's > 7.7, a CM-Sephadex in 50 mM Na-citrate, pH 6.0, is utilized for the electrophoretic retrieval (cathodic migration). The surrounding agarose layers are equilibrated in the corresponding buffers. Electrophoretic elution lasts in general 700 V x h. Right side: after electrophoresis, the resins are transferred to short columns or to plastic syringes and the protein eluted with 200 mM Tris-Gly, pH 9.5 + 200 mM NaCl (anionic species) or with 200 mM Na-formate, pH 4 + 200 mM NaCl (cationic species)(Unpublished).

than 90%) is accomplished by titrating the exchanger at pH 4.0, in 200 mM formate buffer, containing 200 mM NaCl. It has been demonstrated that Immobiline gels, even when incorporating 5 times the standard amount of buffer (75 mM Immobiline at pH=pK, i.e. 50 mM buffering ion and 25 mM titrant) exhibit, under the electric field, negligible ion-exchange properties, thus behaving as ideal supports for isoelectric focusing.

c) In Sephadex-filled channels. For small-scale protein loads in IPGs, an interesting method was described by Bartels and Bock (15) who recovered the protein of interest focused in the Immobiline gel directly into gel filtration media: this was performed by collecting the protein of choice into a layer of Sephadex G-200, inserted into a channel cut into the IPG matrix. How this is done is shown in Figure 9. The IPG plate is first run by applying the sample only in two lateral tracks; after reaching equilibrium conditions, these two zones are cut away and stained for proteins. The developed analytical strips are then aligned back into their original position and, in the middle preparative area of the IPG gel, zones are selected where the protein bands of interest would focus; a channel is cut away with the aid of a scalpel and a spatula in these areas and the trench filled with a slurry of Sephadex G-200 equilibrated in distilled water. Then the sample for the preparative run is applied in tracks corresponding to the different channels cut out, and the IPG run is performed under the same conditions used for the analytical pre-run. Upon reaching equilibrium, the desired protein will collect in the Sephadex-filled channel, and will be forced to stay there by the electric field. At the termination of the run, the Sephadex grains are quickly removed from the different channels, individually transferred into suitable micro-columns and the different protein zones are recovered by gel filtration. The difference between the present method and the systems of Ek et al. (11) and Casero et al. (14) is that both the electrophoretic fractionation and the protein recovery are performed simultaneously in the same mixed Immobiline-Sephadex gel, rather than sequentially in two different gel layers by two separate experiments.

d) In dialysis bags. The elution systems described so far have as a drawback the fact that the proteins are transferred from matrix to matrix, i.e. from an Immobiline gel into granulated, ion-exchange or gel-filtration beads, from which they have to be recovered by an additional elution step, usually in presence of high salt levels. It would be desirable to be able to retrieve the purified protein directly in a liquid phase. This can be achieved if, after the first IPG run, a different electrophoresis chamber is used for recovery, as described by Righetti and Drysdale (47). The set-up is illustrated in Figure 10: the IPG gel strip, containing the focused protein zone, is chopped into pieces and loaded on top of a stacking gel, consisting of a 5 mm thick, 5%T polyacrylamide gel disc, which is in fact the ceiling of an elution chamber containing a few ml (1-3 ml) of a 20% sucrose, closed at the bottom by a dialysis tube (Figure 10). Upon zone electrophoresis, the protein band is removed from the IPG matrix, crosses the stacking gel and is collected in the liquid phase; the recoveries are very high, usually 90-95%, in a quite concentrated solution (up to 50 mg/ml) (48). There is no universal system: the first three described have the advantage of

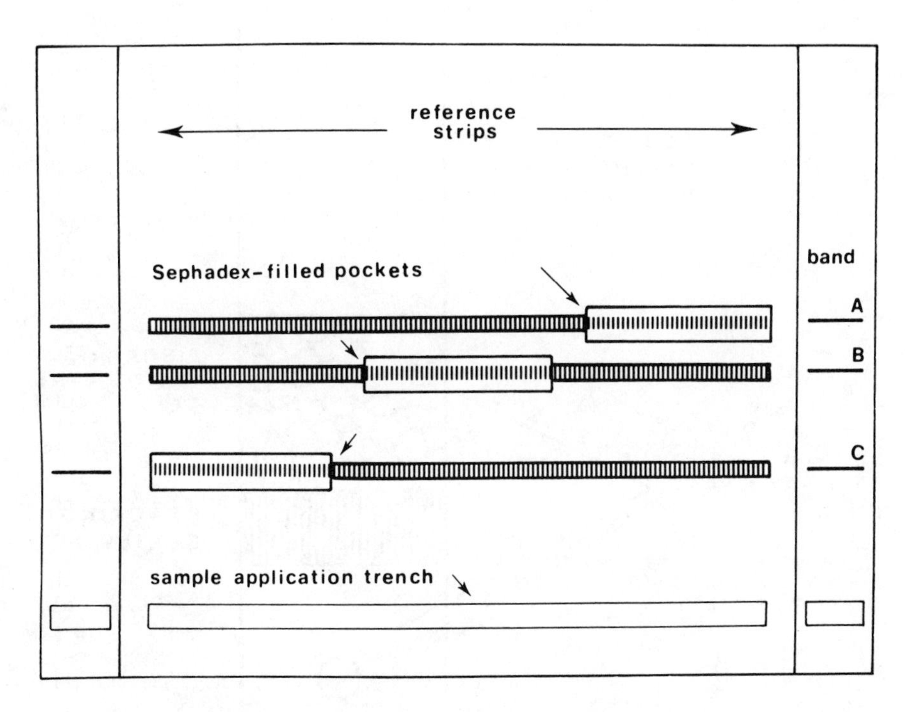


Figure 9. Recovery of proteins in gel filtration media (Immobiline - Canal technique). After preparing the IPG gel, the sample to be purified is applied only in two lateral, reference strips and focused. The strips are stained and re-aligned with the intact IPG gel. In the latter, in correspondence with the focusing positions of the desired proteins (here three major bands, A, B and C are considered), trenches are dug, scraped free of Immobiline matrix and filled with Sephadex G-200. The sample for the preparative run is now loaded into the application trench. Upon completion of the IEF step, the proteins of interest are eluted from the Sephadex grains (modified from Bartels and Bock, see ref. 15; unpublished).

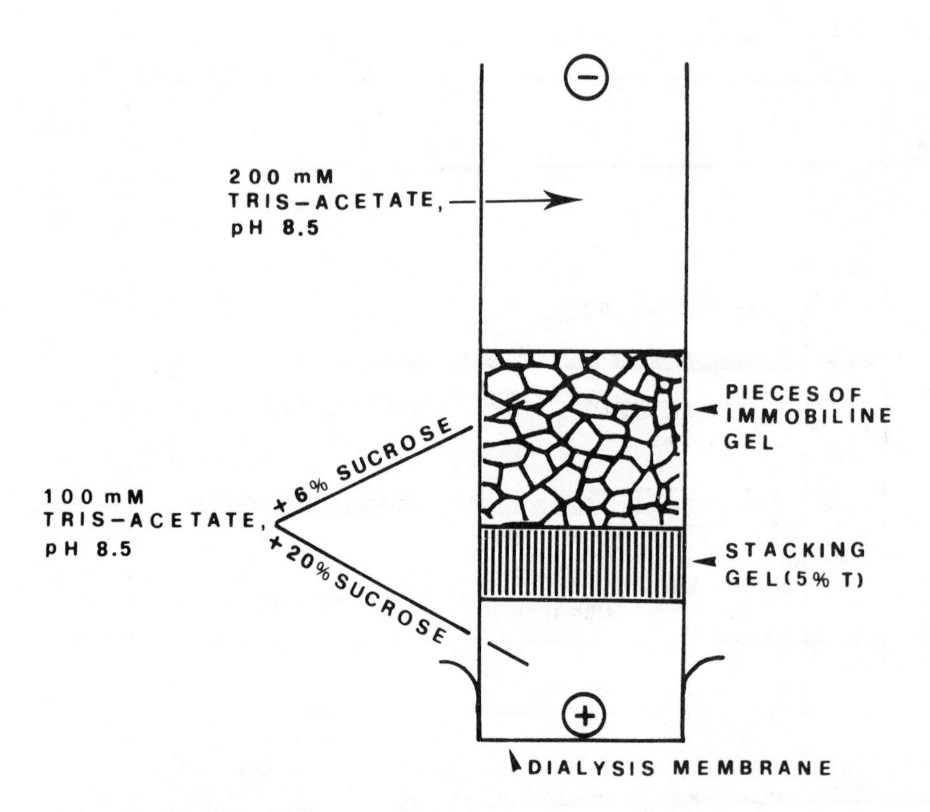


Figure 10. Recovery of proteins into a dialysis bag. Upon termination of the IPG run, the gel strip containing the protein of interest is cut along the contours, chopped to pieces and loaded on top of a stacking gel in a preparative disc electrophoresis apparatus (here the glass tube has an inner diameter of 1.5 cm). After zone electrophoresis (usually 30-45 min at 4*C and 250 V), the protein is collected into the chamber having as a floor the dialysis membrane and as a ceiling the 5%T stacking gel, in a free liquid phase (20% sucrose in 100 mM Tris-acetate, pH 8.5) (Reproduced with permission from Ref. 48. Copyright 1986, from Elsevier).

using the same chamber for both electrophoretic steps, but the obvious disadvantage of having to perform a second elution step from the granulated resins; the latter system allows direct recovery in a liquid phase, but requires use of two different electrophoretic chambers.

## Conclusions

Clearly, the horizon of IPGs is rapidly expanding: a novel case of the Big Bang? Analytically, with the latest development of mixed Ampholine-Immobiline gels, even the remaining difficult cases of poor focusing have been overcome. Preparatively, in addition to the high loading capability and the high-resolving power, many different retrieval techniques are avalaible for protein recovery. One of the major problems with IPG gels is still the sieving of polyacrylamide matrices, substantial even at high dilutions (e.g. 2.5%T). An agarose gelatin would be ideal, but so far attempts at grafting pH gradients to agarose have proven unproductive.

## Acknowledgments

Our research, for the development of IPGs, has been supported by two, five-year grants from Consiglio Nazionale delle Ricerche (Roma), Progetti Finalizzati 'Salute dell'Uomo' and 'Chimica Fine'.

## Legend of symbols

IEF: isoelectric focusing; IPG: immobilized pH gradients; CA: carrier ampholytes; 2-D maps: two dimensional separations based on charge (isoelectric focusing) coupled to orthogonal mass fractionation (SDS-electrophoresis); HbA: adult hemoglobin; HbF: fetal hemoglobin.

## Literature Cited

1. Righetti, P.G. "Isoelectric Focusing: Theory, Methodology and Applications"; Elsevier Biomedical Press: Amsterdam,1983.
2. Righetti, P.G.; Drysdale, J.W. J. Chromatogr. 1974, 98, 271-321.
3. Bjellqvist, B.; Ek, K.; Righetti, P.G.; Gianazza, E.; Görg, A.; Westermeier, R.; Postel, W. J. Biochem. Biophys. Methods 1982, 6, 317-339.
4. Dossi, G.; Celentano, F.; Gianazza, E.; Righetti, P.G. J. Biochem. Biophys. Methods 1983, 7, 123-142.
5. Gianazza, E.; Dossi, G.; Celentano, F.; Righetti, P.G. J. Biochem. Biophys. Methods1983, 8, 109-133.
6. Gianazza, E.; Celentano, F.; Dossi, G.; Bjellqvist, B.; Righetti, P.G. Electrophoresis 1984, 5, 88-97.
7. Gianazza, E.; Astrua-Testori, S.; Righetti, P.G. Electrophoresis 1985, 6, 113-117.
8. Gianazza, E.; Giacon, P.; Sahlin, B.; Righetti, P.G. Electrophoresis 1985, 6, 53-56.
9. Righetti, P.G.; Gianazza, E.; Celentano, F. J. Chromatogr. 1985, 356, 9-14.

10. Righetti, P.G.; Gianazza, E. J. Chromatogr. 1985, 334, 71-82.
11. Ek, K.; Bjellqvist, B.; Righetti, P.G. J. Biochem. Biophys. Methods 1983, 8, 134-155.
12. Gelfi, C.; Righetti, P.G. J. Biochem. Biophys. Methods 1983, 8, 156-171.
13. Righetti, P.G.; Gelfi, C. J. Biochem. Biophys. Methods 1984, 9, 103-119.
14. Casero, P.; Gelfi, C.; Righetti, P.G. Electrophoresis 1985, 6, 59-69.
15. Bartels, R.; Bock, L. In "Electrophoresis '84"; Neuhoff, V., Ed.; Verlag Chemie: Weinheim, 1984; pp. 103-106.
16. Righetti, P.G.; Ek, K.; Bjellqvist, B. J. Chromatogr. 1984, 291, 31-42.
17. Gelfi, C.; Righetti, P.G. Electrophoresis 1984, 5, 257-262.
18. Pietta, P.G.; Pocaterra, E.; Fiorino, A.; Gianazza, E.; Righetti, P.G. Electrophoresis 1985, 6, 162-170.
19. Westermeier, R.; Postel, W.; Weser, J.; Görg, A. J. Biochem. Biophys. Methods 1983, 8, 321-330.
20. Gianazza, E.; Artoni, F.; Righetti, P.G. Electrophoresis 1983, 4, 321-326.
21. Gianazza, E.; Frigerio, A.; Tagliabue, A.; Righetti, P.G. Electrophoresis 1984, 5, 209-216.
22. Gianazza, E.; Giacon, P.; Astrua-Testori, S.; Righetti, P.G. Electrophoresis 1985, 6, 326-331.
23. Gianazza, E.; Astrua Testori, S.; Giacon, P.; Righetti, P.G. Electrophoresis 1985, 6, 332-339.
24. Righetti, P.G.; Gianazza, E.; Bjellqvist, B. J. Biochem. Biophys. Methods 1983, 8, 89-108.
25. Righetti, P.G. J. Chromatogr. 1984, 300, 165-223.
26. Righetti, P.G.; Gianazza, E.; Gelfi, C. In "Electrophoresis '84"; Neuhoff, V., Ed.; Verlag Chemie: Weinheim, 1984; pp. 29-48.
27. Righetti, P.G. Trends Anal. Chem.1983, 2, 193-196.
28. Righetti, P.G.; Delpech, M.; Moisand, F.; Kruh, J.; Labie, D. Electrophoresis 1983, 4, 393-398.
29. Görg, A.; Postel, W.; Johann, P. J. Biochem. Biophys. Methods 1985, 10, 341-350.
30. Sinha, P.K.; Righetti, P.G. J. Biochem. Biophys. Methods 1986, 12, 289-297.
31. Sinha, P.K.; Bianchi Bosisio, A.; Meyer-Sabellek, W.; Righetti, P.G. Clin. Chem. 1986, 31, in press.
32. Bianchi Bosisio, A.; Righetti, P.G.; Egen, N.B.; Bier, M. Electrophoresis 1986, 7, 128-133.
33. Mosher, R.A.; Bier, M.; Righetti, P.G. Electrophoresis 1986, 7, 59-66.
34. Righetti, P.G.; Macelloni, C. J. Biochem. Biophys. Methods 1982, 6, 1-15.
35. Rimpilainen, M.A.; Righetti, P.G. Electrophoresis 1985, 6, 419-422.
36. Gianazza, E.; Astorri, C.; Righetti, P.G. J. Chromatogr. 1979, 10, 161-169.
37. Righetti, P.G.; Morelli, A.; Gelfi, C. J. Chromatogr. 1986, 359, 339-349.
38. Gelfi, C.; Morelli, A.; Rovida, E.; Righetti, P.G. J. Biochem. Biophys. Methods 1986, 12, in press.

39. Gianazza, E.; Quaglia, L.; Caccia, P.; Righetti, P.G. J. Biochem. Biophys. Methods 1986, 12, 227-237.
40. Gahne, B.; Juneja, R.K. In "Electrophoresis '84"; Neuhoff, V., Ed.; Verlag Chemie: Weinheim,1984; pp. 285-288.
41. Gianazza, E.; Astrua-Testori, S.; Caccia, P.; Giacon, P.; Qaglia, L.; Righetti, P.G. Electrophoresis 1986, 7, 76-83.
42. Gianazza, E.; Quaglia, L.; Caccia, P.; Righetti, P.G.; Rimpilainen, M.A.; Forsen, R.I. Electrophoresis 1986, 7, in press.
43. Righetti, P.G.; Tudor, G.; Gianazza, E. J. Biochem. Biophys. Methods 1982, 6, 219-227.
44. Bernardi, G. In "Methods in Enzymology"; 1971; vol. 21, part D, pp. 95-140.
45. Ziola, B.R.; Scraba, D. Anal. Biochem. 1976, 72, 366-371.
46. Guevara, J., Jr.; Johnston, D.A.; Ramagli, L.S.; Martin, B.,A.; Capetillo, S.; Rodriguez, L.W. Electrophoresis 1982, 4, 197-200.
47. Righetti, P.G.; Drysdale, J.W. Ann. N.Y. Acad. Sci. 1973, 209, 163-186.
48. Righetti, P.G.; Morelli, A.; Gelfi, C.; Westermeier, R. J. Biochem. Biophys. Methods 1986, 12, in press.

RECEIVED November 12, 1986

# Chapter 4

# Rehydratable Polyacrylamide Gels for Ultrathin-Layer Isoelectric Focusing

**Bertold J. Radola**

**Technical University of Munich, D-8050 Freising-Weihenstephan, Federal Republic of Germany**

A new approach to isoelectric focusing in 60-240 µm polyacrylamide gels is based on the use of rehydratable gels which in dry form may be stored for extended periods and which prior to use are rehydrated with solutions of any composition. The polymerized gels are washed with distilled water and before drying impregnated with suitable additives to preserve gel functionality on storage. Polyol compounds such as glycerol, sorbitol and dextran, as well as synthetic polymers like polyethylene glycol and polyvinylpyrrolidone are the most efficient additives when incorporated into the gel in a concentration of 1-10 %, either as single substances or in different combinations. Rehydratable gels excel over the traditional wet gels by better standardized properties, convenient handling and flexibility.

In the current practice of polyacrylamide gel electrophoresis wet gels are employed and, usually, gels with a specific composition for each application are prepared just prior to an electrophoretic run. For some applications prefabricated wet gels are available but they do not seem to be widely used. There are many drawbacks to the preparation and use of wet polyacrylamide gels. (i) Polymerization conditions are often poorly standardized in the presence of additives and/or electrolytes with undefined composition or extreme pH values (1,2). Preparation of gels for isoelectric focusing is a typical example where a mixture of carrier ampholytes with practically unknown composition and sometimes extreme pH values is used. (ii) Depending on the gel formulation, particularly on the crosslinker employed, the wet polyacrylamide gel contains variable amounts of unpolymerized monomer (3-5), with potential health risks, and linear polymers (1,6). (iii) Residual amounts of ammonium persulfate or other reagents necessary for free radical generation are present in wet gels and may cause artefacts, inactivate enzymes or interfere with the formation of uniform pH gradients in isoelectric focusing (7-9). (iv) Some chemicals at high concentration inhibit gel poly-

0097-6156/87/0335-0054$06.00/0

merization (10-11). (v) Incorporation of all components into the polymerization mixture is laborious, inefficient and time-consuming because for any change in gel formulation separate gels must be polymerized. This chapter outlines a new approach to polyacrylamide gel isoelectric focusing based on the use of rehydratable gels which in dry form may be stored for extended periods and which prior to use are rehydrated, with solutions of any composition. The gels are polymerized under well standardized conditions and, after polymerization, washed exhaustively to remove any unreacted monomers, catalysts or soluble polymers. Before drying the washed gels are impregnated with suitable additives to preserve their functionality. Experimental details are presented in a recent publication (12).

## Genealogy of Rehydratable Gels for Electrophoresis

The idea of prefabricated rehydratable gels for electrophoresis is not new and has been repeatedly alluded to in the literature (2, 12) but in some recent monographs on electrophoresis rehydratable gels are not even mentioned. Precast wet gels are commercially available for some applications, e.g. isoelectric focusing and gradient gel electrophoresis. These gels save the labor of gel casting but they are offered only for specific applications and they have not been washed for removal of undesirable contaminants. Rehydratable polyacrylamide-agarose gels, washed, dehydrated and preserved between two cellophane films (13), have been commercialized temporarily (Indubiose Plates, Industrie Biologique Française, Villeneuve-la Garenne, France) but they do not seem to have found wide dissemination. One reason might be the long rehydration time of several hours necessary for the 2 mm thick gels.

For many years rehydratable granulated dextran (Sephadex) or polyacrylamide (Bio-Gel) gels have been used for gel chromatography and in the form of hydrated gel layers for analytical and preparative isoelectric focusing (14,15). Recently, rehydratable layers of granulated gels on a plastic film have been developed (16). The presence of carrier ampholytes or addition of 1-2 % glycerol was necessary for retaining functionality of the gels on rehydration by spraying. Some granulated gels (agarose polyacrylamide-agarose) for chromatographic applications are available only in hydrated form indicating that rehydration of gels with resultant functionality may not always be feasible.

Rehydration is now gaining importance in the preparation of gels with immobilized pH gradients (IPG) (17-22). The high proportion of unpolymerized reagents and catalyst present in these gels must be removed by extensive washing. Initially, the gels swollen during washing were partially dried to the original gel weight to avoid liquid exudation during isoelectric focusing. Later, the potential of dry gels, rehydrated prior to use, was recognized. Two protocols for drying IPG gels have been reported differing with respect to an important detail, namely presence of an additive. In the first protocol (20,22) it is stressed that glycerol should be eliminated from the washing solution to prevent the hydrolysis of amide bonds on storage. According to the other protocol (21) the IPGs were equilibrated, before drying, with a solution containing 1 % glycerol. In both protocols 0,5 mm gels were dried with a fan at room temperature but no data were reported about the residual water content of the gels. From the results in the following section it is apparent that these gels will retain considerable amounts of water and owing to the residual moisture might be expected to retain their functionality over a period of several weeks. Under these conditions both gel types remain functional but on continued storage the gels with an additive will prove superior. Data on long term storage of dry IPG gels have not been reported (22).

Gel Polymerization, Washing and Drying. Polyacrylamide gels, cross-linked with N,N'-methylenebisacrylamide (Bis) or in a few experiments N,N'-(1,2-dihydroxyethylene)bisacrylamide (DHEBA) or AcrylAid usually with a composition of 5 % T and 3 % C, were prepared with the flap technique (23). The polymerization mixture contained a 5mM Tris-HCl buffer, pH 8.0, for better pH control during the polymerization step but neither carrier ampholytes nor additives were present at this stage of gel preparation. After polymerization the gels were placed in water to wash out any soluble substances from the gel. Based on experiments with two marker substances, Ponceau S ($M_r$ 760) and horse myoglobin ($M_r$ 17 800) a washing period of 10 min at room temperature was routinely adopted. After washing, the gels were dried at room temperature in a horizontal position.

Preliminary experiments have shown that dry gels prepared by the procedure described above could not be stored. The patterns obtained on isoelectric focusing deteriorated with increasing duration of storage, and depending on thickness, the gels either cracked or locally detached from the supporting film. Both processes were strongly influenced by residual water content of the gel and were always apparent earlier for the 60-120 μm than for the 240 um gels. Additives were necessary to retain the functional properties of the gels on storage and to prevent gel cracking.

Additives. Polyol compounds, such as glycerol, sorbitol and dextran, as well as synthetic polymers, for instance polyethylene glycol, polyvinyl alcohol and polyvinylpyrrolidone, were efficient additives when incorporated into the gel at different concentrations, either as single substances or in combinations. Functionality on isoelectric focusing and gel adherence to plastic supports, under different storage conditions, were the criteria in optimization experiments (Table I). Glycerol, a mixture of dextran with glycerol, and mixtures of polyethylene glycol and polyvinylpyrrolidone with sorbitol were the optimal additives (Figure 1).

The retention of functional properties, particularly on storage for extended periods, is a desirable attribute of rehydratable gels. Heating of rehydratable gels, either open or protected by a plastic or aluminium foil, at 80-100 °C for 1-20 h, strongly affected the functional properties. Although this "forced aging" cannot be strictly compared with long term storage, the approach proved useful for rapid screening of a great variety of additives. All heated gels could be easily rehydrated but, depending on the extent of heat treatment, some of the gels had poor functional properties as a result of increased restrictiveness, most evident for ferritin. Dilution series of different proteins, applied as droplets and allowed to diffuse into the gel for various periods, revealed that penetra-

Table I. Effect of Additives on Different Properties of Rehydratable Polyacrylamide Gels [a)]

| Additive w/v | Gel surface | Adherence to polyester support[b)] | Coalescence on isoelectric focusing [c)] | | | | |
|---|---|---|---|---|---|---|---|
| | | | Marker proteins[d)] | | Ferritin | | |
| | | | 3 h | 20 h | 3 h | 20 h | 100°C |
| Glycerol 10 % | Sticky | ++ | + | + | + | + | |
| Dextran 35 | | | | | | | |
| 5 % | Dry | + | + | + | – | – | |
| 10 % | Dry | + | + | + | + | + | |
| Dextran 35 plus 3 % glycerol | | | | | | | |
| 5 % | Dry | ++ | + | + | + | – | |
| 10 % | Dry | ++ | + | + | + | + | |
| Polyethylene glycol 200[e)] | | | | | | | |
| 1 % | Dry | + | + | + | – | – | |
| 2 % | Dry, oily | + | + | + | – | – | |
| Polyethylene glycol 200 plus 2 % sorbit 1-2% | Dry | ++ | + | + | – | – | |
| Polyvinylpyrrolidone 10 or 25 1-2% | Dry | + | + | + | + | – | |
| Polyvinylpyrrolidone 10 or 25 plus 3% sorbit[f)] 1 - 3 % | Dry | ++ | + | + | + | – | |

Continued on next page

Table I. Continued

a) All gels 5 % T, 3 % C on Gel-Fix support, thickness 120 µm. Washing for 2 x 5 min in distilled water. Equilibration with an excess of additive solution for 10-20 min. Drying overnight at room temperature.

b) + Dry gel detaching slightly from the support within the cutting area. Good adherence during staining and destaining.

++ Good adherence throughout.

c) Focusing temperature for heated gels was usually 15°C. Depending on additive and humidity liquid exudation was observed on isoelectric focusing at 4°C. Freshly prepared gels showed coalescence for marker proteins and ferritin.

d) Marker proteins (14) except ferritin.

e) Sticky gel surface at higher concentrations ($\gtrsim$ 3 %) of sorbitol or glycerol. Polyethylene glycol 200, 3 % : turbid, oily surface. Polyethylene glycol 1000, $\gtrsim$ 1 % : turbid, - oily surface and turbid in 20 % trichloroacetic acid.

f) Polyvinylpyrrolidone plus 3 % glycerol: sticky gel surface. Gels are turbid in the fixative solution (20 % trichloroacetic acid) but become clear on destaining in water/methanol/acetic acid (23).

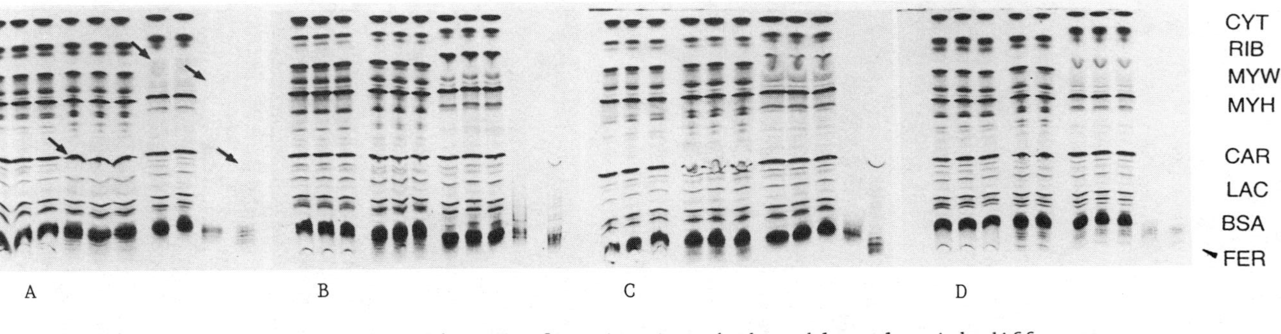


Figure 1. Ultrathin-layer isoelectric focusing in rehydratable gels with different additives. Gels equilibrated before drying with (A) 10% glycerol, (B) 10% Dextran 35 and 3% glycerol, (C) 2% Polyvinylpyrrolidone 25, and (D) 1% Polyvinylpyrrolidone 10 and 3% sorbitol. All gels heated at 100 °C for 3 h ("forced aging"). Rehydration with a 3% solution of Servalyt carrier ampholytes, pH 4–9/pH 3–7 (1:1) with 5% glycerol. Gel format 5 x 5.5 cm, thickness 120 μm. Marker proteins (0.6 μl) applied at different positions (arrows). FER-ferritin, BSA-bovine serum albumin, LAC-β-lactoglobulin, CAR-carbonic anhydrase, MYH-horse myoglobin, MYW-sperm whale myoglobin, RIB-ribonuclease, CYT-cytochrome c (for pI values see Ref. 14). Focusing conditions: 600 Vh, final field strength 400 V/cm; staining with Serva Violet 49 (23, 24). Reproduced with permission from Ref. 12. Copyright 1986 VCH Verlagsgesellschaft.

tion of larger proteins by diffusion was retarded in the rehydratable gels (12). Liquid exudation during isoelectric focusing was observed in heated or stored gels for some additives or higher concentrations of additives (Table I). At low concentrations of some additives, the gels cracked on heating or showed dimished adherence to a variety of polyester supports on cutting the dry gels as well as on staining and destaining. In addition to forced aging, gels with different additives were stored at room temperature, wrapped but not sealed in a plastic cover, for periods up to 12 months. The focusing patterns and coalescence of marker proteins were comparable to those in fresh gels proving that, under these conditions of storage, gel functionality is preserved. We conclude from our experiments that gels retaining functionality after forced aging (100 °C, 8-20 h) will retain also their functional properties after long-term storage at room temperature.

Residual Moisture. Depending on the additive, gel thickness and conditions of drying, the residual moisture of gels varied (Table II). Gels dried at room temperature, although sensorially dry, contain much more residual moisture than gels heated at 100 °C. All gels asymptotically approached constant relative weights after extended heating, the 240 µm gels slower than the 60-120 µm gels. Whereas the relative weight of the dextran gels changed only slightly, under different conditions of drying, the glycerol gels showed great differences, ranging from 17-21 % (gels dried at room temperature) to 5-6 % (100 °C, 24 h). The residual moisture may have profound influence on the functional properties of the dry gels.

## Rehydration

Rehydration Techniques. Prior to isoelectric focusing the dry gels must be rehydrated with a solution of carrier ampholytes. Four rehydration tecniques have been investigated. (i) Flap technique. The technique described for preparation of ultrathin gels has been adapted to rehydration by placing the dry gel with a drop of water on the base plate (cf. Figure 1 in (23)), with two spacers on both edges corresponding in thickness to the original gel layer. Using this technique a perfectly uniform rehydration for all gel formats, up to 26 x 12 cm could be achieved. (ii) Rolling technique. The film with the dry gel was rolled onto a calculated amount of rehydrating solution, care being taken to uniformly spread the solution over the entire gel surface. The technique is economical, because the solution is quantitatively soaked up by the gel layer, and convenient for small gel formats, e.g. 5 x 5 cm. Experiments with dye solutions have shown that with increasing size the polyester support is sagging towards the middle displacing the solution towards the edges. Thus, more of the solution is taken up by the periphery than the gel center and the non-uniform distribution of carrier ampholytes may result in irregular patterns on isoelectric focusing. (iii) Floating technique. The dry gel was rehydrated by placing the film with the gel on the surface of the rehydration solution. The technique requires large volumes of rehydration solution and may be afflicted by non-uniform diffusion of different components from the rehydration solution into the gel (17). (iv) Mold technique. Gels prepared in vertical cassettes, may be conveniently rehydrated in the mold (17). The

Table II. Relative Weight of 5 % T, 3 % C Polyacrylamide Gels after Drying

| Additive | Gel thickness µm | Swelling a) on equilibration: Time min | Swelling a) on equilibration: Weight increase % | Drying at room temperature b) 20 h | Drying at 100°C, Oven b) 1 h | 3 h | 5 h | 20 h |
|---|---|---|---|---|---|---|---|---|
| Glycerol (10 %) | 60 | 20 | 148.9 | 21.1 | 14.1 | 8.2 | 5.9 | 4.9 |
| | 120 | 20 | 137.0 | 17.4 | 12.0 | 8.2 | 5.0 | 5.0 |
| | 240 | 20 | 132.9 | 19.0 | 15.1 | 13.5 | 12.0 | 6.0 |
| Dextran 35 (10 %) | 60 | 20 | 157.7 | | | | | |
| plus glycerol (3 %) | | 30 | 156.9 | 17.8 | 14.5 | 14.5 | 14.5 | 14.5 |
| | 120 | 20 | 124.5 | | | | | |
| | | 30 | 131.9 | 19.1 | 16.7 | 16.3 | 16.3 | 16.2 |
| | 240 | 20 | 107.4 | | | | | |
| | | 30 | 111.8 | 17.8 | 15.6 | 15.6 | 15.6 | 15.5 |

a) After washing the gels were equilibrated in an excess of solution with additives. Swelling did not change if the glycerol-containing gels were incubated for 30 min. The 120 and 240 µm dextran gels continued to swell with but equilibration longer than 30 min was not studied. Weight increase (%) is expressed relative to the weight of the wet gel after polymerization (100 %).

b) All gels were dried uncovered. Weight (%) is relative to the weight of the same gel but in wet form after polymerization (100 %). Parallel determination for two gels.

functional properties of the rehydrated gels strongly depended on the degree of rehydration. Distorted patterns were observed on partial rehydration (50-75 %) and coalescence of ferritin could not be attained. In gels rehydrated to the original volume regular patterns and coalescence of marker proteins, including ferritin, were obtained. Rehydration with an excess of solution (150-200 %) did not offer any advantage.

Kinetics. If rehydratable gels are to replace wet gels their rehydration should be fast, preferably requiring just a fraction of the time necessary for gel polymerization. The rehydration kinetics of rehydratable gels, heated at 100 °C for different periods without or with additives, depended strongly on gel thickness (Figure 2). The 60-120 µm gels were rehydrated in only 20-30 s to 115-155 %, relative to the weight of the gel after polymerization, and rehydration reached a maximum after 2-3 min. Although the initial kinetics were steeper for the 60 µm than for 120 µm gels, both gels were similar in many other respects. For gels containing 10 % glycerol, rehydration decreased after heating and dropped to only 80 % relative weight after heating for 20 h at 100 °C. Rehydration of 60 µm gels containing 10 % Dextran 35 and 3 % glycerol did not change on heating for 1-5 h at 100 °C (relative weight ~ 160-165 %), and was still high after 20 h at 100 °C (relative weight ~ 130 %). The 120 and 240 µm dextran gels could be rehydrated roughly to the same extent (relative weight ~ 120-125 %), irrespective of heating time. Also the gels without additive showed only small changes after heating for different periods and could be rehydrated to 85-100 % of the original gel weight. By contrast, rehydration of the 240 µm gels containing either 10 % glycerol or 10 % Dextran 35 with 3 % glycerol, was much slower than for the 60-120 µm gels, and even after 12 min the gels continued to swell following moderate heating (1-5 h, 100 °C). After more intensive heating (20 h, 100 °C) or in the absence of an additive the rehydration of the 240 µm attained a plateau within 5-10 min and a low extent of rehydration (70-90 %). The 240 µm gels retained their functionality, on isoelectric focusing, better than the 60-120 µm gels.

Functionality of rehydratable gels

Dry ultrathin polyacrylamide gels can be easily rehydrated, though to a different extent, depending on gel thickness, pretreatment and storage. Of course, their important property as an electrophoretic matrix is not swelling but functionality. Dry polyacrylamide gels are irreversibly damaged on storage at room temperature or, much faster, at elevated temperatures. Such gels, even if fully rehydrated to the original volume, are unsuitable for isoelectric focusing by a number of criteria. The focusing patterns are distorted, coalescence of proteins applied at different positions cannot be attained, migration of large proteins, e.g. ferritin, is strongly retarded and the gel surface is modified. All above effects result from an increased restrictiveness of the gel on storage or heating. The probable mechanism of gel damage is crosslinking either through free radicals persisting in the gel (9) or generated on storage. Preliminary electron spin resonance studies indicate that free radicals are involved in gel damage. Hydrogen bonding seems not to contribute to gel damage because it could not be reversed by swelling for 24 h in solutions of 8-10 M urea. Irreversible damage was more evident for the 60-120 µm gels and less distinct for the 240 µm gels under similar conditions of storage or heating. The less restrictive 3 % T, 4 % $C_{Bis}$ and 3 % T, 20 % $C_{Bis}$ gels were more susceptible to irreversible damage than the standard 5 % T, 3 % C gels.

Dry polyacrylamide gels can be protected from irreversible damage by incorporating additives into the gel prior to drying. Compounds differing widely with respect to chemical nature and molecular weight such as glycerol, sorbitol, dextran, polyethylene glycol and polyvinylpyrrolidone, were particularly efficient, either alone or in combination (Table II). We assume that the additives protect the gels by a double mechanism: (i) quenching free radicals in the dry gels on storage, and/or (ii) control of residual moisture. Whereas the ultrathin gels, unprotected by an additive, deteriorated within a few days the 0.5 mm IPG gels were functional also in absence of additives if stored for several weeks (22), probably due to residual moisture. If the air-dried gels, with unknown water content, were further desiccated, e.g. in presence of drying agents, the gels cracked and peeled off from the polyester support. We have observed similar effects for ultrathin gesl in the absence of additives. Dry IPG gels (pH 4-7), containing glycerol according to the protocol of (21), retained functionality and adhered firmly to GelBond supports after incubation at 100 °C for up to 20 h (A. Kinzkofer and B.J. Radola, unpublished observations). Although storage of IPG gels is different, mainly owing to the presence of labile functional groups (22), the results with heating demonstrate that at least some effects found for the uncharged polyacrylamide gels may be also operative. The ultrathin rehydratable gels retained functionality in the presence of additives, even after storage for more than one year at room temperature without additional precautions, e.g. storage in sealed bags or/and at low temperatures. It seems probable that the shelf life of rehydratable gels can be improved by storage under better controlled conditions.

Although the properties of polyacrylamide gels are less critical in isoelectric focusing than in gel electrophoresis, particularly when low molecular weight proteins are separated, restrictiveness may become a major factor when isoelectric points under equilibrium conditions are to be determined or in work with larger proteins. In standard 5 % T, 3 % $C_{Bis}$ gels, polymerized in the presence of carrier ampholytes, proteins with molecular weight up to 500,000 could be focused to equilibrium within a few minutes by miniature isoelectric focusing (24). However, we have repeatedly observed that despite rigid adherence to the published protocol, coalescence of proteins in some gels could not be achieved, not even for the low molecular weight marker proteins (A. Kinzkofer and B.J. Radola, unpublished observations). In all rehydratable gels with the appropriate additives, marker proteins, including ferritin, could be focused to coalescence.

Since rehydratable gels are polymerized in the presence of a defined buffer, any effects of composition, pH range, batch variation or storage-induced changes of the carrier ampholytes are eliminated. The problems involved in achieving reproducible polymerization of

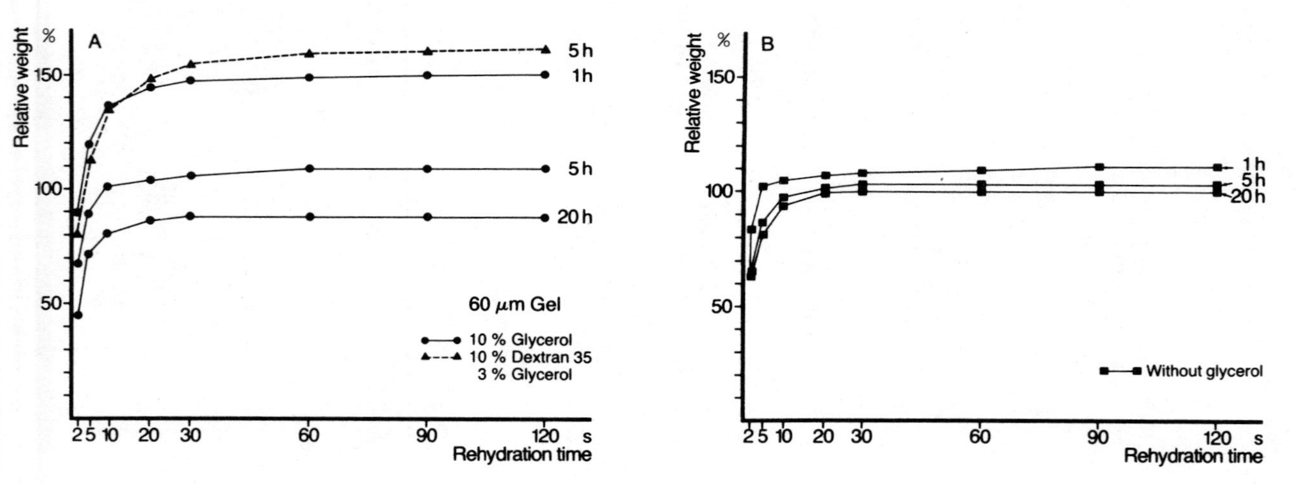


Figure 2. Rehydration kinetics of 60 μm (A,B) and 240 μm (C,D) gels with or without additives. Gels (5 x 5 cm) heated in an oven at 100 °C for the indicated period, without a cover. Rehydration with distilled water, by the floating technique, for the indicated time. Any droplets on the gel surface were vigorously shaken off, and the supporting film was dried with a tissue. The gels were immediately placed in closed Petri dishes for weight determination (double for each point). Results are expressed relative to the weight of the gel after polymerization (100%). Note that the abscissa for (A) and (B) is in seconds but for (C) and (D) is in minutes. Reproduced with permission from Ref. 12. Copyright 1986, VCH Verlagsgesellschaft.
Continued on next page.

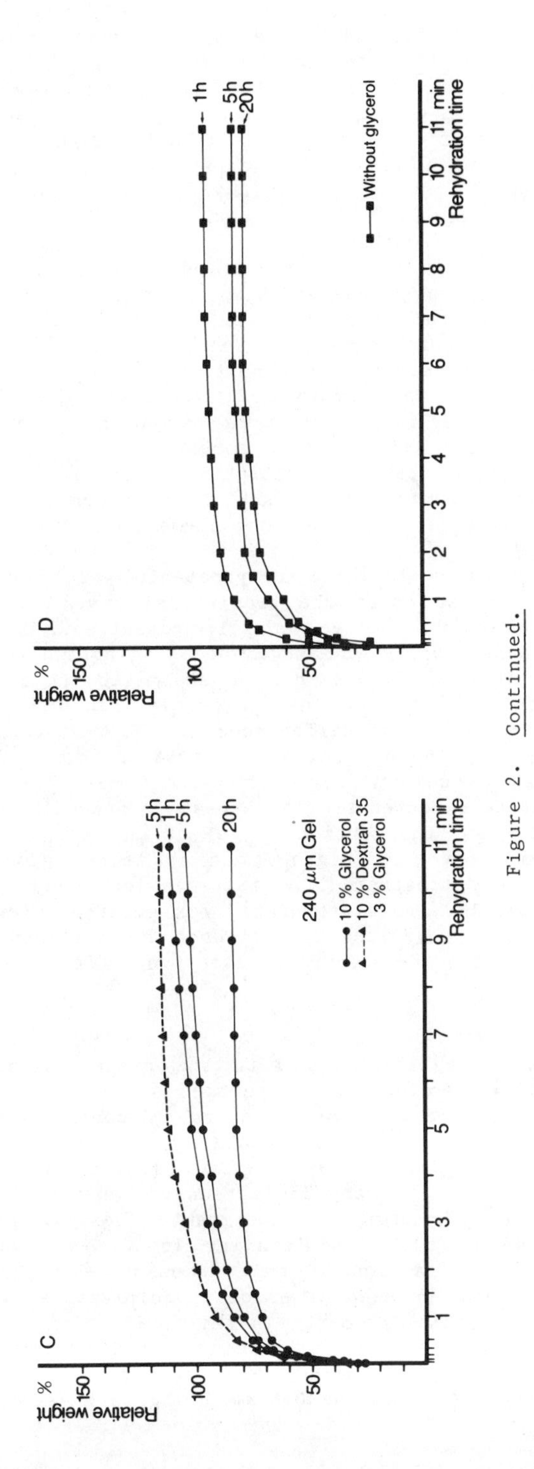


Figure 2. Continued.

acrylamide have been recognized as a major weakness of polyacrylamide gel electrophoresis (1). Instead of optimizing the polymerization conditions for a great number of buffer systems, it has been suggested to prepare a gel under standard polymerization conditions, followed by introduction of the desired buffer either by preelectrophoresis or diffusion (25). This approach proved unpractical for thick gel slabs and cylindrical gels, and thus remained a wishful contemplation but equilibration of ultrathin rehydratable gels with different buffers is feasible.

The washing step in the preparation of rehydratable gels also contributes to better standardization by removing components with potentially detrimental effects on the separation. As a result of washing the initial current is lower, the final field strength higher (Table III) and the electrical conditions in the gel can thus be better standardized. The removal of excessive reagents, catalysts and secondary polymerization products has been described for gels for electrophoresis and isoelectric focusing (26, 27). In the washed gels distortion of patterns on ultrathin-layer isoelectric focusing could be prevented but equilibration of the hydrated ultrathin gels required one week (27), a period that seems to be a serious obstacle to this approach.

Although by many criteria rehydratable gels would appear to be better standardized than the traditional hydrated gels, storage may be anticipated to modify the gel properties. At present this variable is insufficiently examined. Forced aging and storage experiments demonstrate that in presence of additives the restrictiveness of the gels for marker proteins on isoelectric focusing did not change but there were some differences in the rehydration kinetics, extent of water regain and protein penetration. Some of the results with the 60-120 µm gels differed from those with the 240 µm gels, more than might be expected from an linear increase in gel thickness. In the thinner gels, the portion of gel at an interface to the supporting film and cover is higher, and in this region larger pore sizes have been demonstrated by transmission electron microscopy (28). Parenthetically, gels dehydrated for electron microscopic examination have been rehydrated and evidence was obtained that freeze-drying does not alter the electrophoretic properties of the gel (29).

## Applications

Advantages. Rehydratable gels offer two major advantages: (i) convenience, and (ii) better defined properties owing to standardized gel polymerization and removal of undesirable contaminants. Without doubt, for many applications rehydratable gels are more convenient than wet gels. Instead of starting an electrophoretic experiment by gel casting, with its many limitations, a dry gel is rehydrated with any solution depending on experimental design. With ultrathin gels rehydration to full functionality is achieved within a few minutes, in just a fraction of time necessary for the preparation of wet gels and without need of special equipment. We consider fast rehydration an essential property in the application of rehydratable gels and in this respect the ultrathin gels differ decisevely from the early agarose-polyacrylamide gels, requiring rehydration times of several hours (13), and the 0.5 mm IPG gels (19-22). Preparation

Table III. Electrical Conditions on Isoelectric Focusing in Rehydratable Gels (60-120 µm).

| Separation distance | Relative initial current a) | Final field strength (V/cm) | Vh | Total time min |
|---|---|---|---|---|
| 5 cm | ≤ 0.29 | 500 b) | 600 b) | 20 b) |
| | | 800 c) | 800 c) | 22 c) |
| 10 cm | ≤ 0.36 | 400 b) | 2500 b) | 90 b) |
| | | 600 c) | 3300 c) | 120 c) |

a) Initial current of rehydratable gels with 10 % glycerol as additive, relative to initial current of wet gels ( = 1).

b) Wide range carrier ampholytes (Servalyt pH 4-9/3-7, 1:1).

c) Narrow range carrier ampholytes (Servalyt pH 4-6).

of rehydratable gels starts with gel casting, and even necessitates additional steps, namely washing and drying. However, since it is not specific gels but rather multiple gels of the same composition which are prepared, this approach provides a more rational utilization of time. Rehydratable gels will prove on ideal tool for research with its demand for flexibility in handling samples with widely diverging properties. In isoelectric focusing it now seems feasible, with a fraction of efforts necessary so far, to select the optimum pH range of carrier ampholytes, to screen different commercial products and their mixtures, and to investigate the effect of separators or other additives. Also in routine work rehydratable gels might be preferable to wet gels by offering higher consistency of gel properties and operational advantages. By combining gels of different thickness, rehydrated with the same or different carrier ampholytes, an ideal tool will become available for the manipulation of the pH gradient by volume or thickness modification (30,31). Although this report describes only the application of rehydratable gels for analytical isoelectric focusing, the potential for a variety of other applications is apparent.

Ultrathin-Layer Isoelectric Focusing. Ultrathin-layer isoelectric focusing in rehydrated gels differs in several respects from that in wet gels. As a result of extensive washing such impurities as residual monomers, salts and soluble polymers are removed from the gel. The initial current of rehydratable gels is roughly only one third of that in unwashed gels, at the same concentration of carrier ampholytes (Table III). Higher field strengths were tolerated at the final stage of focusing, with typical values of 500-900 V/cm. Equilibrium focusing at a given Vh product can thus be attained in a shorter time, with improved resolution and sharper zones. Prefocusing which is often considered essential for obtaining good patterns in work with wet gels is not necessary with rehydratable gels. The focusing patterns were consistently found to be more regular and better reproducible than in wet gels, both over short (3-5 cm) and long (10-20 cm) separation distances. As added advantage, the washed rehydratable gels proved rather insensitive to high salt concentrations of the sample (12). In wet, unwashed gels distorted patterns were observed on isoelectric focusing of salt-containing samples and focusing hat to be completed at lower final field strength because of local overheating and sparking. Up to 0.5 M salt concentrations were well toleratd by the rehydratable gels and the final field strength could be increased to the same values as for salt-free samples. A potentially important application of rehydratable gels is their use for isoelectric focusing or electrophoresis in presence of high concentrations of urea (12). These gels cannot be stored because of rapid decomposition of urea with the subsequent risk of carbamylation reactions.

Flattening of carrier ampholytes-generated pH gradients with the aid of separators (32) requires at present empirical optimization. Selection of a separator out of a great number of potential spacers, assessment of optimum concentration and polyol additive has so far necessitated preparation of separate gels with specific changes already on gel polymerization (33). Rehydratable gels considerably simplify the screening and optimization work because gels with different formulations are obtained just by changing the composition of the rehydrating solution. Marker proteins and Rohament P, a crude fungal macerating enzyme (23) were focused in a pH 4-6 gradient supplemented with different separators (Figure 3). Depending on added separator, resolution is improved in different parts of the pattern and the position of some of the pI marker proteins permits a rough estimation of the flattening and shift of the pH gradient. By varying the concentation of HEPES from 1-7.5 % the pH gradient is increasingly flattened. From the position of both ß-lactoglobulin components, differing by 0.1 pH, it can be estimated that the pH gradient has been flattened at the highest HEPES concentration, to 0.1 pH/cm, at least over 20 % of the total separation distance.

For the separation of larger proteins ($M_r$ - 400 000 - 500 000) gels less restrictive than the standard 5 % T, 3 % $C_{Bis}$ gels would be desirable. Gels with 3 % T, 4 % $C_{Bis}$ were prepared on commercially available polyester supports but the highly crosslinked 3 % T, 20% $C_{Bis}$ gels could not be backed either to Gel-Fix, GelBond, to silanized polyester supports, at various catalyst/TEMED concentrations, and had therefore to be polymerized on silanized glass (23). The functional properties of these gels were tested with ferritin ($M_r$ 465 000), thyroglobulin ($M_r$ 660 000) and the mixture of marker proteins. In the 3 % T, 4 % $C_{Bis}$ gel with 10 % Dextran 35 and 3 % glycerol, coalescence of anodically and cathodically migrating ferritin and thyroglobulin could be achieved after heating the uncovered gel at 100°C for 4 h and 7 h, and for the gel wrapped in an aluminium foil even after heating for 15 h. By the same criterion of coalescence, gels with 5 % Dextran 35, after heating at the above conditions were suitable only for ferritin. Marker proteins of lower molecular weight coalesced in both gels. Polyvinylpyrrolidone 25 (1-3 %) or polyethylene glycol (1-2 %) both supplemented with sorbitol (Table I) protected the gels less efficiently. The 3 % T, 20 % $C_{Bis}$ gels on glass retained functionality after heating for 4 h at 100°C only in the presence of 10 % Dextran 35 and 3 % glycerol but with other additives heavily distorted patterns were observed.

Other Applications Rehydratable gels offer operational advantages when applied to preparative isoelectric focusing of up to 50-100mg amounts of proteins by improving such important aspects as sample application, non-destructive component location and fast quantitative elution of protein from the gel (F.Kögel and B.J.Radola, in preparation). In addition to isoelectric focusing the rehydratable gels could be applied also for electrophoresis without or in presence of additives, in horizontal and vertical configurations, for the separation of proteins and nucleic acid sequencing. In all above applications the rehydratable gels would serve as a support for the separation. Rehydratable gels in the form of cards or a tape could be used in automated systems for electrophoresis, as described for cellulose acetate on a polyester support (34). In such a system isoelectric focusing over short separation distances, e.g. 3-5 cm (24) and Table III, could be completed within one hour, including all steps from gel rehydration to separation, visualization and densitometric evaluation.

Rehydratable ultrathin gels provide a convenient matrix not

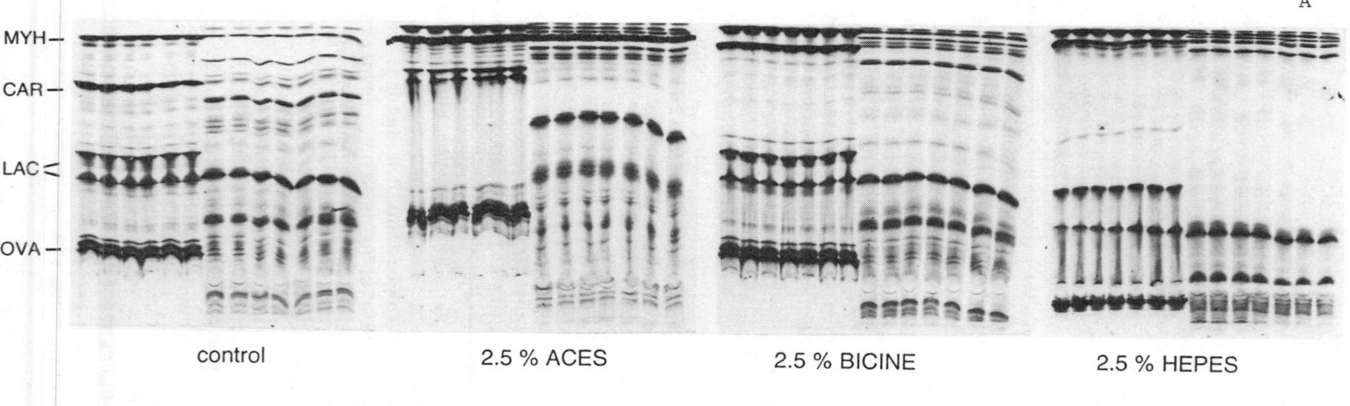


Figure 3. The effect of separators on ultrathin-layer isoelectric focusing in rehydratable gels. Gels (120 μm) containing 10% Dextran 35 plus 3% glycerol were rehydrated with 3% Servalyt pH 4-6 plus 5% glycerol with the indicated concentration of different separators. On the left, a mixture of four marker proteins: MYH-horse myoglobin, CAR-carbonic anhydrase, LAC-β-lactoglobulin, OVA-ovalbumin. On the right, Rohament P. (A) Effect of different separators. Final field strength 500 V/cm, 1000 V x h. (B) Effect of increasing concentrations of HEPES. Final field strength 400 V/cm, 850 Vh (7.5%) or 1000 Vh (1 and 5%). Note increasing distance between the major components of β-lactoglobulin (arrows), corresponding to a flattening of 0.1 pH/cm of the pH gradient. Reproduced with permission from Ref. 12. Copyright 1986 VCH Verlagsgesellschaft.

Continued on next page.

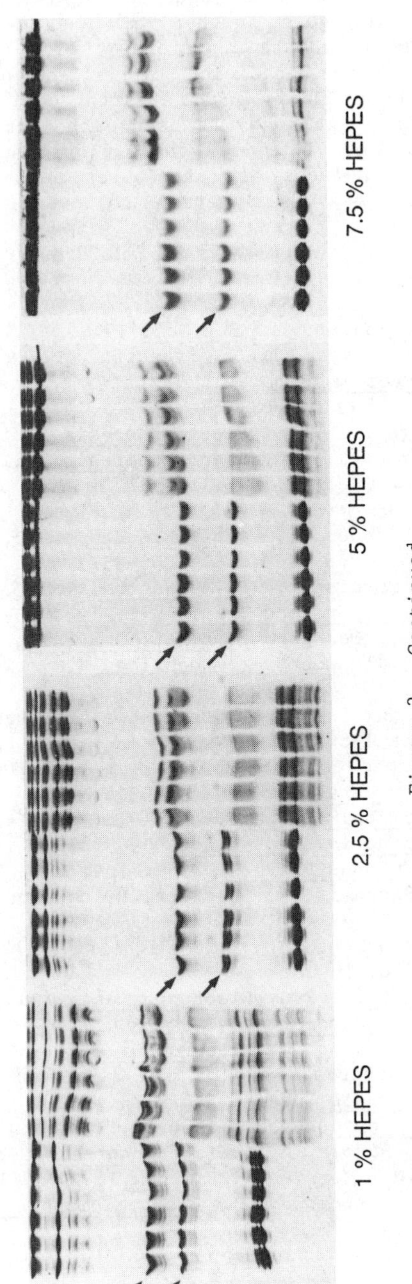


Figure 3. Continued.

only for the separation process but also for subsequent component visualization. In a previous report agarose replicas proved versatile and superior to the traditional agarose overlay technique (35) but ultrathin rehydratable gels further improve enzyme visualization. The dry gels can be rehydrated, preferably with the flap technique, with buffered solutions containing either low or high molecular weight substrates or auxiliary enzymes. After equilibration for defined time the substrate gels may be used either immediately or after storage if this is compatible with the employed substrate. Rehydratable substrate gels possess all attributes necessary for fast and high resolution enzyme visualization following ultrathin-layer isoelectric focusing.

## Literature Cited

1. Chrambach, A.; Rodbard, D. Sep.Sci. 1972, 7, 663-703.
2. Chrambach, A.; Jovin, T.M.; Svendsen, P.J.; Rodbard, D. In "Methods of Protein Separation"; Catsimpoolas, N., Ed.; Plenum Press: New York, 1976; Vol. II, pp. 27-144.
3. Gelfi, C.; Righetti, P.G. Electrophoresis 1981, 2, 213-219.
4. Gelfi, C.; Righetti, P.G. Electrophoresis 1981, 2, 220-228.
5. Righetti, P.G.; Gelfi, C.; Bianchi Boisio, A. Electrophoresis 1981, 2, 291-295.
6. Brooks, K.P.; Sander, E.G. Anal.Biochem. 1980, 107, 182-186.
7. Brewer, J.M. Science 1967, 156, 356-357.
8. Fantes, K.H.; Furminger, I.G.S. Nature 1967, 215, 750-751.
9. Peterson, R.F. J.Agr.Food Chem. 1971, 19, 585-599.
10. Heukeshoven, J.; Dernick, R. Electrophoresis 1981, 2, 91-98.
11. Althaus, H.H.; Klöppner, S.; Poehling, H.M.; Neuhoff, V. Electrophoresis 1983, 4, 347-353.
12. Frey, M.D.; Kinzkofer, A.; Bassim Atta, M.; Radola, B.J. Electrophoresis 1986, 7, 28-40.
13. Dugué, M.; Boschetti, E.; Tixier, R.; Rousselet, F. and Girard, M.L.; Clin.Chim.Acta 1972, 40, 301-304.
14. Radola, B.J. Biochim.Biophys.Acta 1973, 295, 412-428.
15. Radola, B.J. Biochim.Biophys.Acta 1975, 386, 181-195.
16. Radola, B.J. Methods Enzymol. 1984, 104, 256-275.
17. Altland, K.; Banzhoff, A.; Hackler, R. and Rossmann, U. Electrophoresis 1984, 5, 379-381.
18. Gianazza, E.; Artoni, G. and Righetti, P.G. Electrophoresis 1983, 4, 321-326.
19. Gelfi, C.; Righetti, P.G. Electrophoresis 1984, 5, 257-262.
20. Righetti, P.G. J.Chromatogr. 1984, 300, 165-223.
21. Altland, K.; Rossmann, U. Electrophoresis 1985, 6, 314-325.
22. Pietta, P.; Pocaterra, E.; Fiorino, A.; Gianazza, E.; Righetti, P.G. Electrophoresis 1985, 6, 162-170.
23. Radola, B.J. Electrophoresis 1980, 1, 43-56.
24. Kinzkofer, A.; Radola, B.J. Electrophoresis 1981, 2, 174-183.
25. Morris, C.J.O.R.; Morris, P. Biochem.J. 1971, 124, 517-528.
26. Leaback, D.H.; Rutter, A.C. Biochem.Biophys.Res.Commun. 1968, 32, 447-453.
27. Eckersall, P.D.; Conner, J.G. Anal.Biochem. 1984, 138, 52-56.

28. Rüchel, R.; Steere, R.L.; Erbe, E.F. J.Chromatogr. 1978, 166, 563-575.
29. Rüchel, R.; Brager, M.D. Anal.Biochem. 1975, 68, 415-428.
30. Altland, K.; Kaempfer, M. Electrophoresis 1980, 1, 57-62.
31. Låås,T.; Olsson, I. Anal.Biochem. 1981, 114, 167-172.
32. Caspers, M.L.; Possey, Y.; Brown, R.K. Anal.Biochem. 1977, 79, 166-180.
33. Gill, P. Electrophoresis 1985, 6, 282-286.
34. Fosslien, E. Clin.Chem. 1977, 23, 1436-1443.
35. Kinzkofer, A.; Radola, B.J. Electrophoresis 1983, 4, 408-417.

RECEIVED November 25, 1986

# Chapter 5

# Silver-Stain Detection of Proteins Separated by Polyacrylamide Gel Electrophoresis

**Carl R. Merril**

**Section on Biochemical Genetics, Clinical Neurogenetics Branch, National Institutes of Mental Health, Bethesda, MD 20892**

Silver staining now permits visualization of tenths of a nanogram of electrophoretically separated proteins. The mechanisms of protein silver staining depends on the reduction of ionic to metallic silver. Staining properties of individual amino acids, homopolymers, and small peptides, have been used to demonstrate the importance of the basic amino acids, lysine and histidine, and the sulfur containing amino acids in the silver staining process. Many silver stains demonstrate reproducible curvilinear relationships between silver densities and protein concentrations. This sensitivity and reproducibility permits quantitative studies of nanogram amounts of protein. By utilizing sets of operationally constitutive proteins for the normalization of intra-gel stain intensities, quantitative comparisons of protein concentrations may be made from complex biological fluids or cellular extracts.

Development of electrophoretic protein separation techniques have been paralleled by improvements in protein detection methods. Protein detection in early electrophoretic applications, utilizing electrophoretic separations of solutions or colloidal suspensions from about 1816 to 1937, was limited to direct visualization of proteins coated onto microspheres, or studies of naturally colored proteins such as hemoglobin, myoglobin, or ferritin (1-4). An increase in sensitivity and the ability to detect non-colored proteins was achieved by the use of the specific absorption, by proteins, of ultraviolet light. This detection technique permitted Tiselius, in 1937, to demonstrate the quantitative electrophoretic separation of ovalbumin, serum globulin fractions and Bence Jones proteins (5). Tiselius also employed the shadows, or schlieren, created by the boundaries, due to the different concentrations of proteins in the electrophoretic system to detect protein position and concentration (5). These detection methods served as the main methods for protein detection in the liquid electrophoresis systems. However, as the solid support electrophoresis systems supplanted the liquid sustems stains offered a simplier and often more sensitive method for the detection of proteins.

Organic protein stains were the first stains introduced for the detection of proteins along with the development of moist filter paper as an electrophoretic support medium. Many of these stains such as Bromophenol Blue (6), Amido Black (7) and Oil Red (8) had been adapted for the detection of proteins separated by chromatography prior to the introduction of the electrophoretic methods. Coomassie Blue stains, with their capability of detecting as little as half a microgram of protein, are the most sensitive of these organic protein stain (9). This increased sensitivity of the Commassie Blue stains was originally used to detect proteins separated on cellulose acetate. It also became a stain of choice for acrylamide gel electrophoretic methods. Fluorescent protein stains which were introduced by Talbot and Yaphantis in 1971 (10) can now detect as little as one nanogram of protein(11). However, these fluorescent stains usually require reaction conditions that are best performed prior to electrophoresis and their formation of covalent bonds with the protein molecules generally alters the charge of the proteins (12). This charge alteration is not of consequence for electrophoretic techniques that separate proteins on the basis of molecular weight, such as sodium dodecyl sulfate (SDS) electrophoresis, but it can alter separations by isoelectrofocusing (12).

Radioactively labelled proteins may be visualized without staining by autoradiographic methods which were first introduced by Becquerel and Curie in their discovery of the phenomenon of radioactivity (13), or fluorographic techniques for some of the weak beta emittors, such as tritium(14). If the proteins are radioactively labled to a high specific activity, they can be detected with sensitivities equal and often better than those obtained by the most stains. However, the use of radioactively labled proteins is limited as it is difficult to achieve high specific activities in animal studies and unethical in reseach involving humans.

Development Of Silver Stains For Protein Detection The introduction of silver as a general protein stain increased the sensitivity of protein detection 100 fold gain over that attained by the most sensitive commonly used organic stain Coomassie Blue, from a tenth of a microgram to a tenth of a nanogram (15-16). The first silver stains used for the detection of proteins separated by polyacrylamide gel electrophoresis were adapted from histological silver stains and were often tedious, requiring three hours of manipulations and the use of numerous solutions (15-16). In the seven years since the introduction of silver staining as a general method for the detection of proteins in polyacrylamide gel electrophoresis numerous staining protocols have been developed (17). These protocols can be divided into three categories: the diamine or ammoniacal silver stains, the non-diamine stains including stains based on photographic chemistry, and stains based on the photodevelopment or photoreduction of silver ions to form the metallic silver image.

## General Silver Stain Protocols

Diamine Stains Formation of silver diamine complexes, with ammonium hydroxide, offers a means of stabilizing silver ions in an alkaline environment. Diamine silver stains were first developed for the visualization of nerve fibers (18). Silver ion concentration is usually very low in these stains, as most of the silver is bound in diamine complexes (19). The first histological stain to be used for the general detection of proteins separated by polyacrylamide gel electrophoresis was a diamine stain (15-16). Diamine stains tend to become selectively more sensitive for glycoproteins if their concentration of silver ions is decreased. This specificity can be minimized, if the stain is to be used as a general protein stain, by maintaining a sufficient sodium to ammonium ion ratio in the diamine solution (20). However, in some applications, an emphasis of the diamine stains specificity has proven useful, as in the adaptation of a diamine histological silver stain to visualize neurofilament polypeptides in electrophoretic analyses of spinal cord homogenates (21). This stain was saturated with copper ions, which appears to increase its sensitivity. Copper is used in a number of the diamine stains.

In the diamine stains, the ammoniacal silver solution must be acidified, usually with citric acid, for image production to occur. The addition of citric acid lowers the concentration of free ammonium ions, thereby liberating silver ions to a level where their reduction by formaldehyde to metallic silver is possible. The optimal concentration of citric acid also results in a controlled rate of silver ion reduction, preventing a non-selective deposition of silver.

Non-Diamine Chemical Development Stains Most of the non-diamine chemical development silver stains were developed by adapting photographic photochemical protocols (22-26). These stains rely on the reaction of silver nitrate with protein sites in acidic conditions, followed by the selective reduction of ionic silver by formaldehyde in alkaline conditions. Sodium carbonate and/or hydroxide and other bases are used to maintain an alkaline pH during development. Formic acid, produced by the oxidation of formaldehyde, is buffered by the sodium carbonate (27).

Photo-development Silver Stains Photo-development stains utilize energy from photons of light to reduce ionic to metallic silver. Scheele in 1777 recognized that the blackening of ionic forms of silver by light was due to the formation of metallic silver. He demonstrated that silver chloride crystals exposed to light while under water produce a black metallic silver precipitate and hydrochloric acid (28). This ability of light to reduce ionic to metallic silver was adapted by William Fox Talbot, in 1839, as the basis of a photographic processes that dominated photography from its introduction until 1862, when photo-development was replaced by "chemical development" processes (29). The use of photo-reduction provides rapid and simple, yet sensitive silver stain methods for detecting proteins separated by gel electrophoresis (29-30).

Most chemical development stains require a minimum of two solutions, in addition to the fixing solution. This requirement for multiple solutions in the chemical development stains is a result of their use of alkaline solutions for the reduction of silver. The presence of silver ions and an organic reducing agent in an alkaline solution often results in the uncontrolled reduction of silver. However, since light can reduce silver in an acidic solution, a photo-development stain may utilize a fixation solution followed by a single staining solution.

Such single-solution photo-development silver stains have two major advantages over chemical-development silver stains. First, pH gradient effects are eliminated. In chemical development, one solution, containing silver ions, diffuses out of the gel, while the solution containing the reducing agent diffuses into the gel. The interactions of these solutions creates complex pH gradients within the gel. A single-solution photo-development stain reduces such diffusion effects, minimizing staining artifacts due to variations in gel thickness or the use of plastic gel "backings". Proteins on ultra-thin supporting membranes such as cellulose nitrate stain poorly with the "chemical stains" because they retain very little silver nitrate when transferred into alkaline solution for image development. Because the photo-development stain contains the silver ions in the image-developing solution, proteins may be visualized even when bound to thin membranes.

Combination Photo-Development And Chemical-Development Stain By combining silver photo-development and chemical-development methods, a stain has been developed which can detect proteins and nucleic acids in the nanogram range, it can be performed in under fifteen minutes, and it results in minimal background staining (32). This stain utilizes: a silver halide, to provide a light senstive detection medium, and to prevent the loss of silver ions from membranes or thin layer plates; photo-reduction, to initiate the formation of silver nucleation centers; and chemical-development, to provide a high degree of sensitivity by depositing additional silver on the silver nucleation centers (formed by the photo-reduction of the silver halide). This stain displays an average detection sensitivity of 1 ng of protein or 10ng of DNA. The stain's rapidity of action, and its' ablility to stain samples spotted on membranes, such as cellulose nitrate, has afforded the opportunity to investigate some aspects of the mechanism of silver staining.

The first step in this new stain protocol employs copper acetate, a metal salt that is both a good fixative (33) and a silver stain enhancer. The mechanism of copper's stain enhancement, in this and other silver stains, may be similar to its action in the biuret reaction (15), in which a characteristic color shift, from violet to pink, is achieved by titrating peptides in the presence of copper ions. Copper complexes formed with the N-peptide atoms of the peptide bonds are primarily responsible for this reaction. There are also some number of secondary sites which may interact with copper. Any elemental copper formed may displace positive silver ions from solution as copper has a greater tendency to donate electrons than silver, indicated by its position in the electromotive series of the elements. Following the treatment with copper acetate, the membrane is sequentially soaked in a solution containing chloride and citrate ions and then in a solution containing silver nitrate. The membrane is then irradiated with light while it is in the silver nitrate solution. The presence of the resulting silver chloride, in the membrane, produces a significant increase in light sensitivity over that which can be achieved with silver nitrate alone.

Herman Vogel, a 19th century photochemist, postulated that silver nitrate and silver chloride are synergistic in their response to photo-reduction. He reasoned that although silver chloride is more sensitive to the reducing action of light than silver nitrate, it is fixed in position by its insolubility and the potential density of its image would be limited unless the free silver ions supplied by the silver nitrate are present to diffuse into the photo-reduction centers (34). This increase in sensitivity was further enhanced by the presence of acetate and citrate ions (31,34,35). White fluorescent light proved to be the most effective for this photo-reduction. Ultra violet light produced a denser image, but it also produced an unacceptable background stain. Continued irradiation with white light would provide sufficient photo-reduction to produce an image of the protein pattern on the membrane, however, photo-reduction alone usually results in a dense background stain when applied to thin membranes(31). By limiting the light irradiation to a total of four minutes, only enough to initiate the formation of a latent image, formation of a visiable image is achieved by chemical-reduction. The chemical-reduction of ionic to metallic silver was effected by placing the membrane in a solution containing the reducing reagents hydroquinone and formaldehyde. Hydroquinone is known for its ability to produce photographic images of high density with little background fogging. while formaldehyde is a relatively weak reducing agent. Cajal, in 1903, first adapted these photographic reagents for use in silver stains to visualize the histology of the nervous system (36-38). During image formation, ionic silver is reduced to metallic silver, formaldehyde is converted to formic acid (27) and hydroquinone to quinone. Unreacted silver chloride is removed from the membrane, to prevent a grayish cast background, and continued darkening of the membrane as the silver ions in the unreacted silver chloride are photo-reduced, by exposure to light. Removal of the silver chloride is accomplished by complexing the silver chloride with sodium thiosulfate to form a series of complex argentothiosulfate sodium salts, most of which are soluble in water (29). The argentothiosulfate sodium salts, unreacted reagents, and silver grains formed in solution that may have precipitated onto the surface of the membrane are washed away with water.

## Silver Stain Mechanisms

Basic Mechanisms The basic mechanism underlying all protein detection silver stains involves the reduction of ionic to metallic silver. Detection of proteins in the gel or membrane requires a difference in the oxidation-reduction potential between the sites occupied by proteins and adjacent sites of the gel or membrane. If a protein site has a higher reducing potential than the surrounding gel or matrix, then the protein will be positively stained. Conversely, if the protein site has a lower reducing potential than the surrounding gel or matrix, the protein will appear to be negatively stained. These relative oxidation-reduction potentials can be altered by the chemistry of the staining procedure. Proteins separated on polyacrylamide gels have been shown to stain negatively if the gel is soaked in the dark in silver nitrate followed by image development in an alkaline reducing solution (such as Kodak D76 photographic developer). By treating the gel with potassium dichromate prior to the silver nitrate incubation followed by development of the image in an alkaline reducing solution (utilizing formaldehyde as the reducing agent), a positive image is produced (25). Positive images may also by obtained by substituting potassium ferricyanide (22), potassium permanganate (39), or dithiothreitol (26) for potassium dichromate in this stain. Dichromate, permanganate and ferricyanide are thought to enhance the formation of a positive image by converting the protein's hydroxyl and sulfhydryl groups to aldehydes and thiosulfates, thereby altering the oxidation-reduction potential of the protein. Although the formation or presence of aldehydes has often been suggested as essential for silver staining, in certain histological stains neither aldehyde-creating or aldehyde-blocking reagents appreciably affect silver staining (40). Silver staining of fibrils appeared on electron microscopic observation to depend on whether the fibrils were in an ordered or random array, suggesting that, in some cases, tissue silver staining may depend on physical interface phenomena (40).

Dithiothreitol, a reducing agent, also creates a positive image, perhaps by maintaining the proteins in a reduced state. However, other reducing agents, such as beta-mercaptoethanol, do not enhance positive image formation. Alternatively, all of the positive image enhancing compounds may form complexes with the proteins. These complexes may act as nucleation centers for silver reduction (17).

Photo-development Mechanisms In photo-development light photons are utilized to liberate electrons. The mechanism is probably best described by the Gurney-Mott photochemical theory: When a liberated photoelectron combines with a silver ion metallic silver is formed. It is generally accepted that a single silver atom will rapidly undergo oxidation, reverting to a silver ion, unless additional silver atoms are formed nearby. Once a critical number of silver atoms are formed in a local region, they become an autocatalytic center for the reduction of additional silver ions. The rate of silver ion reduction after the photocatalytic formation of stable metallic silver depends on the electron availability or the local oxidation-reduction potential (41).

Prior to silver staining by either the chemical- or photo-development methods, proteins must be "fixed". Fixatives play a dual role in both methods. They retard diffusion of the protein from the gel or membrane, and they elute substances that might interfere with staining (such as reducing agents, detergents, and ampholytes) from the matrix. Electrophoretic systems may, in some cases, require more than one fixation to clear the gel of these substances. In the photo-development stains, the fixative also impregnates the gel with chloride ions. When a gel that has been treated with a fixative, containing a small amount (0.2% w/v) of sodium chloride, is transferred into a silver nitrate solution, a fine bluish-white precipitate of silver chloride forms in the gel. This image can be visualized either with transillumination or incident light. Proteins appear as clear regions, while the rest of the gel contains a fine bluish-white precipitate. This effect is especially noticeable if the band or spot contains one or more micrograms of protein. With illumination by a light source of sufficient intensity, the clear regions containing most proteins will darken (31). The appearance of these clear areas at the sites occupied by the proteins may indicate that proteins exclude silver ions. However this interpretation cannot be correct, since previously clear regions develop the characteristic black or brown images formed by the reduction of silver ions on exposure to light (25). Further evidence that the silver ions are not excluded from gel regions containing electrophoresed proteins was obtained by placing a polyacrylamide gel into a radioactive silver nitrate solution, using $^{110m}Ag$. At very low concentrations of silver nitrate ($10^{-9}M$) sufficient silver bound to the proteins to visualize their positions by autoradiography. At higher concentrations of silver, similar to that used in the silver stains, 0.1M, the distribution of silver was fairly uniform. In no case was there evidence for the exclusion of silver ions (25). Evidence for the presence of chloride ions in the clear zones is more indirect. If chloride ions are excluded from the stain protocol, there is a significant loss of the stain's sensitivity.

Lack of a visible silver chloride precipitate in gel regions containing relatively high concentrations of proteins may be due to altered solute structures affecting interactions between silver and chloride ions. Biological molecules often reorganize solvent molecules in the neighborhood of reactive groups. X-ray diffraction studies of proteins have demonstrated immobilized water oxygen atoms, such that they are visible in specific positions in electron density maps (42). Effects of proteins on surrounding solvents may be augmented by the physical organization of the proteins themselves, as in the previously discussed electron microscopic observations which suggest that the degree of silver staining depends on whether fibrils are arranged in random arrays or in organized bundles (40).

## Protein Silver Stain Reactive Groups

**Staining Of Amino Acids And Homopolymers** A study of amino acid homopolymers and individual amino acids was undertaken with the combination photo-development, chemical-development silver stain to gain information about reactive groups that may be involved in the staining reactions. The only individual amino acids which stained were cysteine and cystine. Poly-methionine and the hydrophilic basic amino acid polymers: poly-lysine, poly-arginine, poly-histidine, and poly-ornithine also stained (32). Staining of the basic amino acids in their homopolymeric form, but not as individual amino acids, may be related to the shift of pKs that is normally associated with the incorporation of amino acids into peptides. This shift in pK toward the neutral range results in an increased presence of ionized amino acid side chains closer to the physiological pH. The ability of reactive group in an amino acid side chains to form complexes with metal ions may be enhanced by such a shift. For example, a shift in the pk of an amino group would reduce the proton competetion that a metal ion must overcome for the amino group's N-atom electron pair. Staining of the basic amino acid and methionine homopolymers, but not their individual amino acids may also indicate the need for cooperative effects of several intramolecular functional groups to form complexes with the silver or copper ions (43).

Heukeshoven and Dernick also observed silver staining of the basic homopolymers of histidine, arginine, and ornithine, although they did not report staining of the basic amino acid homopolymer poly-lysine (44). The role of the basic amino acids in silver staining is further strenghened by the observation by Nielsen and Brown that the basic amino acids: lysine, arginine, and histidine, (in both a free and homopolymeric form) produce colored complexes with silver (45).

Previous studies have reported silver staining with other amino acids. Heukeshoven and Dernick reported silver staining of the homopolymers of glycine, serine, proline and aspartic acid (44) while Nielsen and Brown reported the formation of colored silver complexes with: aspartate, and tyrosine (45). Staining of these homopolymers was not observed in the study of Merril and Pratt (32), and prior metal binding studies failed to demonstrate metal interactions with the side-chain hydroxyl groups of serine, threonine or tyrosine (43).These discrepancies concerning the non-basic amino acids may be due to differences in the staining procedures employed; the Heukeshoven and Dernick study stained homopolymers on polyacrylamide gel, Nielsen and Brown studied formation of silver-amino acid complexes in solution. Both of these studies used formaldehydye in an alkaline sodium carbonate solution for image development, while Merril and Pratt utilized acidic conditions and a combination of light, hydroquinone and formaldehyde for image formation (32).

Staining Of Peptides And Proteins With Known Sequences. The importance of the basic and the sulfur containing amino acids in the current staining protocol was corroborated by observations with purified peptides and proteins of known amino acid sequence. Leucine enkephalin, which has neither sulfur containing nor basic amino acids does not stain with silver, while neurotensin which also has no sulfur containing amino acids but does have three basic amino acid residues (one lysine and two arginines) does stain. Gastrin produced a weak staining reaction. It lacks basic amino acids but it has one sulfur containing amino acid, methionine. Oxytocin stains fairly vigorously. It also has no basic amino acids but it does have two sulfur containing cysteines. The staining reaction of angiotensin II was rather anomalous. It produced a negative stain rather than a positive silver stain despite its two basic amino acids, arginine and histidine. All the other polypeptides; insulin somatostatin, alpha-melanocyte stimulating hormone, thyrocalcitonin, aprotinin, vasoactive intestinal peptide and ACTH, contained both basic and sulfur containing amino acids and they all produce positive silver staining reactions (32).

The importance of the basic amino acids has been further substantiated by evaluations of the relationship between a denatured protein's amino acid mole percentages and its ability to stain with silver. The best correlations were achieved when a comparison was made between the slope of the linear portion of a denatured protein's staining curve and the protein's mole percentages of the basic amino acids, histidine and lysine (32). A similar correlation was observed by Dion and Pomenti(46). Dion and Pomenti suggested that this correlation may be due to an interaction between lysine and glutaraldehyde, which was used in their stain protocol. The bound glutaraldehyde could supply aldehyde groups to facilitate the reduction of ionic silver. While this mechanism may play a role in the stain protocol employed by Dion and Pomenti(46), it is unlikely to be a factor in the Merril and Pratt protocol, since that protocol does not use glutaraldehyde. Dion and Pomenti also suggested that alkaline conditions may be important for the formation of silver complexes with lysine and histidine. However the Merril and Pratt protocol utilizes acidic conditions. No significant correlations were found between a protein's amino acid mole percentages and its ability to stain with silver for native, undenatured, proteins(32). This lack of significant correlation is probably due to the inaccessibility of many of the potentially active amino acid side chains in the undenatured protein structures.

Reactivity Of Amino Groups The significant correlation of silver staining intensity to the mole percent of lysine is most likely due to the reactive "amino group" at the terminus of lysine's side chain. The amino group's metal binding is due to its strong electron-donor qualities and the ligand-field effect of its nitrogen atoms (43). In general, the lower the $pK_a$ of a potential metal binding group, the more likely it is to form a metal-ligand bond. Given this general "rule," one might predict the order of metal binding to be: carboxyl>imidazole>amino groups. However, acid dissociation criterion do not include the role of enthalpy and entropy changes which provide a measure of the relative thermodynamic stabilities of the complex. Reactive group properties, such as the group's electron donor ability and its ligand field effects, must dominate over the acid dissociation constants for the functional groups in the current protocol. This interpetation is supported by the lack of significant correlations between the stain's intensity and the mole percent of the carboxylated side chain amino acids, aspartic or glutamic acid. Furthermore, these carboxylated side chain amino acids do not stain with silver, either as individual amino acids or as homopolymers (32).

Amino groups involved in peptide bonding and N-terminal amino groups are in themselves insufficient for visualization with silver stain. If they were capable of independently reducing silver ions, all peptides, proteins, and amino acids would stain positively. However, the amino groups involved in peptide bonding and N-terminal atoms may be of some importance for the intensity of the stain, as these atoms have been observed to form 13 different complexes with copper between pH 1.5 to pH 11.0 (47). Bound copper may be reduced under the conditions some stain protocols and then be displaced by silver. Alternatively, silver may also interact directly, but weakly, with these groups.

Reactivity Of Histidine Groups. The contribution of histidine to silver staining, as demonstrated in the homopolymer studies and the correlation between the mole percent of histidine and silver staining intensity is not surprising, since the imidazole groups in the histidine side-chains are often important for metal-binding in metalloproteins. The effectiveness of histidine in metal binding is probably due to the fact that imidazole groups are good electron donors (43). The enthalpy changes in the formation of metal-nitrogen(imidazole) bonds are only slightly less than those found with metal-nitrogen(amino) bonds (48). The slightly lowered ability of the imidazole group, relative to the amino group, to donate electrons for the formation of metal complexes may be balanced by imidazole's lower pK. The lower imidazole group's pK, in contrast to the higher pK of an amino group, reduces the metal ion's competition with protons for the imidazole's nitrogen atom's electron-pair (43).

Reactivity Of Guanidine Groups The guanidine group in arginine's side chain proved to be less active than either the amino or the imidazole groups in the side chains of lysine and histidine, respectively. Arginine's correlation coefficient was not found to be significant in studies of comparing staining densities to mole percent of arginine. This lack of activity of the guanidine group may have been, in part, responsible for the negative staining reaction of the peptide angiotensin II which contains the two basic amino acids arginine and histidine (one residue of each). However, neurotensin, which contains two arginine residues and one lysine residue, stained fairly well. Cooperative metal binding effects between active groups may play a role in the staining process. In angiotensin II the arginine residue is separated from the histidine by three residues, while in neurotensin the two arginines are adjacent to each other and only one residue separates them from lysine.

Reactivity Of Nonpolar And Uncharged Polar Groups Of the nonpolar and uncharged polar amino acids, only the sulfur containing amino acids, methionine, cysteine and cystine, showed any silver staining reactivity with the Merril and Pratt Protocol(32). Cysteine and cystine were the only amino acids to stain as individual amino acids and they may account for the silver staining properties of the peptide oxytocin. Oxytocin contains no basic amino acids and its only sulfur containing amino acids are two cysteine residues. The ability of cysteinyl side-chains to form complexes with silver ions is well known. At the low pH utilized in this protocol, the predominant species is $Ag(HCys)_2^+$,(43). It has been suggested that the ability of reducing agents [including: thiosulfates, sulfides, borohydrides, cyanoborohydrides, mercaptoethanol, thioglycolic acid, cysteine, tributylphosphine reducing metal salts (such as $FeCl_2$,$SnCl_2$ and $TiCl_3$) and dithiothreitol] to intensify silver stains may be related to the generation of thiol groups in cysteine residues (49). However, proteins that contain no cysteine or proteins with an alkylated cysteine(s) were also affected by these reducing agents in some stain protocols (44).

Methionine's ability to participate in the silver staining process has been demonstrated by silver staining of methionine homopolymers. Methionine may also be responsible for the staining of the peptide gastrin. Gastrin contains no basic amino acids and only one methionine residue. In general the thioether sulfur atoms in the methionine residues are weaker electron donors than the sulfhydryl sulfur atoms in the cystiene residues. The only metal ions that have been observed to bind to the thioether's sulfur atoms are those with electrons in the $d^8$ and $d^{10}$ configurations ($Pd^{++}$, $Pt^{++}$, $Ag^+$, $Cu^+$, and $Hg^{++}$). The affinity of sulfur ligands for metal ions may be explained by the highly polarized state sulfur atoms achieve during interactions with small metal ions containing high charge densities. Sulfur's electron distributions and energies enhance the enthalpies of metal ion bonding (they have high crystal field stabilization energies). There may also be electron resonance bonding in the metal-sulfur bond (43). Insignificant staining correlations were observed between staining densities and mole percentages of the sulfur containing amino acids methionine and cysteine(32). This observation may indicate a relatively minor silver staining role in proteins containing large numbers of basic amino acids. However, this poor staining correlation is somewhat of a paradox since poly-methionine stained with a higher silver density than the basic amino acid homopolymers (32). This paradox may be explained by a strong requirment for cooperative effects between sulfur atoms and silver atoms which is disrupted in heteropolymers.

Comparison Of Silver Stain Reactive Groups With Commassie Stain Reactive Groups Recent studies concerning the mechanisms of Coomassie dye staining of proteins have indicated a similar importance for the basic amino acids. Righetti and Chillemi (50) noted that polypeptides rich in lysine and arginine were aggregated by Coomassie G dye molecules, suggesting that the dye interacts with the basic groups in the polypeptides. Studies of proteins with known sequences, by Tal et al.,confirmed these observations and demonstrated a significant correlation between the intensity of Coomassie blue staining and the number of lysine, histidine and arginine reidues in the protein (51).

Properties Of Silver Stains

Color Effects Most proteins stain with monochromatic brown or black colors. However, Goldman et al.(1980) noted that certain lipoproteins tend to stain blue while some glycoproteins appear yellow, brown or red in a study of cerebrospinal fluid proteins (52). This color effect is most likely an analogue of a photographic phenomena first described by Herschel in 1840 (25, 53). Herschel noted in 1840, that if he projected the spectrum of visible light obtained by passing sunlight through a prism onto a silver chloride-impregnated paper, the colors of the spectrum appeared on the paper, particularly a "full and fiery red" at the focal point of the red light (53). Since these observations by Herschel it has been found that colored images may be obtained if the particles of metallic silver are small in comparison to the wavelength of light. The color produced depends on three variables: the size of the silver particles, the refractive index of the photographic emusion or electrophoretic gel, and the distribution of the silver particles. In general, studies with photographic emulsions have shown that smaller grains (less than 0.2 microns in diameter) transmit reddish or yellow-red light, while grains above 0.3 microns give bluish colors, and larger grains produce black images (35). Modifications of the silver staining procedures, such as lowering the concentration of reducing agent in the image development solution, prolonging the development time, adding alkali, or elevating the temperature during staining will often enhance color formation. Some silver stain protocols have been developed to produce colors that may aid in identification of certain proteins (45,54,55). Production of color with silver stain depends on many variables. Nielsen and Brown (1984) have shown that charged amino acid side groups play a major role in color formation (45). However, variations in protein concentration and conditions of image development may also produce color shifts, confusing identification. Furthermore, color-producing silver stains tend to become saturated at low protein levels and often produce negatively stained bands or spots. These factors tend to make quantitative analysis more difficult.

Specificity Silver stains can demonstrate considerable specificity. Hubbell et al. stained nucleolar proteins with a histological stain (56), while Gambetti et al. adapted a silver stain specific for neurofilament polypeptides (21). Many silver stain protocols detect not only proteins but also DNA, (57-59), lipopolysaccharides (60), and polysaccharides (61). In a study of erythrocyte membrane proteins, sialoglycoproteins and lipids stained yellow with a silver stain protocol, while the other membrane proteins counterstained with Coomassie Blue (62). All silver-stains do not detect proteins such as calmodulin or troponin C. However, pretreatment with gluteraldehyde often permits positive silver staining of these proteins (63). Histones may also fail to stain with silver. Fixation with formaldehyde coupled with simultaneous prestaining with Coomassie Blue partly alleviates this problem. However, even with this fixation procedure sensitivity for histones is decreased 10-fold compared with detection of neutral proteins (64). Another example of differential sensitivity was demonstrated in a study utilizing four different silver stain protocols to stain salivary proteins. Different protein bands were visualized with each of the stains(65).

Quenching Of Autoradiography Quenching of $^{14}C$-labelled proteins is minimal with most of non-diamine silver stains and even the most intense diamine stained, $^{14}C$-labelled proteins can be detected by autoradiography with only a 50% decrease in image density. This loss of autoradiographic sensitivity can generally be compensated for by longer film exposures. However, detection of $^{3}H$-labelled proteins is severely quenched by all silver stains. Destaining of the silver stained gel with photographic reducing agents can often permit detection of as much as half of the fluorographic density of $^{3}H$-labelled proteins, providing that the initial staining was performed with a non-diamine silver stain. Many diamine stains continue to quench, even after treatment with photographic reducing agents, so that fluorographic detection of $^{3}H$-labelled proteins is not feasable. This impediment to $^{3}H$ detection with diamine stains is likely to be due to a greater amount of residual silver deposited in the gels by the diamine stains, which block the weak-beta emissions from $^{3}H$. Residual silver has been demonstrated in gels that have been cleared by photographic reducing agents by the faint silver image of the protein can be observed after drying the gel with heat. Silver has also been demonstrated in these "cleared" gels by electron beam analysis (66).

Sensitivity Silver stains currently offer the most sensitive non-radioactive method for detecting proteins separated by gel electrophoresis. They are 100-fold more sensitive than the Coomassie stains for most proteins (15-16). Chemical-development silver stains are in general, more sensitive than photo-development silver stains. This loss in senstivity may be compensated for by the ability of photo-development stain to produce an image within 10 to 15 minutes after gel electrophoresis (31). Unfortunately, photodevelopment often produces negatively stained protein bands. Many chemical-development silver stains, which have been modified to enhance color, also display this effect. The presence of negative and positive regions, in photo-developed or color-enhanced silver stained gels makes quantitative analysis difficult (30-31).

Intensification Sensitivity can often be increased by recycling the electrophoretogram through the silver staining procedure. Such recycling often permits visualization of trace proteins that otherwise might not have been detected (24). Image-intensification methods developed for photography may also be used to intensify silver stained gels. Such intensifiers may add more silver to the existing deposits in the image, as in the recycling procedure, or they use other dense metals. Silver intensifying procedures usually exhibit an increase in density that is proportional to the original amount of silver in the gel. Copper, mercury, copper iodide and mercuric chloride intensifiers generally increase contrast by adding additional metal to the heaviest original silver deposits (these are superproportional intensifiers). Intensifiers which preferentially build up the less dense regions, uranium, mercuric iodide and chromium, are subproportional (67).

Destaining Silver stained gels may also be destained in a proportional, subproportional or superproportional manner, with some of the silver stains being more resistant to destaining than others. Farmer's reducer is a subproportional reducer and, if allowed, will remove all the silver from the lighter regions of the gel. Farmer's reducer, a photographic reducing agent, utilizes ferricyanide as the silver solvent and thiosulfate to complex and solubilize silver salts (67). The first photographic reducer used to destain silver-stained polyacrylamide gels employed ammonium hydroxide, copper sulfate and sodium thiosulfate (16). It is also a subproportional reducer. The diamine stains are most resistant, while the photodevelopment stains are most sensitive to destaining.

If quantitation is intended, only proportional processes should be employed. Given the fine balance required to achieve proportional destaining or intensification, caution should be employed in quantitatively analysing gels which have been manipulated by these techniques.

Artifacts Attaining high sensitivities with silver stains requires care in selecting reagents, small traces of contaminants may cause a loss of sensitivity and result in staining artifacts. Artifactual bands with molecular weights ranging from 50 to 68 kiloDaltons have commonly been observed in silver stained gels. Evidence has been presented indicating that these contaminating bands are due to keratin skin proteins (68). The presence of these bands indicates that samples, solution and equiptment must be handled carefully to minimize artifactual bands or spots. The high sensitivity obtained with the silver stains also increases the need to be on guard against bacterial and fungal contamination. Water used to make solutions should have a conductivity of less than 1 mho.

Quantitation With Silver Stains

Relation of Stain Density to Protein Concentration A reproducible relationship between silver stain density and protein concentration has been found with most silver stain protocols. The linear portion of this relationship extended over a 40-fold range in concentration, beginning at 0.02 nanograms per $mm^2$ for most proteins (16,24-25,30). Protein concentrations greater than 2 ng/$mm^2$ generally cause saturation of silver images, resulting in non-linearity above that concentration. Saturation can usually be recognized by bands or spot with centers which are less intensely stained than the regions near the edges. This effect is similar to the "ring-dyeing" noted with some of the organic stains. An often quoted report by Poehling and Neuhoff (69) states that "Silver does not stoichiometrically stain proteins, unlike Coomassie Blue". However, their silver-stain data actually is linear over a 30-fold range in protein concentration, while their Coomassie Blue data is only linearity over a 20-fold range (17,24).

Curve-fitting techniques, such as those described by Coakley and James (70), may be employed for the analysis of the relationship between silver stain densities and protein concentrations. Coakley and James developed these techniques to examine the similar curvilinear relationships which are found in the Folin-Lowry method of protein estimation (71). With careful measurement of total stain densities, estimates of relative protein concentrations have been made over a 220 fold concentration range with six purified proteins(32).

Protein Specific Staining Curves Plots of silver stain densities versus protein concentrations produce different staining curves for each proteins studied (16,24,25,32). Protein specific staining curves have also been observed with the organic stains, including Coomassie Blue (51,72) and with most protein assays such as the commonly used Lowry protein assay (71). These curves are governed by the basic mechanisms underlying the detection and assay methods. The fact that each protein produces a unique density verses concentration curve in these studies, illustates a dependence on specific reactive groups contained in each protein. Furthermore, the occurance of protein-specific curves argues against a stain mechanism that depends on some fundamental subunit common to all proteins, for example the peptide bond, or a unique element in each protein, such as the terminal amino acid. A stain that depended on a subunit, such as the peptide bond, would result in similar staining curves for all proteins, when the density of staining for each of the protein bands or spots was plotted against the mass of protein contained in each of the bands or spots. Similarily, a stain that was based on a reaction with a unique element in each protein, for example the terminal amino group, would produce similar plots for each protein when the stains densities were plotted against the number of molecules contained in each band or spot. It is possible that these protein-specific curves may be utilized to differentiate proteins and to provide insights concerning the reactive groups responsible for the staining reactions. The importance of the basic amino acids, particularly lysine and histidine as discussed in the section "Staining of Peptides and Proteins with Known Sequences" illustrates the use of these proteins specific staining curves. It indicates the need for a careful choice of a "standard protein(s)" if this stain is used quantitatively to estimate protein consentrations. A protein containing an abnormally large number of stain reactive groups would produce a curve which would tend to underestimate the concentration of proteins containing normal numbers of reactive groups. A similar correlation between the intensity of Commassie Blue staining and the number of basic amino acids in proteins (50-51) caused Tal et al. to suggest the use of egg white lysozyme rather than the more commonly used bovine serum albumin as a protein standard. This suggestion is based on their observation that the basic amino acid content of proteins ranges between 10-17 mole percent, with a modal content of 13 mole percent (51). Egg white lysozyme has a basic amino acid mole percent of 13.2 while bovine serum albumin has a basic amino acid content of 16.5 mole percent. For similar reasons, egg white lysozyme may also prove to be an optimal standard for quantitative silver stain applications.

Quantitative Inter-gel Protein Comparisons The occurance of Protein specific staining curves with silver staining requires that quantitative inter-gel comparative studies limit comparisons to homologous protein bands or spots on each gel. For example, the actin spot on one gel can be compared with an actin spot on another gel, but not with a transferrin spot. These limitations to homologous comparisons are also applicable to most of the organic stains, including the Coomassie Blue stains, (72).

Quantitative inter-gel comparisons requires the presence of reference proteins for the normalization of spot or band staining densities. One scheme for normalization utilizes "operationally constitutive proteins", a subset of proteins contained in each gel that have constant intra-gel density ratios to each other in all of the gels in a study. The sum of the densities of the "operationally constitutive proteins" in an arbitrarily designated "standard gel" are compared with the sums of the densities of the constitutive proteins in all other gels, and a specific normalization factor is determined for each gel. These gel specific normalization factors are then utilized to correct the densities of all the proteins on each of the gels to those of the standard gel. This scheme corrects for variations in staining, in image digitization, and initial protein loading; a variation of initial protein loading of up to 10-fold may be tolerated (24,30).

## Literature Cited


1. Porrett, R. Ann. of Phil. 1816, July, 78-83.
2. Abramson, H.A. "Electrokinetic Phenomena and Their Applications To Biology and Medicine"; The Chemical Catalog Co. Inc: New York, 1934; pp 17-104.
3. Picton, H.; Linder, S.E. J. Chem. Soc. 1892, 61, 148-172.
4. Davis, B.D.; Cohn, E.J. Ann. N. Y. Acad. Sci. 1939, 39, 209-212.
5. Tiselius, A. Trans. Faraday Soc. 1937, 33, 524-531.
6. Durrum, E.L.J. Am. Chem. Soc. 1950, 72, 2943-2948.
7. Grassman, W.; Hannig, K. Z. Physiol. Chem. 1952, 290, 1-27.
8. Durrum, E.L.; Paul, M.H.; Smith, E.R.B. Science 1952, 116, 428-430.
9. Fazakas de St. Groth, S.; Webster, R.G.; Datyner, A. Biochem. Biophys. Acta 1963, 71, 377-391.

10. Talbot, D.N.; Yaphantis, D.A. Anal. Biochem. 1971, 44, 246-253.
11. Barger, B.O.; White, F.C.; Pace, J.L.; Kemper, D.L.; Ragland, W.L. Anal. Biochem. 1976, 70, 327-335.
12. Bosshard, H.F.; Datyner, A. Anal. Biochem. 1977, 82, 327-333.
13. Becquerel, A.H. Comp. Rend. Acad. Sci. (Paris) 1896 122, 420-421.
14. Wilson, A.T. Nature 1958, 182, 524.
15. Merril, C.R.; Switzer R.C.; Van Keuren M.L. Proc. Natl. Acad. Sci. USA 1979, 76, 4335-4339.
16. Switzer R.C.; Merril C.R.,; Shifrin S. Anal. Biochem. 1979, 98, 231-237.
17. Dunn, M.J.; Burghes, A.H.M. Electrophoresis 1983, 4, 173-189.
18. Bielschowsky, M. J. Psychol. Neurol. 1904, 3, 169-189.
19. Nauta, W.J.H.; Gygax P.A. Stain Technology 1951, 26, 5-11.
20. Allen, R.C. Electrophoresis 1980, 1, 32-37.
21. Gambetti, P.; Autilio-Gambetti, L.; Papasozomenos S.C.H. Science 1981, 213, 1521-1522.
22. Merril, C.R.; Dunau, M.L.; Goldman, D. Anal. Biochem. 1981, 110, 201-207.
23. Merril, C.R.; Goldman, D.; Sedman, S.A.; Ebert, M.H. Science 1981, 211, 1437-1438.
24. Merril, C.R.; Goldman, D.; Van Keuren, M.L. Electrophoresis 1982, 3, 17-23.
25. Merril, C.R.; Goldman, D. In "Two-Dimensional Gel Electrophoresis of Proteins"; Celis, J.E.; Bravo, R., Eds.; Academic Press: New York, 1984; pp. 93-109.
26. Morrisey, J.H. Anal. Biochem. 1981, 117, 307-310.
27. Ehrenfried, G. Photogr. Sci. Tech. 1952, 18B, 2-6.
28. Eder, J.M. "History of Photography" Translated by Eptean, E., 1945; Columbia University Press: New York, 1932; pp. 96-99.
29. Newhall, B. "Latent Image, the discovery of Photography"; University of New Mexico Press: Albuquerque, New Mexico, 1983; pp. 8-17; 117-118.
30. Merril, C.R.; Harrington, M.G. Clin. Chem. 1984, 30, 1938-1942.
31. Merril, C.R.; Harrington, M.; Alley, V. Electrophoresis 1984, 5, 289-297.
32. Merril, C.R.; Pratt, M.E. Anal. Biochem., In Press.
33. Sheinin, J.J.; Davenport, H.A. Stain Technology, 1931, 6, 131-148.
34. Reilly, J.M. "The Albumin and Salted Paper Book"; Light Impressions Corporation: Rochester, New York, 1980; p. 4.
35. Mees, C.E.K., "The Theory of the Photographic Process, 1st edition";The MacMillan Company: New York, 1952; p. 305 and pp. 563-583.
36. Cajal, S.R. Trab. Lab. Invest. Biol., Univ. Madrid 1903, 2, 129-222.
37. Cajal, S.R. Trab. Lab. Invest. Biol., Univ. Madrid 1904, 3, 1-7.
38. Heimer,L. In "Contemporary Research Methods in Neuroanatomy"; Nauta, W.J.H.; Ebbesson, S.O.E., Eds.; Springer-Verlag: Berlin. 1970; pp. 106-107.
39. Ansorge, W. In: "Electrophoresis '82"; Stathakos, D., Ed.; de Gruyter: Berlin, 1983; pp. 235-242.
40. Thompson, S.W.; Hunt, R.D. "Selected Histochemical and Histopathological Methods"; Thomas: Springfield, Illinois, 1966; pp. 798-802.

41. Hamilton, S.F. In: "The Theory of the Photographic Process, 3rd edition"; James, T.H., Ed.; MacMillian Publishing Co.: New York, 1977; pp. 105-132.
42. Wolfenden, R. Science 1983, 222, 1087-1093.
43. Freeman, H.C. In: "Inorganic Biochemistry, Vol. 1"; Eichhorn, G.L., Ed.; Elsevier: Amsterdam/New York, 1973; pp.121-166.
44. Heukeshoven,J.; Dernick,R. Electrophoresis, 1985, 6, 103-112.
45. Nielsen, B.L.; Brown, L.R. Anal. Biochem. 1984, 144, 311-315.
46. Dion, A.S.; Pomenti, A.A. Anal. Biochem. 1983, 129, 490-496.
47. Osterberg, R.; Sjoberg,B. J. Biol. Chem., 1968, 243, 3038-3042.
48. Meyer,J.L.; Bauman, J.E. J. Am. Chem. Soc., 1970, 92, 4210-4215.
49. Morrissey,J.H. Anal.Biochem. 1981,117, 307-310.
50. Righetti,P.G.; Chillemi,F. J. Chromatogr., 1978, 157, 243-251.
51. Tal, M.; Silberstein, A.,; Nusser, E. J. Biol. Chem., 1985, 260, 9976-9980.
52. Goldman, D.; Merril, C.R.; Ebert, M.H. Clin. Chem. 1980, 26, 1317-1322.
53. Herschel, J.F.W. Phil. Trans. Roy. Soc., London,1840, 131, 1-59.
54. Sammons, D.W.; Adams, L.D.; Nishizawa, E.E. Electrophoresis 1981, 2, 135-141.
55. Sammons, D.W.; Adams, L.D.; Vidmar, T.J.; Hatfield, A.; Jones, D.H.; Chuba, P.J.; Crooks, S.W., In "Two-Dimensional Gel Electrophoresis of Proteins"; Celis, J.E.; Bravo, R., Eds.; Academic Press: New York, 1984; pp. 112-127.
56. Hubbell, H.R.; Rothblum, L.I.; Hsu, T.C. Cell. Biol. Int. Rep. 1979, 3, 615-622.
57. Somerville, L.L.; Wang, K. Biochem. Biophys. Res. Commun. 1981, 10, 53-58.
58. Boulikas, T.; Hancock, R.J. Biochem, Biophys. Methods 1981, 5, 219-222.
59. Goldman, D.; Merril, C.R. Electrophoresis 1982, 3, 24-26.
60. Tsai, C.M.; Frasch, C.E. Anal. Biochem. 1982, 119, 115-119.
61. Dubray, G.; Bezard, G. Anal. Biochem. 1982, 119, 325-329.
62. Dzandu, J.K.; Deh, M.H.; Barratt, D.L.; Wise, G.E. Proc. Natl.Acad. Sci. USA 1984, 81, 1733-1737.
63. Schleicher, M.; Watterson, D.M. Anal. Biochem. 1983, 131, 312-317.
64. Irie, S.; Sezaki, M. Anal. Biochem. 1983, 134, 471-478.
65. Friedman, R.D. Anal. Biochem. 1982, 126, 346-349.
66. Van Keuren, M.L.; Goldman, D.; Merril, C.R. Anal. Biochem. 1981, 116, 248-255.
67. Wall, E.J.; Jordan, F.I. In "Photographic Facts and Formulas"; Carrol, J.S., Ed. American Photographic Book Publishing Co.: New York; 1976, pp. 168-189.
68. Ochs, D. Anal. Biochem. 1983, 135, 470-474.
69. Poehling, H.M.; Neuhoff, V. Electrophoresis 1981, 2, 141-147
70. Coakley, W.T.; James, C.J. Anal. Biochem. 1978, 85, 90-97.
71. Lowry, O.H.; Rosenbrough, N.J.; Farr, A.L.; Randall, R.J., J. Biol. Chem. 1951, 193, 265-275.
72. Fazekas de St.Groth, S.; Webster, R.G.; Datyner, A. Biochim. Biophys. Acta, 1963, 71, 377-391.


RECEIVED July 24, 1986

# Chapter 6

# Color Silver Staining of Polypeptides in Polyacrylamide Gels

**David W. Sammons[1] and Lonnie Adams[2]**

**[1]Center for Separation Science, University of Arizona, Tucson, AZ 85721**
**[2]The Upjohn Company, Kalamazoo, MI 49001**

Silver staining of polypeptides in polyacrylamide gels is one of the methods that can be used in detection of proteins by analytical gel electrophoresis. There are three primary methods commonly used to reduce ionic silver to metallic silver. The ammoniacial method for SDS gels (1) is an adaptation of a histochemical stain for tissue sections in light microscopy (2) and utilizes an acidic environment of citric acid and formalin. The second method is more stable, uses less silver, and utilizes a weak base, sodium carbonate, and formalin together (3). The third method also uses less silver, is the most sensitive of the methods, (4) results in polychromatic staining of proteins, and uses a strong base environment of sodium hydroxide and formalin followed by the weak base sodium carbonate (5,6). The third method is commonly referred to as the GELCODE (Registered trademark, Health Products, Inc.) Color Silver Stain and is the subject of this article.

The GELCODE silver stained proteins are found in four color classes of blue, brown - black, red and yellow. Shades of colors have also been reported with the ammoniacal silver stain (7) and with the weak basic silver nitrate methods (8). Color provides a third dimension (9) to the analysis of proteins by 2-D electrophoresis. Comparison of the method to the other two basic procedures for silver reduction has shown that the GELCODE stain is superior in reliability and reproducibility of the color staining (10).

Color staining of the gel with GELCODE reveals protein patterns which are easily interpreted for qualitative analysis. The color aids in the analysis by distinguishing overlapping spots or bands which inherently have similar or identical isoelectric points or molecular weights. Quantitative analysis of color silver stained gel

0097-6156/87/0335-0091$06.00/0

protein patterns have been reported to be unreliable by the other methods (11,12); however, GELCODE silver stain reacts in a stochiometric manner with proteins of the various color classes (9). Although scanning systems are not currently commerically available, it is likely that data acquisition systems can be configured that will utilize the color information to calculate the absolute protein concentration in protein spots and/or bands of different color classes.

## The Method

The method is designed for use in 10% to 20% polyacrylamide SDS gels and it is necessary to remove the SDS and buffer salts such as glycine and Tris before staining. Generally, these are removed by agitating the submersed gel in the staining solution contained in a tray on a reciprocating laboratory shaker. To effect proper fixation of the protein and removal of the bound SDS, the polyacrylamide gel is agitated in a solution of 50% ethanol and 5% glacial acidic acid. The SDS is primarily removed in the first fixative solution and the rest removed in several subsequent water washes. Fixation is somewhat of a misnomer for this step since some polypeptides may diffuse from the gel and are lost. The rate of loss is dependent on the solubility in acid or alcohol solutions and the amount of cross linking in the polyacrylamide gel. It has been shown that alcohol, acid, and fixation times are critical variables affecting staining sensitivity (9). Protein loss is less in the ethanol acetic acid than the commonly recommended methanol acetic acid and some researchers have suggested that trichloroacetic acid is superior to the acetic alcohols (13,14). Another method of fixation that has been effectively used with GELCODE for the fixation of acid soluble peptides is gluteraldehyde (15). Reaction of gluteraldehye with protein in gels results in increased sensitivity for most proteins and without loss of its characteristic color.

Equilibration of silver nitrate into the gel is the first step of GELCODE silver staining. It is accomplished by soaking the gel in sufficient volume of silver nitrate for it to be adequately submersed and for an adequate time for equilibration throughout the gel.

The second step is a brief rinse with water. This rapid rinse of about 10 seconds removes the excess silver nitrate from the immediate surface of the gel and from the wall and sides of the tray or container.

The third step is a critical one and is responsible for the reduction of silver nitrate to colloidal silver. Formaldehyde is mixed with the strong base, sodium hydroxide, and the gel is floated within the solution while agitating. As the silver nitrate diffuses from the gel the reducing reactants diffuse into the gel, usually diffusing from two directions. A double layer of silver is deposited as shown in Figure 1. The colloidal silver, deposited laterally throughout the protein precipitate, has a characteristic of being deposited internally and not on the surface of the gel. It has been reported that some proteins fail to attract colloidal silver and thereby result in negative staining. This phenomenon also occurs with GELCODE, however because of the amber background the negatively stained protein appears as a yellow spot and therefore can be detected.

The fourth and final step involves enhancement of the sensitivity by placing the gel in a solution of carbonate. Although it has

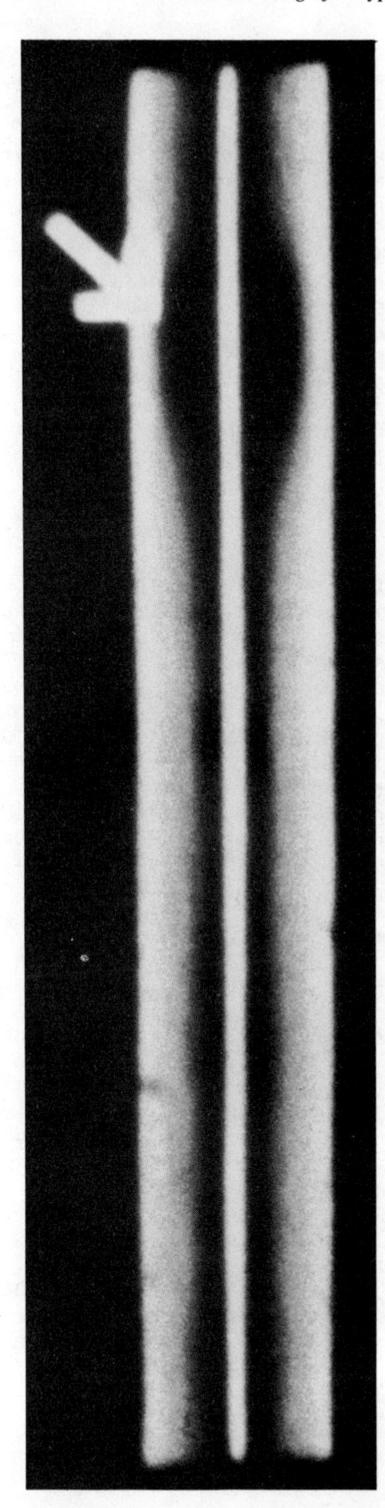

Figure 1. Cross section of protein stained with GELCODE. A 2-D protein pattern of a 1.5 mm gel was stained and a spot was cut through its center. The gel cross section was placed on its side and viewed on a light box.

not been demonstrated experimentally, it is surmised that a silver carbonate complex is formed with the collodial particles that are already formed within protein precipitate within the gel, thereby increasing the grain density within the protein precipitate. The carbonate serves two other important functions. The dilute carbonate solution neutralizes the basic environment of the gel and diminishes the swelling which is caused by the sodium hydroxide hydrolysis of the bis-acrylamide crosslinkages. Finally, the excess carbonate stabilizes the collodial particle on the protein precipitate in a manner similar to the electrolyte stabilization of AgBr by KBr (16). The silver stain development ceases as the sites are saturated and the reactants are depleted from the gel by diffusion. Overstaining, so characteristic of the other silver stain methods, is not normally observed since a final stable endpoint is reached by exhaustion of reactants. Whenever an unacceptable dark green or black background is observed impure water or chemicals, improper reagent concentrations or incorrect reaction times have been used.

Possible Mechanism

It is known that coloration of a given protein varies depending on the basic method of staining i.e. citric acid, weak carbonate, or in the strong base sodium hydroxide thereby supporting the idea that the environment of the staining reaction must also be considered in the evaluation of the mechanism of color silver stains. Proteins are negatively charged in the strong basic environment of the GELCODE reduction step. The negative surface charge on the protein precipitate allows binding of the colloidal silver and results in nuclei formation. Addition of the gel to a solution of sodium carbonate would therefore cause coagulation of nuclei and growth of particle size due to formation of protein-silver-carbonate complexes. Early work with colloidal systems showed that the initial size of the colloidal particles are small during the nucleation stage and the nuclei grow through reaction in a supersaturated solution and by aggregation of the small nuclei by the process of coagulation (16). The mechanism of GELCODE silver staining may be analogous to this process, however, the reader should be reminded that this is a working hypothesis and more data are needed.

At the time of initial nucleation most spots appear yellow to brown. As the reaction continues in the carbonate solution some proteins stain red or blue, or black while others remain yellow or brown. It is generally accepted that the smaller silver colloidal particle (approximately $\leq$ .2 um) are seen as yellow - red coloration and that larger particles (above $\geq$ .3um) are blue to black in coloration. The mechanism of different coloration of proteins may be dependent on protein sequence or structure. Thus coalescence of differing amounts of silver colloidal particles onto precipitated proteins is at least one plausible explanation of the colors seen in the gels. Therefore, colors observed in the pro tein pattern are influenced by the chemical environment as well as the charge and steric characteristics of the precipitated protein.

## Uses of Color

Color furnishes a third dimension to the electrophoresis analysis of proteins even without knowing the mechanism. In fact, color of the GELCODE stained protein can be used to characterize a polypeptide much the same way as molecular weight and isoelectric points are used. In a two dimensional gel, isoelectric point and molecular weight are inferred by measuring the x and y coordinate positions and extrapolating their apparent size and charge characteristics from an appropriate standard reference curve. The color of the stained protein can be determined by direct vision or automatically by computer controlled scanners and multispectral image analysis.

Utilization of the color parameter enables one to confirm color assignments of unknown proteins separated in gels. Table I lists the visual assignment colors to known proteins stained with GELCODE. These were run on an SDS gel and visually scored by viewing on a fluoresent light box with a color temperature of 5000 degrees Kelvin. This survey represents only a fraction of the purified proteins that are commercially available and even less of the total proteins that have been isolated by scientists around the world. The color data in the survey of Table I is a representative sampling of the color classes expected when the other proteins are characterized. Use of a color assignment eliminates some of the ambiguities which occur when isoelectric points and molecular weights are used to characterize and identify proteins in 2-D electrophoresis gels.

Figure 2 illustrates the potential of using multispectral image analysis to assign color classes and to confirm the observation that proteins remain the same color throughout the linear range of the Integrated Intensity vs protein concentration. Multivariate color discrimination analysis of the spectral signatures of spots from the linear range (see Figure 3) of three replicate 2-D gel was performed. Each spot was assigned a color by visual inspection and confirmed on the color monitor. The data was collected by scanning each gel with a different filter, a blue, green, and red, as described by Vincent et.al. (17). The spectral signatures of spots from each assigned color class were grouped and ratios of the values were utilized in the calculation of the color variable 1 and variable 2. Plots of the two variables gave almost perfect separation of the color classes; however, two spots designated by the asterisk in Figure 2 did fall outside their expected color class. After manual reexamination of the original gel images, it was seen that the misclassification occurred because of local discolorations of the gel backgrounds, hence giving false spectral values. Black spots were not included in this study, however Adams (18) has shown that black spots fall within the brown class. He also independently has shown that GELCODE stained spots can be classified into four color classes.

## Color and Quantitation

GELCODE stained proteins in polyacrylamide gels can be quantitated because there is a linear relationship between concentration and integrated intensity. However, the slope of a line formed by plotting the intensity versus the concentration becomes nonlinear as the concentration increases beyond a certain concentration. The concentration at which the slope deviates from linearity varies depending upon

Table I Mapping of Polypeptides by Color

Proteins were obtained from various vendors. Stock solutions of 1 mg/ml were freshly prepared in 2% SDS + 2% B-mercaptoethanol. Appropriate dilutions of the stock were made and then diluted 1:1 with a 1.5% agarose buffer composed of 27mM Tris adjusted to pH 8.0 with HCl and 0.1% SDS. The protein-agarose mixture was heated to 95$^{o}$ for 5 minutes and drawn into a capillary tube with an internal diameter of 1.5 mm. After the agarose solidified it was extruded and a piece was run on a one dimension SDS 10-20% gradient gel as previously described.

| Polypeptide | Source | Color |
|---|---|---|
| Acetylcholinesterase | Electrophorus eel | Brown |
| Actin | Rabbit | Black |
| Actomysin | Rabbit | Brown |
| N-acyl-Neuraminic Acid | Bovine | Brown |
| Adenosine Deaminase | Bovine | Red |
| Albumin | Rat | Brown |
| Alcohol Dehydrogenase | Horse | Black |
| Aprotinin | Beef | Brown |
| (beta)-Amylase | Sweet potato | Brown |
| Amylglucosidase | A. niger G4 | Yellow |
| Alpha-1-Antitrypsin | Human | Red |
| Carbonic Anhydrase | Bovine | Brown |
| Citrate Synthase | Porcine | Blue |
| Citrate Synthase | Pigeon | Black |
| Creatine Phosphokinase | Rabbit | Blue |
| Cytochrome C | Horse | Brown |
| Diaphorase | Porcine | Brown |
| Enolase | Rabbit | Red |
| Fructose-1, 6-diphosphatase | Rabbit | Black |
| Fumarase | Porcine | Brown |
| Glutamate Dehydrogenase | Beef | Brown |
| Glucose-6-phosphate Dehydro | Beef | Brown |
| Glutathion Reductase | Yeast | Red |
| Glycerinaldehyde-3-phosphate Dehydrogenase | Rabbit | Black |
| Glycogen Synthase | Rabbit | Black |
| Growth Hormone | Bovine | Blue |
| Haptoglobin Beta Chain | Rat | Red |
| Hemoglobin | Beef | Blue |
| Hexokinase | Yeast | Brown |
| Histone H4 | Bovine | Blue |
| (beta)-Hydroxyacyl-COA-Dehydrogenase | Porcine | Blue |
| Immunoglobulin light Chain | Rat | Yellow |
| Isocitrate Dehydrogenase | Porcine | Black |
| (alpha)-Lactoalbumin | Bovine | Black |
| Lactic Dehydrogenase | Porcine M4 | Black |
| Myokinase | Rabbit muscle | Yellow |
| Myosin | Rabbit | Red |

Table I. Continued

| Polypeptide | Source | Color |
|---|---|---|
| NAD-Pyrophosphorylase | Porcine | Brown |
| Neuraminidase | Clostridium | Black |
| Phosphodiesterase | Beef | Brown |
| Phosphoglucose Isomerase | Rabbit | Black |
| Phosphoglucose Isomerase | Yeast | Red |
| Phosphoglucomutase | Rabbit | Black |
| 3-phosphoglycerate Kinase | Yeast | Brown |
| Phosphorylase a | Rabbit | Brown |
| Phosphorylase b | Rabbit | Brown |
| Prolidase | Porcine | Black |
| Pyrophosphotase | Yeast | Brown |
| Pyruvate Decarboxylase | Yeast | Brown |
| Pyruvate Kinase | Rabbit | Red |
| RNA Polymerase Subunits | E. coli | |
| 185K | " | Red |
| 155K | " | Black |
| 57K | " | Black |
| 38K | " | Brown |
| Transaldolase | Yeast | Red |
| Transferrin | Human | Red |
| Triosephosphate Isomerase | Rabbit | Brown |

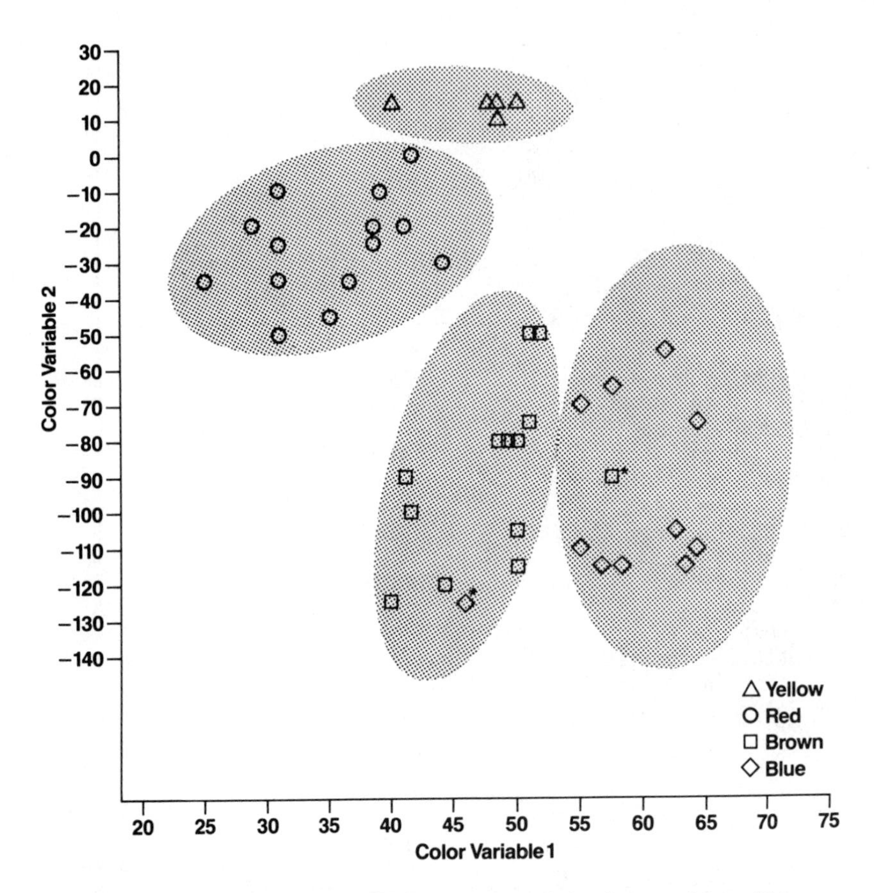


Figure 2. Grouping of proteins into color classes by multispectral image analysis. A liver extract was made in iso-urea sample buffer (20) and serial dilutions made from 1/16 to 1/1024. Gels were run on each sample dilution and the linear range was separately determined for blue, brown, red and yellow spots (see Figure 3). Spots from each color class were selected from the linear range and the intensity level from blue, green and red channels of a multispectral scan were used to calculate ratios as previously described (17). The ratios for each spot within a color class were plotted.

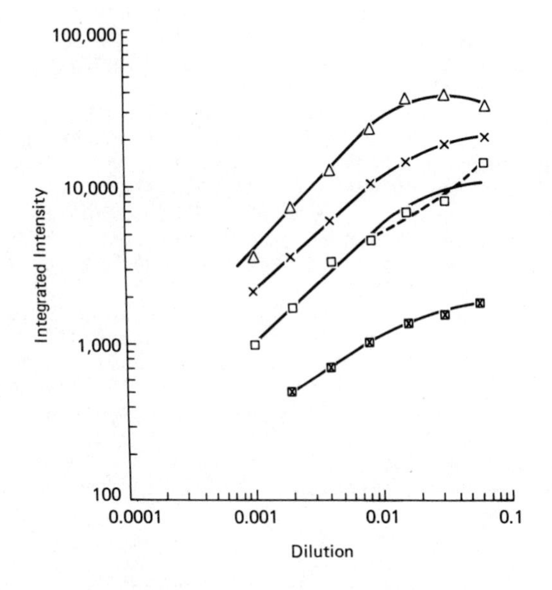


Figure 3. Linear range and intensity saturation for each color class. Integrated intensity values described in Figure 2 were plotted versus dilutions.

the color class of the protein. For example, the linear range of stained proteins are 100, 85, 78, and 4 fold for blue, brown, red and yellow respectively (18). The order of decreasing slopes of the plots are blue, brown, red, yellow, and it naturally follows that the order of decreasing plateau values are blue, brown, red and yellow (Fig. 3).

The color classification fails to hold as the linear region of the curve is exceeded (9). For example, glutamate dehydrogenase is brown in linear range and red in plateau region. Hemoglobin is blue in the linear range and is red as the concentration in the gel reaches a certain limit. Thus, as illustrated by these examples, erroneous color judgement can be made if concentrations of the proteins exceed the linear range of the stain. This limitation of saturation is true for all of the silver staining methods.

It has been repeatedly reported that silver stained methods are not suitable for computerized quantitation because of their capriciousness and nonlinearity. This is apparently true of the stains based on reduction with citric acid and weak carbonate because there is no predicting the slope of a plot of integrated intensity versus protein concentration without the use of a reliable and reproducible color. Thus, in order to use quantitation with these two methods one must perform a standard curve with each protein as its own standard or accept some relative standard for normalization. The relative approach has been successfully used with GELCODE and allowed measurement of protein changes within an experimental protocol (19). These disadvantages have discouraged the acceptance of silver staining to its full potential application.

In GELCODE stained gels there is predictable relationships between a protein's integrated intensity and concentration within a color class (9). This feature allows one to realize that a standard concentration curve of any protein of similar color is sufficient to determine an unknown protein's actual concentration within a gel, if it is of similar color. It will be necessary to experimentally confirm, at least once, that a given integrated intensity for a silver scanning system and for a given color is within the linear range.

## Conclusion

I would like to submit that the elements required for the characterization of all proteins according to their color is presently available. This can be accomplished with GELCODE Color Silver Staining of 2-D gels, digitizing with appropriate multispectral cameras, and analyzing the resulting data with computer programs.

For instance, the reliable and reproducible stain can be used to classify proteins according to color. Computerized scanning cameras are now available. Proteins belonging to the same color class have the same slope characteristics when intensity and concentrations are plotted. Thus, proteins can be automatically assigned to their color classes and the concentration of each protein determined by extrapolating from the appropriate standard color curve.

## Acknowledgments

Special acknowledgments are made to Donald Jones, and Tom Vidmar of the Upjohn Company and to John Hartman of BioImage Corporation for technical assistance and discussions, and to Carol Strong of the

Center for Separation Science for typing the manuscript. The work was partially supported by NASA Grant NAGW-693, The Upjohn Company and Upjohn Diagnostics.

Literature Cited

1. Schwitzer, R. C. III; Merril, C. R.; Shifrin, S. Anal. Biochem. 1979, 98, 231-237.
2. Nauta, W. J. H.; Gygax, P. A. Stain Technology 1951, 26, 5-11.
3. Merril, C. R.; Goldman, D.; Sedman, S. A.; Ebert, M. H. Science, 1981, 211, 1437-1438.
4. Ochs, D. C.; McConkey, E. H.; Sammons, D. W. Electrophoresis 1981, 2, 304-307.
5. Adams, L. D.; Sammons, D. W. "Electrophoresis '81"; Allen, R. C.; Arnaud, P., Eds.; Walter DeGruyter, Berlin, 1981, pp. 155-166.
6. Sammons, D. W.; Adams, L. D.; Nishizawa, E. E. Electrophoresis 1981, 2, 135-140.
7. Goldman, D.; Merril, C. R.; Ebert, M. H. Clin. Chem. 1980, 26, 1317-1322.
8. Dzandu, J. K.; Deh, M. E.; Barrett, D. L.; Wise, G. E., Proc. Natl. Acad. Sci. USA 81, 1984, 1733-1737.
9. Sammons, D. W.; Adams, L. D.; Vidmar, T. J.; Hatfield, C. A.; Jones, D. H.; Chuba, P. J.; Crooks, S. W. in "Two-Dimensional Gel Electrophoresis of Proteins"; Celis, J. E.; Bravo, R., Eds.; 1984, pp. 112-125.
10. Bladon, P. T.; Cooper, N. F.; Wright, R. M.; Forster, R. A.; Wood, E. J.; Cunliffe, W. J. Clin. Chim. Acta. 1983, 127, 403-406.
11. Rochette-Egly, C.; Stussi-Garaud, C. Electrophoresis 1984, 5, 285-288.
12. Giulian, G. G.; Moss, R. L.; Greaser, M. Anal. Biochem. 1983, 129, 277-278.
13. Frey, M.; Radola, B. J. Electrophoresis 1982, 3, 27-32.
14. Allen, R. C. Electrophoresis 1980, 1, 32-37.
15. Slisz, M.; Van Frank, R. Electrophoresis 1985, 6, 405-408.
16. Sheludko, A. "Colloid Chemistry" Sheludko, A. Ed.; Elsevier Publishing Co.: New York, 1966; p. 4.
17. Vincent, R. K.; Hartman, J.; Barrett, A. S.; Sammons, D. W. In "Electrophoresis '81"; Allen, R. C; Arnaud, P., Eds.; De Gruyter, Berlin, 1981; pp. 371-381.
18. Adams, L. D.; Electrophoresis '84 Abstract, 1984, Tucson, Az.
19. Schoenle, F. J.; Adams, L. D. Sammons, D. W. J. Biol. Chem. 1984, 259, p. 12112-12116.
20. Anderson, N. G.; Anderson, N. L.; Tollaksen, S. L. ANL-BIM-79-82 1979, Argonne, IL.

RECEIVED December 2, 1986

# Chapter 7

# Development of Electrophoresis and Electrofocusing Standards

**Dennis J. Reeder**

**Organic Analytical Research Division, Center for Analytical Chemistry, National Bureau of Standards, Gaithersburg, MD 20899**

This work reviews some of the approaches to standardization in several different areas of electrophoretic separations. While no definitive standards have been established, some practical standards have been reported and are being used by researchers. Standards usage is part of quality assurance programs and is necessary for interlaboratory comparability.

It is known that conventions begin to exist from the time that a technology first originates. As the conventions begin to become widespread, changes are introduced. As changes are advanced, chaos often results and recovery often takes a long time. A good analogy of the process of standardization is detailed in the description of the standardization of railroad track gauges (1). The introduction of various gauges of railroad tracks in the early days of railroad development led to many problems in rail transportation that have continued until recent standardization activities.

In the field of electrophoresis, standardization is yet to be achieved in many areas. In laboratories throughout the world, variations of techniques exist. Interlaboratory intercomparisons of results are not widespread and there are apparent needs to have better control of electrophoretic results. Also, many techniques need to be validated for experimental appropriateness. Taylor (2) expressed the need for test validation: "A plethora of methods, procedures, and protocols based on the same measurement principle can arise for a given analytical determination. Usually, they are worded differently, and they may contain subtle or major differences in technical details. The extent to which each needs to be validated is a matter of professional judgment. It is evident some validation tests could be merely a matter of experimentally testing the clarity of the written word."

This chapter will describe the efforts taken to attempt to bring order to various electrophoretic methods by means of physical standards or standards activities. The discussion will not extend to electrophoresis of nucleic acids, which because of the nature of relative comparisons, needs to be addressed independently.

Because electrophoresis is a separations technique, it is especially useful in studying complex processes. It has the further advantage of requiring only small amounts of proteins for analysis. At first glance, electrophoresis gives the impression of being a very simple technique. A macromolecule such as a protein or nucleic acid that has a net charge in an electric field is subjected to a force which will accelerate it until the opposing frictional force is equal in magnitude. This results in a steady-state mobility of the macromolecule which is directly proportional to its net charge and inversely proportional to its frictional coefficient, which reflects both its size and shape. However, standardization is made difficult by many subtle differences in the chemistry of the system. Situations that must be considered include the cloud of counterions around the macromolecule, interactions between the macromolecule and the components of the solvent that affect both the apparent net charge and the hydrodynamic properties, and distortions of the counterion cloud by the electric field.

Alterations of net charge, size, or shape of macromolecules can be interpreted by their changes in mobility, either by a shift in their measured isoelectric points, or by a change in their apparent molecular weight. In order to interpret the changes correctly, adequate standards must be employed. To date, there has not been a systematic pursuit of standardization in many areas of electrophoresis.

## Historical Perspective

Electrophoresis was first put into practical use by Tiselius (3,4). A number of years later, Smithies introduced starch gels in the 1950's (5,6). His technique of zone electrophoresis with starch gel as the supporting medium gave a resolving power equal, if not greater, to that of the Tiselius method. Shortly thereafter, Raymond and Weintraub (7) introduced the use of acrylamide as an anticonvective medium. In the mid 1960's, Ornstein (8) and Davis (9) popularized the technique of gel electrophoresis on acrylamide gels in tubes. The original technique of "disc" gel electrophoresis consisted of three separate layers, each containing a different buffer and a different concentration of acrylamide. However, this makes the technique rather time consuming and present-day practice is to dispense with the upper sample gel and the middle spacer gel and to use a single small pore gel for the separation. The Ornstein and Davis techniques elicited a great deal of excitement in the resolving power of electrophoresis and the use of the technique became more widely used as the methods of Weber and Osborne (10) and Laemelli (11), despite their deficiencies, provided researchers with a convenient tool for determining molecular weights of proteins. Hjerten (12), published one of the first papers that detailed the use of agarose as an anticonvection agent in electrophoresis. By 1979, the laboratory use of agarose had progressed to the point where a "Proposed Selected Method" was published in *Clinical Chemistry* (13). The usefulness of agarose continues to expand, especially for larger molecular weight species and for cellular particles such as viruses.

Isoelectric focusing, another powerful method of electrophoresis, had its beginnings with the introduction of carrier ampholytes by Svensson in density gradient focusing (14) and the subsequent introduction of synthetic carrier ampholytes for the production of stable pH gradients (15). The techniques at first were crude, with electrofocusing being performed in liquid columns. These columns had the deficiency of requiring large amounts of reagents, and of being slow to resolve proteins. Once resolved, proteins would concentrate at their isoelectric points in the liquid sucrose gradient and then disrupt the gradients by sinking to the bottom because of the mass that was concentrated at that point, thus degrading the resolution. Improvements in isoelectric focusing evolved as solid supports were promoted. The current use of thin-layer and ultra-thin layer electrofocusing has opened many potential applications. However, there is a paucity of data that relates isoelectric focusing patterns to specific proteins.

Instead, effort has been directed to the information that is derived from patterns resulting from two-dimensional electrophoresis (2-DE). O'Farrell (16), and independently, Scheele (17) introduced the practical aspects of 2-DE. In the first dimensional separation, the proteins are separated according to their isoelectric points, on isoelectric focusing gels made with polyacrylamide. In the second dimension the proteins are further separated according to their molecular weights by means of a discontinuous polyacrylamide gel system containing the anionic detergent sodium dodecyl sulfate (SDS). With this technique, O'Farrell was able to resolve 1,100 different components from *Escherichia coli*. Proteins differing by only a single charge could be resolved and proteins comprising $10^{-4}$ to $10^{-5}$% of the total protein content could be detected. The Andersons (18) standardized a number of the potential variables in the O'Farrell technique thus rendering it reproducible and transferable. This was done by taking special precautions to standardize a number of the experimental variables. Detailed procedures were published describing the operations necessary to achieve reproducibility (19). Many researchers found these procedures to be very useful because they required only a few modifications to function in most situations. In 1978, a procedure (20) was developed for the determination of amino-acid composition among similar cellular proteins separated by two-dimensional gel electrophoresis.

In spite of the progress brought about by the Andersons, there are many standardization tasks to be applied to two-dimensional electrophoresis. Providing standards for this technique may turn out to be a difficult, but important task. For example, use of the same internal marker proteins would facilitate inter- as well as intra-laboratory comparisons, especially for computerized matching systems. Investigations in the newer areas of biotechnology may determine that these electrophoretic techniques are valuable for examining many proteinaceous materials produced by bioengineering techniques.

Standardization Needs

With the rapid proliferation of electrophoretic data, often on closely related systematic groups, it is becoming increasingly desirable to find a standard means of permitting objective comparisons of data from diverse sources and quantifying the various levels of experimental error involved in the different sets of data (21).

Uriano and Gravatt (22) have expounded the need for reliable data. Daufeldt and Harrison (23), have pointed out that "... adequate quality control measures and clear quality assurance criteria with respect to both analytical performance and experimental design are needed for clinical 2-D investigations. Currently, neither defined performance criteria nor external proficiency testing is available for 2-D laboratories. The absence of such criteria results in an overall lack of standardization that adversely affects interlaboratory data comparison, which is essential to conduct a large multicenter project such as the Human Protein Index."

In 1984, the Electrophoresis Society of the Americas and the National Bureau of Standards co-sponsored a workshop entitled Electrophoresis Standardization: Approaches and Needs. Several approaches to standardization in electrophoresis were presented and a number of manufacturers expressed interest in providing materials that will meet the needs of researchers who are extending the capabilities of electrophoresis into analysis of smaller samples with higher precision (24). The overall direction of future efforts will be towards gel evaluation and criteria for electrophoretic performance.

Quality Assurance and Quality Control. The terms "quality assurance"(QA) and "quality control"(QC) need to be defined. They are often used interchangeably, but to the professional they refer to two different activities. Quality control refers to those actions taken in the laboratory in an attempt to keep the measurement system in control. Examples would be running reference standards, calibrating instruments, keeping quality control charts, etc. Quality assurance refers to the system or program whereby management assures itself (and its clients) that the quality control measures are being applied, and that the results reported do, in fact, refer to the sample that was submitted or collected by the laboratory.

In many laboratories quality control is limited to the occasional analysis of a particular control material and plots of the results on conventional control charts. Such repetitive activity carried out without understanding or thought may give false comfort and assurance rather than truly objective assessment of the variability in results.

In electrophoresis, good reproducibility depends upon close adherence to quality control protocols. Details must be closely monitored for all phases of the process, including sample preparation, storage conditions, denaturation techniques, the volume of denaturant relative to the volume of the specimen, the source and pH range of the carrier ampholyte for isoelectric focusing, as well as the manufacturer; lot numbers, volumes, source and purity of the acrylamide and bisacrylamide; the temperature of the electrophoretic run; types of staining techniques used; and the proper statistical control of the experiment by use of adequate standards, numbers of samples, and replications to assure continued high quality and freedom from artifacts in the measurement process.

Training. Generally, a laboratory using published procedures needs more experimental details than are contained in published research reports of a method. With widespread usage of a newly-introduced technique, efforts to develop standardized procedures become attractive. Commercial workshops, especially in the clinical area, have been useful to bring laboratories to a common state-of-the-art. Valuable training workshops in two-dimensional electrophoretic techniques were held for a number of years by the Molecular Anatomy program at Argonne National Laboratory.

Contamination Problems. As analytical techniques become more sensitive, contamination becomes more of a problem. For example, Ochs (25) reported that many erroneous bands in polyacrylamide gels run with SDS were from proteins found on the skin of the researchers handling the gels. Her work indicates the need to exercise extreme care in sample preparation. In addition, suitable sample blanks handled in a parallel fashion would enable the researcher to identify contamination problems. Secondary checking of materials by two-dimensional electrophoresis would also help identify the nature and type of contamination.

Material Variability. Very little has been published that has addressed specifications or tests for chemicals used in electrophoresis. Too often, researchers depend on word of mouth, or empirically-derived tests to determine the source and/or lot number of the particular chemical to be used in their particular electrophoretic method. There has been but a smattering of work publicizing the differences in homogeneity of electrophoretic chemical preparations, stability of the specific compounds, and lot-to-lot differences of a particular manufacturer's chemicals. As analyses for constituents of electrophoretic gels become available and become part of the labeling process, specifications or tests for chemicals used in electrophoresis will be more widely promulgated.

The separation of the protein tubulin exemplifies the problems caused by material variability. Baxandall et al. (26) reported that two different batches of SDS obtained from the same supplier resulted in reversal of the order in which tubulin subunits were separated. Sullivan and Wilson (27) reported that different laboratories achieved different degrees of separation of alpha and beta tubulins and suggested subtle variations in techniques and reagent quality probably explained the variabilities.

As most researchers are aware, manufacturer differences in reagent production often result in noticeable differences in electrophoretic patterns. Recent advertising literature by many companies has emphasized advances in reagent purity.

Staining Problems. For a detailed treatise on a number of staining methods, the reader is referred to a series of chapters in Methods in Enzymology, vol 104. (28). Staining methods are also contained in the recent book of Allen and Saravis (29).

There appear to be problems of comparing protein patterns from gels stained by different techniques. This is probable because proteins stain to a different degree depending upon which staining procedure is employed. In addition, diverse dynamic ranges are often observed. Also, the pattern from any electrophoresis separation of blood or blood serum may contain in each zone several protein constituents including isoenzymes, lipoproteins, hemoglobins and haptoglobins, among other potential analytes. The proper selection of the fixation-staining technique used to visualize the individual components of each family of constituents to be considered depends upon the type of analyte being investigated (30).

## Approaches to Internal Standardization

Internal markers or standard proteins have most often been used by researchers to assure themselves that separations by electrophoresis are consistent and reproducible (21). In addition, charge markers in isoelectric focusing are also used, although standard preparations of these materials are less reliable, especially under denaturing conditions. Other internal standards that are often reported are materials for assuring radiolabel activity, enzyme activity, or other materials to monitor repeatability of measurements. In all cases, the stability of the standard material is the key to long-term quality control. Large batches of standards that are reproducible from lot to lot would also be useful.

Molecular Weight Markers. Johnson (21), was one of the first to describe adequately the value of internal marker proteins added to each biological preparation prior to the electrophoretic run. Their value derives from the fact that the added markers travel through the gel in the same path as the samples of interest during the course of the run. These markers, usually commercially obtained proteins, are chosen so as to be different in physical characteristics; any change in experimental conditions is intended to affect the migration of one more than that of the other. The ratio of the mobilities of the markers then provides a sensitive standardization of the run, as any experimental error is reflected in an alteration in that ratio. Once it has been established that a standard ratio is within the range of values normally obtained, and thus that running conditions for a particular experiment are normal, the mobility of a "variant" protein may be expressed relative to that of one of the internal standards. For many polymorphic variant proteins under study, standardized ratios have proven highly reproducible by a variety of investigators over periods of years.

Giometti et al. (31) examined a number of potential molecular weight standards for 2-DE ranging over serum, red cell lysates, urinary proteins, seed proteins, and a variety of rabbit, rat and mouse tissues. They found that the most satisfactory material was a molecular weight standard. By including the homogenate in the agarose overlay used in holding the isofocusing gel, 80 horizontal lines could be observed. Since the exact molecular weight of many of these heart muscle proteins are known (provided no alterations have occurred in the preparation and storage steps), a calibration curve may be drawn for determination of apparent SDS molecular weights. Rabbit psoas muscle homogenate has also been used as a molecular weight standard (32). The preparation may be modified by the addition of known proteins such as human serum albumin and equine heart cytochrome c for additional bands. Phage proteins have also been proposed as a source of proteins to cover a wide span of molecular weights (33). However, these proteins are not readily available and do not cover the spectrum of protein molecular weights found in tissues. Johnston, et al (34) examined various methods for standardization of protein patterns in 2-DE and recognized that the marker protein approach was preferred, but had shortcomings. However, the definition for standardization that Johnston, et al used, referred to the matching of selected polypeptide spot positions as a result of two-dimensional electrophoresis of complex mixtures of proteins and did not refer to standardization in terms of position in the gels; i.e., known molecular weight or isoelectric point. At this time, most standards approaches for the commonly used molecular weight ranges are still being directed towards obtaining stable, pure, and well-characterized proteins.

There is a particular need for size standards for large molecular weight proteins. In her search for a high molecular weight standard, Quittner (35) described a molluscan hemocyanin polypeptide with a molecular weight of 300,000 in its subunit-SDS complex. However, when this material was measured by electrophoretic techniques, it appeared to behave differently. The value given by an electrophoretic technique, based on a linear extrapolation from lower molecular weight standards, was 240,000; the actual value was 300,000. Likewise, Gower and Rodnight (36) found that their high molecular weight standard (apoferritin) showed a higher free mobility than expected, reflecting radically different mobility characteristics from the other standards. Thus, the complex subunit structure of high-molecular weight standards may cause them to behave in a manner that precludes them from being used as reliable standards. As an approach to standardization, these researchers ran standards on gels of different acrylamide concentration and constructed Ferguson plots [(log RF/%T)] for their standards and for the polypeptides that they had isolated. This approach appears to be the most rigorous and accurate.

In our laboratory experience, we have found that many commercially available proteins are not as pure as they are purported to be. Using sensitive silver staining processes and the resolving power of the 2-dimensional technique to examine purchased standards, we have observed multiple forms of purchased proteins. It appears that many companies simply blend proteins to provide major bands for molecular weight controls. Unfortunately, breakdown products and protein impurities may be easily observed. Care must be taken in using these proteins as standards for comparison purposes.

Sometimes there is a lack of uniformity in the stated molecular weights of proteins in purchased standards. The following molecular weights of commonly used proteins have been suggested as accurate in the biochemical handbooks and from manufacturers selling molecular weight proteins for electrophoresis (28,37):

| Protein | Most Commonly Reported Mol. Wt. | Other Variations Reported | | |
|---|---|---|---|---|
| Apoferritin | 450,000 | 440,000 | | |
| Catalase (beef liver) | 240,000 | 232,000 | | |
| Myosin | 200,000 | 212,000 | 205,000 | |
| Aldolase | 149,100 | 160,000 | | |
| Lactate Dehydrogenase | 133,000 | 140,000 | | |
| Beta-galactosidase (subunit) | 116,000 | | | |
| Phosphorylase b (subunit) | 97,000 | 97,400 | 94,000 | |
| Phosphorylase a | 94,000 | 92,000 | | |
| Albumin, Bovine Serum | 68,000 | 68,500 | 67,000 | 66,000 |
| Hemoglobin | 64,450 | | | |
| Catalase | 60,000 | 57,500 | | |
| Ovalbumin | 45,000 | 43,000 | | |
| Actin | 43,000 | | | |
| Glyceraldehyde-3-PDH | 36,000 | | | |
| Pepsin | 34,700 | | | |
| Carbonic Anhydrase | 29,000 | 30,000 | | |
| Chymotrypsinogen A | 25,000 | | | |
| Trypsinogen | 24,000 | | | |
| Lactalbumin | 23,000 | | | |
| Trypsin Inhibitor (soybean) | 20,100 | | | |
| Beta-lactoglobulin subunit | 18,400 | 14,400 | 14,200 | |
| Myoglobin (horse) | 17,800 | | | |
| Hemoglobin | 16,000 | | | |
| alpha-Lactalbumin subunit | 14,400 | | | |
| Lysozyme | 14,300 | | | |
| Ribonuclease a | 13,700 | | | |
| Cytochrome c (horse heart) | 12,500 | 12,300 | | |

Charge Standards for Isoelectric Focusing. Isoelectric focusing has made substantial progress towards reproducibility since the early days of focusing in sucrose gradients. The use of thin or ultrathin gels is beginning to result in better reproducibility between laboratories. For standardization of isoelectric focusing, the use of "charge trains", or single proteins with artificially-induced charges, seems to be the direction that researchers have taken. Anderson and Hickman (38) produced internal isoelectric point standards for 2-DE by carbamylation of rabbit muscle creatine kinase (EC 2.7.3.2; Sigma Chemical Co., St.Louis, MO 63178). These standards are visualized by Coomassie Blue or silver stain and have been widely used in 2-DE. Carbamylation produces shifts toward more acidic isoelectric points and higher apparent molecular weights. Carbamylation of the free amino groups of proteins occurs as the proteins are heated in a solution of urea, producing cyanate. The extent of carbamylation can be varied by heating the protein for results in the loss of a positive charge on the protein, thus shifting the isoelectric point of the protein to a more acidic pH. A unit shift in the isoelectric point is obtained with each amino group carbamylated. A "carbamylation train" can be established by mixing a number of differentially carbamylated samples of the protein, followed by isoelectric focusing in urea and SDS-electrophoresis. Carbamylation trains have been extensively utilized as standards for isoelectric points, but a universally accepted protein and assignment of pH values is not a reality. This may be a result of the observation that unit differences in charge show different relative migration distances in gels depending on the molecular weight of the proteins that have been modified. In addition, there have been problems in stability and reproducibility with the charge trains that are commercially available. As more experience is gained in the manufacturing of large lots of these standards, we can expect that better control of pH ranges and better reproducibility will follow.

Different proteins have been carbamylated to be used as charge standards. Carbamylated beta chains of hemoglobin (38) were reported to be stable standards, while Fawcett (39) took the approach of acetylation of myoglobin to produce charge standards for isoelectric focusing. Building on these standards, Tollaksen et al. (40) used carbamylated charge standards for testing batches of ampholytes used in 2-DE. This work was preceded by an earlier report of the use of simple isoelectric focusing to characterize ampholytes (41). Whatever standard that is developed for isoelectric point measurements, it must be evaluated in a urea system (the usual first-dimension condition in a two-dimensional separation) so that it will be of use in assigning pH values to unknown proteins.

Use of Complex Standards for Cell Proteins and Two-Dimensional Electrophoresis. One of the more widely-used cell lines chosen as a reference standard is the lymphoblastoid cell line GM607, derived from a normal individual and available from the Human Genetic Mutant Cell Depository, Camden, NJ 08103. This cell line may be grown in defined media, labeled with a radioactive tracer, and reproducibly separated in a 2-DE system. Heat shock proteins may readily be isolated and visualized from this cell line, as shown by Anderson et al. (42). For serum, a reference preparation for serum proteins is available as a certified reference material prepared and assayed by the College of American Pathologists (CAP) and by the U. S. Centers for Disease Control. A widely available human serum standard is that provided by the National Bureau of Standards as SRM 909. If sufficient interest from the user community is evident, a full electrophoretic characterization of this material can be included in the documentation. If the amount of selected standard proteins loaded on a gel is known, "relative" quantification of similar proteins could be obtained. In addition, the National Bureau of Standards could serve as an impartial evaluator of potential national standards (e.g. molecular weight standards, "tie-point" proteins, and isoelectric point standards) to assess suitability and stability.

Techniques for optimal acrylamide concentration. To find the optimal acrylamide concentration to use in a gel for the separation of proteins, a set of computer programs, "PAGE-PACK" of Rodbard (43) are convenient. One enters the relative mobility (RF) of selected proteins run at several gel concentrations, the RF of molecular weight standards run in the same gels, and other information based on the pH of the buffer system and slope/intercepts of Ferguson plots. The resulting computations give the desired optimally resolving gel concentration for selected proteins. Further information, such as ellipsoidal "confidence envelopes" for the various zones of the electropherogram are easily obtained.

Controversy still exists over the effects of acrylamide purity on electrophoretic separations. Clarity and polymerization properties are obviously related to acrylamide reagent purity. Most researchers realized that unwanted polymers, acrylic acid, and other unwanted breakdown products develop in acrylamide on long storage. Careful workers circumvent these problems by procuring fresh lots of highly pure, twice-crystallized acrylamide, or by recrystallizing laboratory-stored acrylamide and keeping the freshly made materials frozen at -20 degrees C under desiccation.

Standards for Publishing Electrophoretic Results. In 1971, the IUPAC-IUB Commission recommended that standardized rules for the systematic orientation of electrophoretic photographs and diagrams be adopted (44). These recommendations have been augmented and illustrated (45).

Reference Materials. Standards are only part of the total QA picture. In order to have an analytical system completely under control, other variables need to be taken into consideration. At the present time, there are no defined reference materials for electrophoresis in the form of pure chemicals certified for use as electrophoretic standards. Manufacturers recognize the value of pure reagents and are increasingly advertising ultra-pure reagents for electrophoresis. Until reference procedures are devised for the many electrophoretic methods available, researchers will have to rely on the manufacturer to provide consistently pure materials for electrophoresis that will assure reproducibility as new lots of material are ordered.

Enzyme Standards. Enzyme electrophoresis has proliferated with increased genetic profiling for medical and forensic use. Standards for enzyme analyses in the form of defined kits containing banks of enzymes of different isoenzyme patterns are not commercially available. For the most part, forensic laboratories use individuals from within their own laboratories who have known phenotypes for the enzymes of interest. Such donors become de facto standards. The interlaboratory exchange of samples and rigorous continued training in enzyme phenotype identification will improve this standards base.

Water Standards. Standardization of the grade of water required for electrophoresis has not been thoroughly researched. Investigators have informally commented that irreproducibility of some protein quality. When in doubt as to the type to use, the researcher should use the highest grade water available. Water purity has been defined by several standards-setting organizations. The American Society for Testing and Materials (ASTM), the College of American Pathologists (CAP) and the National Committee for Clinical Laboratory Standards (NCCLS) have defined water purity according to four major classes of use (46,47,48): 1) Type I: Reagent grade water which is suitable for precision work requiring maximum accuracy and freedom from background impurities, 2) Type II: Analytical grade water which is suitable for all but the most critical analytical work, and 3) Types III and IV: General laboratory water which is suitable for most qualitative chemistry, glassware rinsing, or as feedwater to a reagent grade polishing system.

Detergents. Sodium dodecyl sulfate, the most extensively studied detergent for use as a dissociating agent, binds to most water-soluble proteins to induce conformation changes. In the presence of a reducing agent, the proteins are dissociated to their constituent polypeptides and SDS is bound at a level of approximately 1.4 g SDS per gram of protein. Subsequent electrophoresis of the polypeptides allows determination of molecular weight based on the relative mobilities of the proteins and standards. However, several criteria must be met: unknown proteins and standards must bind the same amount of SDS; the conformational changes in the proteins and standards must be the same, and the proteins and standards must be subjected to the same electrophoretic conditions. A recent review of SDS-polyacrylamide gel electrophoresis details many SDS applications (49).

Detergents such as SDS are used to solubilize proteins to make them easy to separate in an electrophoretic system. However, the exchange among different detergent moieties, that is, substitution of a neutral or zwitterionic detergent for SDS, or vice versa, is often a difficult and purely empirical procedure. The popular series of nonionic detergents, tert-butoxy-phenyl polyoxyethylene, and others (e.g.Triton X-100 and Nonidet-P40), are plagued by a high absorption at 280 nm which hinders ultraviolet monitoring of chromatographic separations.

Staining. Visualization of proteins by dye-binding has continued to be the subject of many papers in the biochemical journals and recently has been reviewed (50). This chapter will not attempt to elaborate on the relative merits of each dye other than comment that adequate standards have not yet been established. It is should be noted, however, that different proteins react to stains in different ways. Certain proteins have an unusually low ability to bind dyes because of their molecular structure and their content of carbohydrates. Zak (30) described a number of problems associated with separations, chemical reactions in which a visual pattern of proteins could be observed, and quantification by densitometry. Included in their observations were problems associated with albumin trail, resolution, unequivalent staining, prestaining, and the densitometry problems associated with band widths, opacity effects and polychromaticities.

The chemistry of silver staining is not entirely known but a recent review, (50), reaffirms that basic and sulfur-containing amino acids contribute in a substantial way to the staining reaction. The exact nature of the chemistry of formation of nucleation centers for silver reduction in proteins is still being investigated.

Of several newer methods for visualizing proteins, DABITC (dimethylaminoazobenzene isothiocyanate) derivatized proteins have been used to evoke discernible bands on SDS-polyacrylamide gels (51). With this technique, colored proteins can be directly detected in the picomole range as yellow-colored bands. The sensitivity is further enhanced by exposing the gel to an acidic solution which turns the yellow bands into red bands. This method avoids the use of radioisotopes and time-consuming staining and destaining procedures.

Data Handling and Data Presentation. Because of the considerable number of proteins that can be separated in one analysis, data handling becomes complex as huge files have to be stored and monitored. The use of acrylamide gradients in 2-DE adds more complexity to the analysis problem. The gradient-making process creates a wedge-shaped gel that spreads low molecular weight proteins further apart than high-molecular weight proteins. Software to correct for inhomogeneous and linear distorted spot patterns in such gels has been developed by several laboratories: by Anderson et al. (52), by Lester et al. (53,54) by Tracy (55), by Garrels (56), and by Hruschka, Massie and Anderson (57). A discussion of the problems of spot detection, segmentation, integration and pairing is beyond the scope of this paper. It is sufficient to report that most digitization programs show between-gel errors of about 20% and digitization errors of about 15%. Estimation of gel-preparation errors range about 12-15%. As automated digitization procedures become more sophisticated, errors may decrease. At this time, national programs to allow intercomparison of data are not available.

Use of Interlaboratory Round Robin Procedures. Interlaboratory exchanges of samples and results can be part of formal external mechanisms for quality assurance programs. Interlaboratory testing can narrow the dispersion among results by directly improving the work of participating laboratories. For example, cumulative comparisons provide an excellent method for demonstrating long-term bias in individual laboratories. Data from interlaboratory exchanges can be used to identify imprecision and inaccurate methods. Evaluation of such survey data identifies problems and suggests priorities. Survey data may also provide a ready means of following trends in method popularity. There is now general agreement that interlaboratory testing can identify a continuous pattern of poor performance, but this objective is minor in comparison to the use of interlaboratory testing as an integral component of the quality control program of each laboratory. The ASTM has published a Standard Practice for conducting an interlaboratory test program to determine the precision of test planning, conducting, analyzing, and interpreting results of an interlaboratory study of a test method. Electrophoretic methods are now becoming sufficiently controlled that interlaboratory round robins are possible. By following the guidelines set forth in this ASTM publication, practitioners will have a rational basis for judging their performance against a national average of performance. Laboratories conducting such tests will be assured that precision statements can be adequately described for the test methods being studied.

## Summary

Over the last three decades, electrophoresis has developed into a technique which allows detailed examination of small amounts of biological materials. Subtle differences in composition and properties may be observed. Though several components of the technique need more rigorous standardization, many new innovations are being advanced. In any developed technology, the timing of the introduction of the proper standards is critical. If rigid protocols are introduced too early, innovation may be stifled. If calibration and material standards are advanced too late, they may not be useful in pulling together divergent methodologies. However, it is apparent that useful standards for electrophoresis are being developed to promote timely and useful techniques for bioanalytical research. As improvements in methodology are introduced, the need for calibration standards will be even more important for the researcher to assure compatibility between previously accepted techniques and newer procedures.

## Literature Cited

1. Cropper, W. V. Chemtech 1979, September, 550-9.
2. Taylor, J. Analy. Chem. 1983, 55, 600-8A.
3. Tiselius, A. Biochem. J. 1937, 31, 313-7.
4. Tiselius, A. Trans. Faraday Soc. 1937, 33, 524.
5. Smithies, O. Biochem. J. 1959, 71, 585-7.
6. Smithies, O. Biochem. J. 1955, 61, 629-41.
7. Raymond, S.; Weintraub, L. Science 1955, 130, 711-2.
8. Ornstein, L. Ann. N. Y. Acad. Sci. 1964, 121, 321-49.
9. Davis, B. Ann. N.Y. Acad. Sci. 1964, 121, 404-27.
10. Weber, K.; Osborn, M. In "The Proteins"; Neurath, H.; Hill, R. L., Eds.; Academic Press. New York, 1969; Vol. I, p. 179.
11. Laemelli, U. K. Nature (Lond.) 1970, 227, 680-5.
12. Hjerten, S. Biochem. Biophys. Acta 1961, 53, 514-17.
13. Jeppsson, J.-O.; Laurell, C. B.; Franzen, B. Clin. Chem. 1979, 25, 629-38.
14. Svensson, H. Acta Chem. Scand. 1961, 15, 325-41.
15. Vesterberg, O. Acta Chem. Scand. 1969, 23, 2653-66.
16. O'Farrell, P.H. J. Biol. Chem. 1975, 250, 4007-21.
17. Scheele, G. A. J. Biol. Chem. 1975, 250, 5375-85.
18. Anderson, N. L.; Anderson, N. G. Proc. Natl. Acad. Sci. USA 1977, 74, 5421-5.

19. Tollaksen, S. L.; Anderson, N.L.; Anderson, N. G. Operation of the ISO-DALT System, 7th ed., US Dept. Energy Publ. ANL-BIM-84-1, Argonne, IL, 1984
20. Cabral, F.; Gottesman, M. M. Anal. Biochem. 1978, 91, 548-56.
21. Johnson, G. Biochem. Genetics. 1975, 13, 833-47.
22. Uriano, G.; Gravatt, C. C. CRC Crit. Rev. in Analyt. Chem. 1977, 6, 361-411.
23. Daufeldt, J. A.; Harrison, H. H. Clin. Chem. 1984, 30, 1972-80.
24 Reeder, D. J. J. Res. of N.B.S. 1985, 90, 259-62.
25. Ochs, D. Anal. Biochem. 1983, 135, 470-4.
26. Baxandall, J.; Forsman, R. A.; Bibring, T. J. Cell Biol. 1979, 83, 339-43.
27. Sullivan, K. F.; Wilson, L. J. Chromatog. Library 1983, 18B, 185-193. ed. Z. Deyl. Elsevier Sci. Pub. Co. NY
28. Blackshear, P. J. Methods in Enzymology 1984, 104, 237-55.
29. Allen, R. C.; Saravis, C. A.; Maurer, H. R. "Gel Electrophoresis and Isoelectric Focusing of Proteins"; Walter de Gruyter: Berlin/New York, 1984; p. 181-240.
30. Zak, B.; Baginski, E. S.; Epstein, E. Annals Clin. Lab. Sci. 1978, 8, 385-95.
31. Giometti, C. S.; Anderson, N. G.; Tollaksen, S. L.; Edwards, J. J.; Anderson, N. L. Anal. Biochem. 1980, 102, 47-58.
32. Edwards, J. J.; Tollaksen, S. L.; Anderson, N. G. Clin.Chem. 1982, 28, 941-48.
33. Gersten, D. M.; Kurian, P.; Ledley, G.; Park, C. M.; Suhocki, P. V. Electrophoresis 1981, 2, 123-25.
34. Johnston, D. A.; Capetillo, S.; Ramagli, L. S.; Guevara Jr., J.; Gersten, D. M.; Rodriguez, L. V. Electrophoresis 1984, 5, 110-6.
35. Quittner, S.; Watts, L. A.; Roxby, R. Anal. Biochem. 1978, 89, 187-95.
36. Gower, H.; Rodnight, R. Biochem. Biophys. Acta 1982, 716, 45-52.
37. Hames, B.D. in "Gel Electrophoresis of Proteins: A Practical Approach" (B. D. Hames and D. Rickwood, eds.), p. 39. IRL Press, Oxford and Washington D.C., 1981.
38. Anderson, N. L.; Hickman, B. J. Anal. Biochem. 1979, 93, 312-20.
39. Fawcett, J. S. in "Isoelectric Focusing" (J. P. Arbuthnott and J. A. Beesley, eds.), ch. 2, Butterworths, London, 1975.
40. Tollaksen, S. L.; Edwards, J. J.; Anderson, N. G. Electrophoresis 1981, 2, 155-60.
41. Gelsema, W. J.; De Ligny, C. L. J. Chromatog. 1977, 130, 41-50.
42. Anderson, N. L.; Giometti, C. S.; Gemmell, M. A.; Nance, S. L.; Anderson, N. G. Clin. Chem. 1982, 28, 1084-92.
43. Rodbard, D. (author) Biomedical Computing Technology Information Center, Room-1302, Vanderbilt Medical Center, Nashville, TN, program identification MED-34 PAGE-PACK.
44. IUPAC-IUB Commission on Biological Nomenclature. J. Biol. Chem. 1971, 6127-8.

45. Inoue, T. In "Electrophoresis '83"; Hirai, H. Ed.; Walter de Gruyter. Berlin/New York, 1984, p. 125-8.
46. ASTM standard specification for reagent water, designation D 1193-77 (reapproved 1983). American Society for Testing Materials, Philadelphia.
47. Hamlin, W. B. "Reagent Water Specifications". 1978, College of American Pathologists, Commission of Laboratory Inspection and Accreditation, Chicago.
48. NCCLS Approved Standard ASC-3. 1980, National Committee for Clinical Laboratory Standards, Villanova, PA.
49. Sherman, L. R.; Goodrich, J. A. Chem. Soc. Reviews 1985, 14, 225-36.
50. Merril, C. R. in "Proceedings of the 1986 Meeting of the Americas Branch of the Electrophoresis Society" (D. J. Reeder, ed.) p. 36. NBSIR 86-3345, available from NTIS, Washington, D.C.
51. Aebersold, R.; Ledermann, F.; Braun, D. G.; Chang, J-Y. Anal. Biochem. 1984, 136, 465-9.
52. Anderson, N. L.; Taylor, J.; Scandora, A. E.; Coulter, B. P.; Anderson, N. G. Clin. Chem. 1981, 27, 1807-20.
53. Lester, E. P.; Lemkin, P. F.; Lipkin, L. E. Anal. Biochem. 1981, 53, 390A-404A.
54. Lemkin, P. F.; Lipkin, L. E. Electrophoresis 1983, 4, 71-81.
55. Tracy, R. P.; Currie, R. M.; Young, D. S. Clin. Chem. 1982, 28, 908-14.
56. Garrels, J. I. J. Biol. Chem. 1979, 254, 7961-77.
57. Hruschka, W. R.; Massie, D. R.; Anderson, J. D. Anal. Chem. 1983, 55, 2345-8.
58. ASTM publication: ASTM Designation E 691-79. ASTM Manual for Conducting an Interlaboratory Study of a Test Method. ASTM STP 691, Am. Soc. Testing Mats., 1979, 706-739

RECEIVED January 23, 1987

# Chapter 8

# Standardization in Isoelectric Focusing on Ultrathin-Layer Rehydratable Polyacrylamide Gels

**Robert C. Allen and Peter M. Lack**

**Department of Pathology, Medical University of South Carolina, Charleston, SC 29425**

Rehydratable, stabilized, dry polyacrylamide gels (250-275μ thick) were compared to those made conventionally, both fresh and stored in sealed aluminized bags, for their characteristics in isoelectric focusing. Removal of the impurities and unknown polymerization products by washing, following the polymerization procedure, produced a gel, that on rehydration with ampholytes in the presence of glycerol had a conductance five-fold or more lower than that of a conventionally prepared gels of the same volume and ampholyte concentration, as marked by a lower milliamperage at the same voltage gradient. The lower conductance allowed higher voltage gradients to be applied at completely controllable Joule heat loads, which increases the resolution of the system. Comparative studies of the separation of hemoglobins on Pharmalyte™ narrow range pH gradients indicated that with conventionally cast gels, either fresh or stored, there were additional peroxidase positive hemoglobin-like bands present in both fresh standards and in patient whole blood samples soaked into PKU filter paper test strips. These bands were presumably due to artifactual binding products of the polymerization process with ampholyte and or the hemoglobins. The rehydratable gels were found to be stable on storage, giving reproducible patterns after over one year of storage at room temperature.

Isoelectric focusing on polyacrylamide gels, freshly prepared, or on those stored in sealed bags obtained commercially, or on in house prepared, contain a number of characteristics that are undesirable, or even unknown: A) Gel polymerization products remain in the gel. These consist of unpolymerized acrylamide monomers, linear polymers, breakdown products of acrylamide such as B',B",B"' Nitrilotrispropionamide (a strong base), ammonium persulfate break down products, acrylic acid and TEMED to name the best known [1]. B) Polymerization kinetics in the presence of synthetic carrier ampholytes are also dependent on the ampholyte and the pH range chosen. The requirement for more TEMED in acid range gels and less in basic pH range gels

0097-6156/87/0335-0117$06.00/0

further complicates a set of conditions that already are characterized poorly [2]. C) Incorporation of additives, such as urea, during the polymerization process may also affect the polymerization kinetics, and eventual pore size and structure. D) Reproducibility between gels and gel types is difficult, even when care is taken to use purified reagents and freshly recrystallized monomer [3]. In any attempt to standardize conditions for isoelectric focusing, a defined gel product is a primary requirement. The foregoing reasons suggest that polymerization should be carried out using a standard buffer and pH under standardized conditions of temperature and time for the polymerization process. Ampholytes, other buffers, or reagents should be added after the polymerization step [4, 5]. One requirement to have a support medium that can be dehydrated and then rehydrated requires that the gel be covalently bonded to a support medium to control initial dehydration and rehydration in two of the three dimensions of the gel structure. Ultrathin-layer gels are desirable, since rehydration equilibration is contingent on the gel thickness and gels ( greater than 500 μ ) require a longer time to reach equilibrium [6-8]. In addition, ultrathin-layer gels have a higher resolution potential than those over 500μ thick due to the fact that higher voltage gradients may be used with production of less Joule heat at a given voltage gradient [9].

Morris and Morris solved the problem of the effects of by-products of polymerization in the gels by soaking their gels in repeated changes of distilled water and then placing them in the buffer desired prior to carrying out their electrophoretic separation. [10-11]. However, this approach was limited to certain apparatus and generally to continuous buffer systems rather than isoelectric focusing.

The development of a washed gel free of unreacted reagents and then dried for storage prior to later rehydration for use in electrophoresis was first accomplished by Uriel [12-13] with an acrylamide-agarose composite gel. These gels were washed for at least 16 hours in distilled water and then kept in an appropriate running buffer for up to several weeks before use. Alternatively, the gels were treated with glycerol and dried. Such gels were backed on an appropriate support, and rehydrated in the desired buffer before use. Wren and Mueller produced a similar material, but with agarose alone backed on mylar [14]. In the Uriel composite gels, proteins and nucleic acids could be separated by conventional electrophoresis. However, isoelectric focusing was impossible in such a system because of the residual charges on the agarose.

In 1972 Robinson [15] polymerized polyacrylamide gels by either photopolymerization or with persulfate in the absence of ampholytes, washed the gels against extensive changes of distilled water , dried them and then reconstituted the dried gels in the presence of 2 per cent Ampholine™. He did not add any stabilizer to these gels, nor attempt to store them. Unfortunately there was no hint of this treatment in the title of the original article and this work went largely unnoticed. Thus, for over the last 15 years, the only gels available for isoelectric focusing have been those cast directly in house or obtained commercially packaged in special sealed foil, with all of the contaminants mentioned above remaining therein.

Frey *et al.* [6,7] developed a method for the treatment of polyacrylamide similar to that of Uriel using a washed ultrathin-layer polyacrylamide gel backed on mylar film (Gel Fix, Serva or Silanized polyester films) stabilized with 10 per cent glycerol. This procedure produced a partially dehydrated gel that could be fully rehydrated with synthetic carrier ampholytes thus, producing a gel devoid of the previously encountered contaminants. Such gels could be stored only by laying a plastic sheet with repellent surface properties on top of the partially dehydrated gel. Unfortunately, the surface of such gels on rehydration was often flawed due to the plastic sticking to the gel and caused severe surface loading problems where samples run and smear into one another. Forced aging studies by heating the gels to 100 °C, indicated a shelf life of at least six months. However, this procedure may bear little relation to practical shelf life studies.

Allen [16] has produced storable, dried gels without glycerol that may be stored for over a year at room temperature without apparent loss of structural functionality or deterioration in pattern quality using a standard test protein and enzymes. Recently, Gelfi and Righetti [8] have speculated also on the desirability of rehydratable gels and have stored, over a short term, immobilized ampholytes gels. Altland, more recently, has successfully dehydrated and rehydrated Immobiline™ gels in a cassette with other additives such as urea gradients [17]. Assessment of rehydrated polyacrylamide gels backed on glass and polyester film, both for separation qualities on a series of proteins and enzymes and their potential use for a standardized reference base for isoelectric focusing is reported here. In addition to a defined and reproducible gel, quality assurance on all reagents employed in the system, as well as precise control of the temperature, time, power and separation conditions of any given apparatus are defined. These parameters also served as the basis of a round robin initiated by the National Bureau of Standards, Analytical Division, in order to define the necessary parameters for standardization of polyacrylamide gel isoelectric focusing (PAGIF), and indeed, to determine if such standardization was feasible.

## Methods and materials

Rehydratable gels were obtained from Micro-Map, Inc., Boca Rotan, FL, at both 3 per cent T, 3.5 per cent C and 5 per cent ,T 3.5 per cent C. (T refers to the per cent of monomer present, while C is defined as the per cent of cross-linker relative to T. The per cent of the gel is the total of these two values) For comparative studies conventional gels were prepared at the same acryl amide concentration from the same bottle of two times recrystallized acrylamide (Serva). Acrylamide was recrystallized also from chloroform as previously described [2] . These gels were 250 or 375μ thick and were bonded to silanized glass plates, or covalently bonded to GelBond ™ Pag 7 mil mylar sheets (Marine Colloids, Rockland ME ) using previously published methods [18-19] and were randomized between the two laboratories. Single lots of ampholytes from various manufacturers were used for the comparative studies in each laboratory to eliminate lot variation.

## Gel Rehydration

Gels were rehydrated either in an aqueous buffer or in one that contained 10 per cent glycerol. The latter conditions the surface of the gel and allows increased surface loading of the sample to be accomplished without sample smearing or surface running. If tabs or a mask were used as the sample application vehicle, the glycerol could be left out. Rehydration for isoelectric focusing was performed using a solution of 3.6 ml of 10 per cent glycerol and 0.4 ml of the desired ampholyte per 100 $cm^2$ gel surface area. Gels up to 150 $cm^2$ covalently bonded to mylar were rehydrated by floating the gel on the rehydration fluid placed on a glass plate as shown in Figure 1. Larger gels and gels covalently bonded to glass are rehydrated as shown in Figure 2.

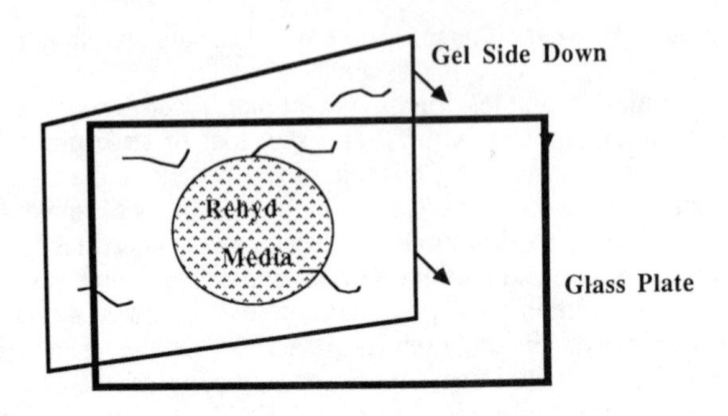


Figure 1. Rehydration of washed and dried gels covalently bonded to GelBond Pag.

## Sample Loading

When separations are carried out under conditions where high humidity is present, gels must be loaded at ambient room temperature, the electrode wicks and electrode cover placed on the gel and then focusing initiated at a low voltage prior to turning on a cooling device. Over cooling of the gel during loading, or in the initial phases of the separation, causes moisture to condense on the surface of the gel which will, at best, decrease resolution and at worst, ruin the separation from smearing . Gels were rehydrated either in an aqueous buffer that contained 10 per cent glycerol. The latter conditions the surface of the gel and allows increased surface loading of the sample to be accomplished without sample smearing . Rehydration for isoelectric focusing was performed using a solution of 3.6 ml of 10 per cent glycerol and 0.4 ml of the desired ampholyte per 100 $cm^2$ gel surface area of gel as shown in Figure 1 above.

## Visualization of Hemoglobin

Hemoglobin was stained for its peroxidase activity using 27 mg of 3,3' diaminobenzidine·HCL per 30 ml of 0.05 **M** sodium citrate. Hydrogen peroxide was added in an amount of 0.5 ml to the foregoing mixture and the TCA fixed hemoglobin gel reacted for 15 min. at room temperature.

## Run Conditions

Comparative run conditions for conventional and rehydratable gels of the same volumes for pH 6-8 Pharmalyte™ and 3-7 Servalyte™ at a four per cent concentration are given in Tables I-IV. The power settings and the temperatures shown were used on an MRA COLD FOCUS© apparatus employing a Pharmacia 300 volt power supply with a volt / hour integrator. In Tables I and III it should be noted also that the initial conductance of these gels was 4-5 fold less than their conventionally cast counterparts, as indicated by the initial milliamperage and voltage.

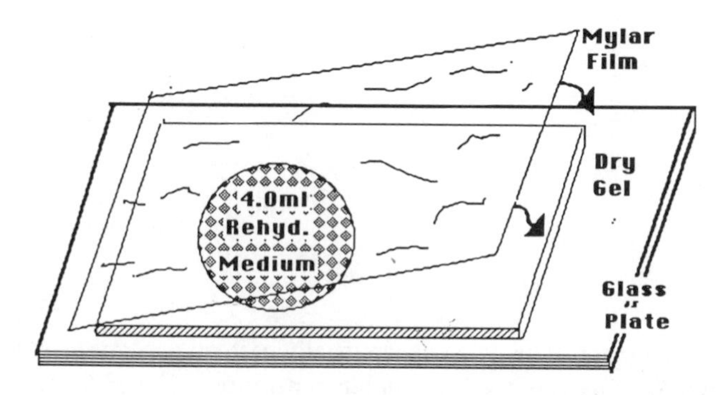


Figure 2. Rehydration from the top surface of a gel covalently bonded to glass or one covalently bonded to mylar greater than 150 cm$^2$. The volume of rehydration medium shown is for a gel of 2.4 cm$^3$. A gasket of any desired thickness may be placed on the lower support glass and the upper film substituted with a glass plate to form a cassette for precise rehydration volumes or to add gradient solutions to the gel.

**Table I.** Run parameters on a rehydrated gel separated on an **MRA Cold Focus©** app hemoglobin separation. Gel ,5 % T, 3.5 % C, 250µ thick and ambient temperature Ampholyte 4 per cent **Pharmalyte** ™ pH 6-8 with a distance of 5.4 cm between electr edges. Catholyte 1.0 M NaOH, anolyte 1.0 M $H_3PO_4$. Time given in minutes.

| Time | Volts | Ma | Watts | Plate Temp | Volt/Hr | Volt/cm |
|---|---|---|---|---|---|---|
| 0 | 170 | 2 | 0.36 | 16.5 | 0 | 31 |
| 6 | 200 | 2 | 0.4 | - | 20 | 37 |
| 6 | 420 | 1 | 0.4 | 16.5 | - | 78 |
| 12 | 520 | 1 | 0.5 | - | 67 | 96 |
| 12 | 850 | 3 | 2.5 | 15 | - | 157 |
| 18 | 1030 | 3 | 3 | - | 160 | 190 |
| 18 | 1300 | 4 | 5 | 16 | - | 240 |
| 24 | 1500 | 3 | 5 | - | 297 | 277 |
| 24 | 1860 | 5 | 8 | 16.5 | - | 344 |
| 30 | 2060 | 6 | 8 | - | 494 | 381 |
| 30 | 2540 | 5 | 12 | 16.5 | - | 470 |
| 36 | 3020 | 4 | 12 | 17 | 750 | 559 |

**Table II.** Run parameters on a conventionally cast gel separated on an **MRA Cold Focus©** apparatus for hemoglobin separation. Gel ,5 % T, 3 .5% C, 250μ thick and ambient temperature 17° C. Ampholyte 4 per cent **Pharmacia** ™ pH 6-8 with a distance of 5.4 cm between electrode wick edges. Catholyte 1.0 **M** NaOH, anolyte 1.0 **M** $H_3PO_4$. Time given in minutes.

| Time | Volts | Ma | Watts | Plate Temp | Volt/Hr | Volt/cm |
|---|---|---|---|---|---|---|
| 0 | 170 | 8 | 1.3 | 16.5 | 0 | 26 |
| 6 | 240 | 4 | 1 | 15.5 | 18 | 44 |
| 6 | 460 | 8 | 3 | - | - | 85 |
| 12 | 800 | 5 | 3 | 16.5 | 83 | 148 |
| 12 | 970 | 6 | 5 | - | - | 179 |
| 18 | 1250 | 4 | 5 | 16.5 | 193 | 231 |
| 18 | 1590 | 6 | 8 | - |  | 294 |
| 24 | 1800 | 5 | 8 | 16 | 360 | 333 |
| 24 | 2050 | 6 | 12 | - | - | 379 |
| 30 | 2160 | 5 | 11 | 15.5 | 625 | 400 |
| 30 | 2320 | 6 | 14 | - | - | 428 |
| 38 | 2370 | 6 | 14 | 16.5 | 750 | 438 |

**Table III.** Run parameters for a rehydratable gel separated on an MRA Cold Focus™ apparatus. Gel, 3 % T,3.5 % C, 250μ thick, ambient temperature 23° C, Ampholyte 4 per cent Servalyte pH 3-7 in 10 per cent aqueous glycerol with a distance of 7 cm between electrode wick edges. Catholyte, 1.0 M NaOH, anolyte 1.0 M $H_3PO_4$. Time given in minutes.

| Time | Volts | Ma | Watts | Plate Temp | Volt/Hr | Volt/cm |
|---|---|---|---|---|---|---|
| 0 | 280 | 3 | 1 | 25 | 0 | 40 |
| 10 | 350 | 2 | 1 | 18.5 | 57 | 50 |
| 10 | 780 | 4.5 | 3 | 18.5 | | 101 |
| 20 | 1000 | 3 | 3 | 18 | 194 | 143 |
| 20 | 1310 | 4 | 5 | - | - | 187 |
| 30 | 1740 | 3 | 5 | 17.5 | 440 | 250 |
| 30 | 2100 | 4 | 8 | - | - | 300 |
| 40 | 2430 | 4 | 8 | 18 | 815 | 347 |
| 40 | 2870 | 4 | 11 | - | - | 410 |
| 45 | 2980 | 4 | 11 | 18.5 | 1070 | 426 |
| 45 | 3000 | 4 | 12 | - | - | 428 |
| 47 | 3020 | 4 | 12 | 18 | 1250 | 431 |

**Table IV.** Run parameters on a conventionally cast gel separated on an MRA Cold Focus™ apparatus. Gel ,3 % T, 3.5 % C, 250μ thick , ambient temperature 26° C, Ampholyte 4 per cent Servalyte pH 3-7 with a distance of 7 cm between electrode wick edges. Catholyte 1.0 M NaOH, anolyte 1.0 M $H_3PO_4$. Time given in minutes.

| Time | Volts | Ma | Watts | Plate Temp | Volt/Hr | Volt/cm |
|---|---|---|---|---|---|---|
| 0 | 180 | 12 | 2 | 23 | 0 | 26 |
| 5 | 270 | 10 | 2 | 21 | 20 | 39 |
| | | | Load Samples Cooling Shut off | | | |
| 5 | 190 | 7 | 1 | 19 | | 27 |
| 15 | 270 | 5 | 1 | 18 | 61 | 39 |
| 15 | 450 | 9 | 3 | 18 | | 64 |
| 25 | 740 | 5.5 | 3 | 17 | 162 | 106 |
| 25 | 890 | 6 | 5 | 17 | | 127 |
| 35 | 1170 | 5 | 5 | 16 | 335 | 167 |
| 35 | 1410 | 6 | 8 | 16 | | 201 |
| 45 | 1530 | 6 | 8 | 16 | 565 | 219 |
| 45 | 1750 | 7 | 12 | 16 | | 250 |
| 55 | 1800 | 6 | 12 | 17 | 856 | 257 |
| 55 | 1950 | 7 | 14 | 18 | | 280 |
| 65 | 2030 | 7 | 14 | 18 | 1169 | 290 |
| 65 | 2170 | 7 | 15 | 18 | | 310 |
| 68 | 2170 | 6 | 13 | 19 | 1260 | 310 |

Densitometry

Densitometry was performed on a Biomed SL-2D Densitometer using their 2D-stepover program with a slow scan rate and a soft laser source for the silver stained cell extracts. The separated standard cell extract proteins on three per cent T gels and the hemoglobin separations on five per cent T gels were fixed in 20 per cent trichloro acetic acid. The cell extracts were stained with diammine silver [20-21] and hemoglobin separations were reacted for peroxidase activity with diamino benzidine and hydrogen peroxide in a citric acid buffer and scanned with a tungsten lamp light source at 420nm. All data was then handled for presentation and quantification using a Videophoresis II (Biomed Inst.) program with superimposition capability.

Results

As may be seen in Tables I and II, using narrow range Pharmalyte and in tables 3 and 4 using a wider pH range Servalyte, the rehydratable gels had a markedly lower conductance than those cast and run in the conventional manner. The reduction in conductance is evidenced by the lower milliamperage at the same voltage gradient in the initial phase of the separation. Thus, there is no need to perform a pre-run before the application of the sample in order to remove the by-products of polymerization by electrophoresis in the rehydratable gels. It is obvious, that some products of the polymerization are not removed from the gel during the pre-focus or during the run as indicated also by the higher conductance at the end of the gel separation, which in turn, limits the final voltage gradient that can be applied for sharpening the bands at the end of the run. This is further apparent in the hemoglobin separations shown in Figures 3-5, where the appearance of additional peroxidase-positive hemoglobin-like bands were found in the conventionally cast gels. In Figure 3A, the gel is made conventionally and was cast on silanized glass. In Figure 3B, a rehydratable gel of 5 per cent T, 3.5 per cent C was used for the same standard hemoglobin A, F, S and C (Isolab) and stained similarly for peroxidase activity of the hemoglobins with benzidine. Additional cathodal bands are apparent in the conventionally cast gel which are absent in the washed and rehydrated gel. The relative amounts of hemoglobin A,F, S and C determined by quantitative microdensitometry of the benzidine-reacted bands, indicated that in the conventionally cast pH 6-8 polyacrylamide gel that the ratios were nowhere near those given by the manufacturer. On the other hand, rehydratable gels on polyester film gave values close to those given for the fresh Isolab standard used. In Figures 4A and B, hemoglobin separations blocked with 10 per cent pH 3-10 (of the total ampholyte) are shown to illustrate the difference between gels bounded to glass conventionally and rehydratable gels bonded to polyester film . In Figure 5 the scans of the standard S, A, F, S and C from Figure 4A and B along with an infant patient A, F hemoglobin, are superimposed (using the computer in the Biomed system) to show graphically the differences between conventionally cast and rehydratable gels. The reason for the increase in HbF on conventionally cast gels, benzidine stained or only acid fixed, and then scanned at 420mu, is not clear from these preliminary studies.
The higher voltage gradients used in Tables I and III with the rehydratable gels would result in excessive heating and burning if applied to conventionally cast gels, even employing the highly efficient Peltier cooling devices. Lower voltage gradients must be employed in focusing on conventionally cast gels which limits their resolution

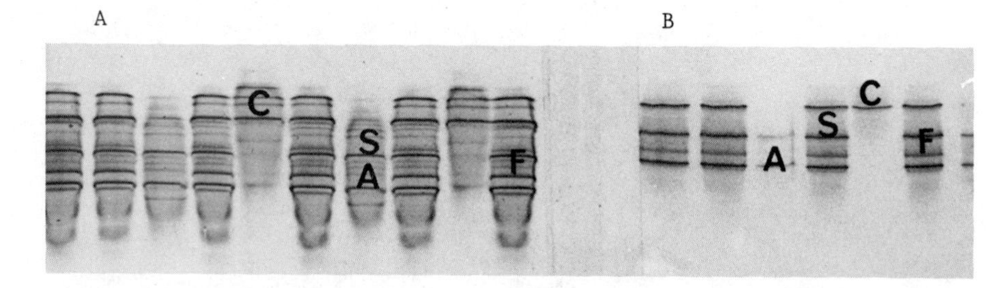


Figure 3. A) Hemoglobin standards A, F, S and C, AS and C separated on a five per cent T, pH 6-8 gradient of Pharmalyte . B) Same samples on a similar rehydratable gel. Both A and B were made with the same base reagents at the same time and were bonded to silanized glass plates.

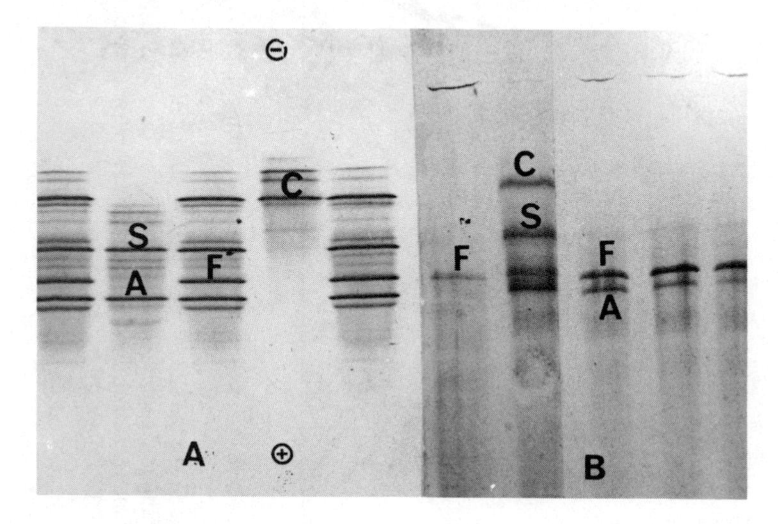


Figure 4. A) Samples the same as in 4A above. B) Standard A, F, S and C and infant patient samples with A and F. Gel in A backed on glass and in B on GelBond Pag..

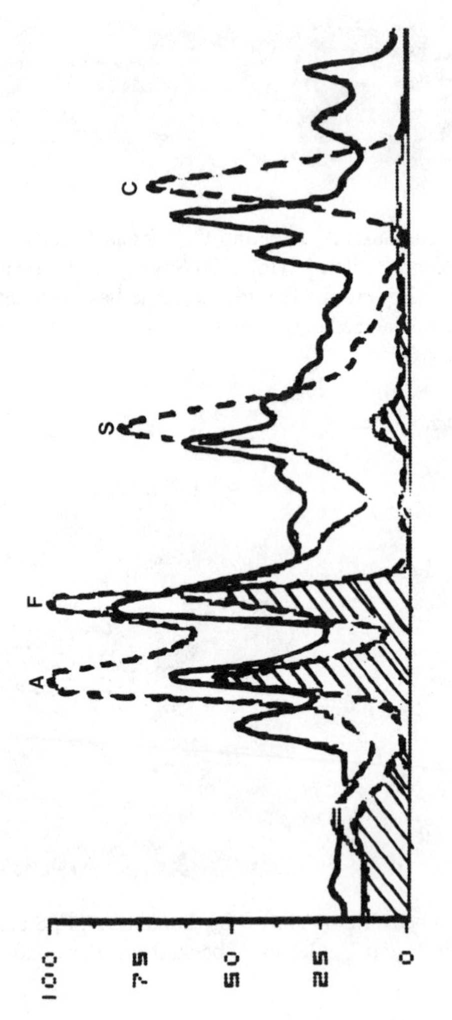


Figure 5. Densitometric scans of the hemoglobin A, F, S and C standard from Figure 5A (solid line) and 5B (dashed line) superimposed along with an infant blood sample showing only hemoglobins A and F (filled with slanted lines). The computer generated scan was traced over and filled by hand to improve the visualization of the three scans, which were presented in different colors on the CRT display.

potential [22]. Not only is the voltage gradient that may be applied limited, but also the reproducibility of conventional gel lots cast from a single batch of reagents may vary with storage. Thie latter appears to be due, in part, to the inability to maintain such gels at an equivalent degree of hydration from day to day during the storage period. It is of further interest to find that in using narrow pH range ampholytes, a small amount of pH 3-5 with the pH 4.2--4.9 Pharmalyte employed in the phenotyping of alpha 1-antitrypsin, not shown, must be employed to block the acid region of the gel. Otherwise, the acid end of the gel will not come to equilibrium. This is not a required procedure in rehydratable gels and suggests that additional acidic material remains in the gel, even after a pre-run and the focusing procedure.
Additionally, as may be seen in Figures 6A and B, where similar run conditions with pore sizes obtained in three per cent T and five per cent T gels show a generally similar pattern, but the more open pore size of the three per cent T gel allows for better entry of a sample, such as the cell extracts used here, into the gel when surface loading is used as readily is apparent in region A. Where resolution might be expected to be superior in a narrower pH range gradient, the effect of pore size appears to play an important role.In the five per cent T, pH 3-7 gels as opposed to the three per cent T, pH 3-10 gel in Figure 4, greater resolution is apparent in the latter gel. Previous studies have always suggested that a five per cent T gel be used for isoelectric focusing due to its minimal sieving effects on the macromolecules. However, it is apparent that the more open pore three per cent T gel is desirable for focusing, although it is more difficult to cast.

## Discussion

The use of rehydratable gels eliminates possible interfering affects from extraneous mat - erials that can complex or bind with the ampholytes or the sample components in ways that are as yet poorly understood. In addition, the necessity of blocking the narrow range pH rehydrated gels at the acid end indicates that acidic impurities are present in the conventionally cast gels that may be interfering with the pH gradient formation, in as yet, an undefined manner. As can be seen from Tables I and III, there is a much higher conductance in the conventional gel shown in Tables II and IV both at the beginning and the end of the separation; thus, some product, or products, of the polymerization procedure in the presence of ampholyte appear to remain in the gel adding to the electrical conductance.
The differences in the hemoglobin patterns between the conventionally prepared gels and the rehydratable gels indicate that there are different conditions present in the two media as evidenced both by unreacted hemoglobin and that reacted for peroxidase activity. The additional bands in the conventionally cast gels may represent complexes with the hemoglobin. However, the patterns appear similar to those that might be obtained when spacers are added to the ampholytes. Thus, the possibility that this difference is due, at least in part, to polymerization products complexing to the ampholyte and producing compounds with isoelectric points similar to the hemoglobins may not be ruled out. The findings here are in direct contrast to those described by Altland and Rossmann [23] on the oxidation effects of gels occurring during the drying process and their subsequent effect of inducing band-splitting which occurs even in the presence of reducing agents such as mercaptoethanol and di-thiothreitol in the dried Immobiline gels. Gels in their studies did not have any protective surrogate, such described by Frey *et al.* [7]. These were added before the

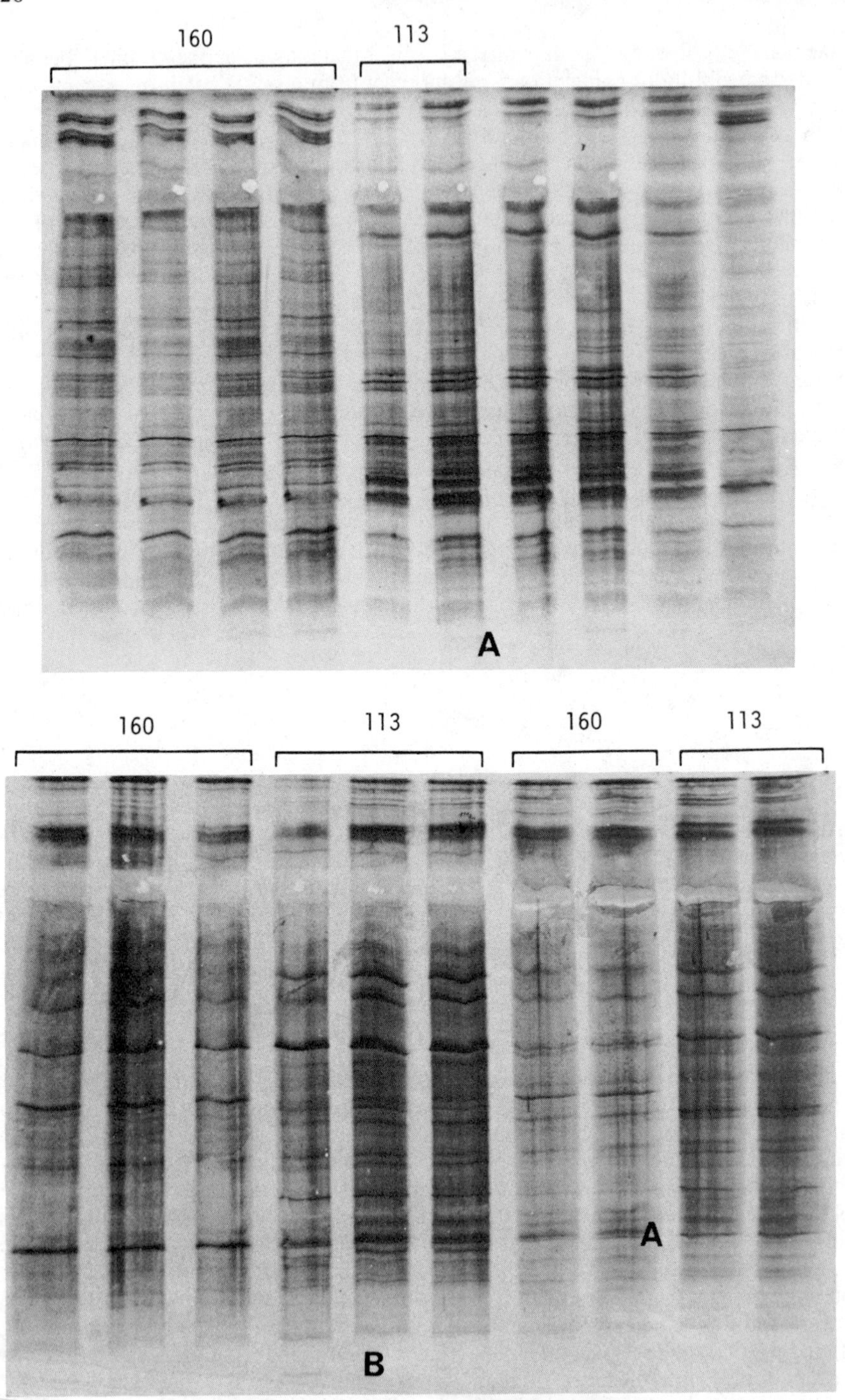


Figure 6. A) Separation of cell extracts 113 and 160 on a three per cent T rehydratable gel with 4 per cent pH 3-10 Servalyte and 7 cm between wick edges. B) Same samples separated on a five per cent T gel.

drying process and the Immobilines were in the gel during the polymerization process. The addition of reducing agents to isoelectric focusing gels containing synthetic carrier ampholytes, as opposed to immobilized ampholytes, has been discussed by Righetti *et al.* [25] who have shown that 2-mercaptoethanol, which is a buffer with a pK of 9.5, ionizes at the basic end of a gel and sweeps away focused carrier ampholytes. The phenomenon described in the present study is more likely indicative of a complexing of the hemoglobin and possibly unexpected production of spacers by as yet undefined reactions taking place during the polymerization process. It is unlikely that the hemoglobin results experienced in this study were due to incomplete absorption of the ampholyte from the rehydration medium into the gel during the rehydration step, since doubling the amount of ampholyte doubled the milliamperage in a similar volume gels, prepared from either the same or different lots of reagents.

The studies of Rüchel on the effect of increasing pore size in polyacrylamide gels at the junction of two dissimilar surfaces [24] and the rehydration overshoot, indicating even greater pore distension, as reported by Frey *et al.* [7], is readily demonstrated in the rehydratable gel systems where ultrathin-layer polyacrylamide gels are used. This is demonstrated in Figure 6, where the entry of the sample into the gel was greatly facilitated in the three per cent T gels as contrasted to the five per cent T gels. Thus, the common use of five per cent T gels as minimally sieving gels for isoelectric focusing may be of dubious utility and more open pore size gels probably should be used for optimal results. Such gels must be backed on a support material as they are much too soft to be handled as a free standing gel. The stabilization of pore size after storage in the dried state is of critical importance to reproduce results between laboratories and is basic in attempting meaningful standardization studies.

The results shown in Figure 7 from initial round robin studies with the rehydratable gels indicates that a 95 per cent correlation resulted between two laboratories as measured on the standard sample, run under similar conditions, on similar instruments, but qualitatively assayed by two different microdensitometers. This interlaboratory comparison, indicates that randomly selected rehydratable gels made at different times and stored for a period, here seven months, are reproducible and may well serve as a beginning for meaningful standardization in the field of protein separation using isoelectric focusing. The apparent differences between gels convalently bounded to glass and those backed on polyester film are, perhaps in part, due to the silanization process used with glass as opposed to the very different processes presently used to bond gels to polyester films. Also the two presently commercially available films Gel Fix (Serva) and GelBond Pag (Marine Colloids), use entirely different processes to convalently bind the polyacrylamide to the film. Thus, these various parameters also require further ellucidation and understanding before one support material can be recommended over another.

The use of the rehydratable gels obviates the previously nagging problem of having to prepare new gels for every pH range perturbation that one wishes to submit a sample to and allows one simply to take a film or glass-backed, dried gel off the shelf at will and rehydrate it with the desired ampholyte, or buffer. The preliminary results in this study indicate that rehydratable polyacrylamide gels offer a new approach to the standardization of a technique that has both delighted and frustrated users over the past 19 years.

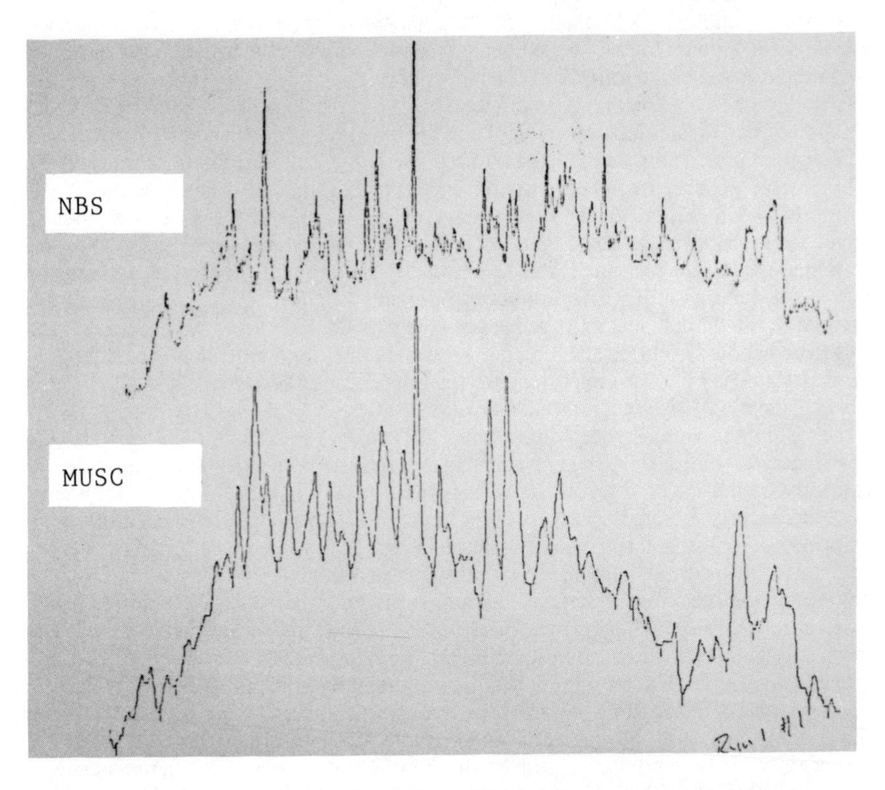


Figure 7. Comparison of densitometric results on cell extract #160 separations on randomized seven month old rehydratable gels in two different laboratories. The NBS separation was scanned on an LKB laser densitometer and MUSC gel was scanned on a Biomed SL-2D soft laser densitometer.

References


[1] Allen, R. C. In , Electrophoresis and Isoelectric Focusing in Polyacrylamide Gel, Allen, R. C. and Maurer, H. R., (Eds.), Walter deGruyter, Berlin, 1974; P. 220.

[2] Chrambach, A., Jovin, T. M., Svendsen, P. J. Rodbard, D., In: . Methods of protein Separation, Catsimpoolas, N. Ed. Plenum Press, New York, Vol. 2, 1976; p.27.

[3] Allen, R. C., Saravis, C. A. and Maurer, H. R. Gel Electrophoresis and Isoelectric Focusing of Proteins: Selected Techniques Walter deGruyter, Berlin, 1984; p. 2.

[4] Radola, B. J., In : Elektrophorese Forum '85 , Radola, B. J. (Ed.), Technische Universität, München , 1985, p. 21.

[5] Kinzkofer, A. and Radola, B. J. Electrophoresis , 1983, 4, 408.

[6] Frey, M. D., Kinzkofer, A., Bassim Atta, M. and Radola, B. J., Electrophoresis, 1986, 7, 28.

[7] Frey, M., Bassim Atta, M. and Radola, B. J. In Electrophoresis '84, Nuehoff, V., (Ed.) Verlag Chemie, Weinheim, 1984; p. 123.

[8] Gelphi, C. and Righetti, P. G. Electrophoresis 1984, 5, 257.

[9] Allen, R. C. and Arnaud, P. Electrophoresis , 1983,4, 205 .

[10]. Morris, C. J. O. R. and Morris, P., Biochem J. ,1971; 40, 43.

[11] Morris, C. J. O. R. and Morris, P.,In: Electrophoresis and Isoelectric Focusing in Polyacrylamide Gel , Allen, R. C. and Maurer, H. R., Eds., Walter deGruyter, Berlin, 1974; p. 10.

[12] Uriel, J. French Patent No. 1,483,742 17 May 1968 an addendum 91,262.

[13] Uriel, J., US Patent No. 3,578,604, May 11, 1971.

[14] Wren, D. W. and Mueller, G. P. US Patent No. 3,527,712, Sept. 8, 1970

[15] Robinson, H. K. Anal. Biochem. 1972, 4 353-366.

[16] Allen, R. C. Abs., 190th ACS Meeting, Chicago, 1985, AGFD 0017.

[17] Altland, K., Banzhoff, A., Hackler, R. and Rossmann, U., Electrophoresis, 1984, 5, 379.

[18] Radola, B. J., In : Electrophoresis '79, Radola, B. J. (Ed.), Walter deGruyter, Berlin, 1980, p. 79.

[19] Nochmuson, S. and Wood, H. US Patent No. 4,415,428, Nov. 15, 1983.

[20] Allen, R. C., Electrophoresis , 1980; **1** , 32.

[21] Allen , R. C. In Electrophoresis '82 Stathakos, D. Ed., Walter deGruyter, Berlin , 1982; p. 835.

[22] Allen, R. C., Saravis, C. A. and Maurer, H. R. Gel Electrophoresis and Isoelectric Focusing of Proteins: Selected Techniques Walter deGruyter, Berlin, 1984; pp. 72-75.

[23]. Altland, K. and Rossmann, U. Electrophoresis, 1985; 6, 314.

[24] Rüchel. R., Steere, R. L. and Erbe, E. F., J. Chromatogr., 1978, 166, 314.

[25] Righetti, P.-G., Tudor, G. and Gianazza, E., J. Biochem. and Biophys. Methods, 1982; 6, 219.


RECEIVED August 11, 1986

Chapter 9

# Application of High-Resolution, Two-Dimensional Electrophoresis to the Analysis of Wheat, Milk, and Other Agricultural Products

Norman G. Anderson and N. Leigh Anderson

Proteus Technologies, Inc., 12301 Parklawn Drive, Rockville, MD 20852

High resolution two-dimensional electrophoresis allows hundreds of proteins to be separated and characterized in submilligram samples of complex protein mixtures. Applications of this method to the analysis of agriculturally important products, including milk, meat, and wheat are reviewed. In a model study we analyzed 100 individual kernels of the wheat cultivar Newton (*Triticum aestivum L.*) for electrophoretic variants. One variant protein was found in 47 kernels, while three variant proteins occurred together in two of the kernels. The implications of two-dimensional electrophoresis for cultivar identification and the problem of relating electrophoretic protein variants to genetic variants are discussed.

High resolution two-dimensional electrophoresis with computerized image analysis and data reduction is the highest resolution method currently available for the analysis of complex protein mixture. In this discussion we review the potential of the method for the analysis of agricultural food products including milk, meat products, and wheat, and present one representative study on the analysis of Newton wheat for electrophoretic variants.

While 2-D electrophoresis is fundamentally a very simple technique, numerous technical problems arise if the method is to give high resolution results which are reproducible and intercomparable, and if quantitative results are to be obtained. The interpretation of data posses additional challenges because the vast majority of proteins in both plant and animal tissues have been neither described nor named, and therefore have functions which remain to be discovered.

Our objective is to develop analytical systems and protein data bases for agricultural products which allow proteins to be identified with confidence by position, to identify as many "known" proteins as possible in the patterns, and to develop methods for quickly and reliably finding both quantitative and qualitative differences

0097-6156/87/0335-0132$06.00/0

between samples. The data bases are linked electronically to the patterns displayed in color on a CRT. This arrangement allows rapid recovery and display of all available information relative to each protein resolved, and differences between one or more patterns may be discovered. New multi-windowing techniques allow the same section of several dozen patterns to be seen and intercompared in one image.

## Fundamentals of 2-D Electrophoresis

High resolution 2-D electrophoresis is done in acrylamide gels using either isoelectric focussing or non-equilibrium pH gradient electrophoresis in the first dimension, and electrophoresis in the presence of sodium dodecyl sulfate in the second (1-4). Denaturing and reducing conditions are used in both dimensions so that multimeric proteins are dissociated into subunits, and the tertiary structure of subunits is largely lost. The resulting separation is based on charge (which reflects the ratio between acid and basic amino acids in each protein), and mass. Since the proteins are unfolded, all amino acids contribute to the overall charge.

The resolving power of 2-D electrophoresis is very high. When complex mixtures of proteins are carbamylated to produce charge shift trains for proteins having a large range of molecular weights, it is found that single charge differences can be easily detected over the mass range of approximately 10,000 daltons to over 100,000 daltons (5). Carbamylated proteins have therefore been used to produce charge trains which appear on 2-D patterns as horizontal rows of spots which are useful for calibration in the first dimension (6,7).

For calibration in the second dimension, use is made of proteins of known molecular mass which give horizontal streaks across the final gel which each define a different and known mass (8). Mass differences of greater than 2% may be detected over the entire mass range resolved, which means that single amino acid deletions cannot ordinarily be seen (5).

Resolution may be defined in terms of the total number of spots of the sizes seen on a particular gel which could be packed into a 2-D pattern and still be resolved (9). This is a theoretical number, and it is independent of the actual number seen in a gel. It accurately reflects differences in resolution seen between batches of gels. Experimentally we have produced gels with theoretical resolutions as high as 38,000 using our ISO-DALT system.

Recently the ISO-DALT system has been extensively modified so that 40 isoelectric focussing gels may be run in parallel, the slab gels are cast using a computerized gel casting system, and larger second dimension slab gels are used (N.L.Anderson, unpublished studies).

A number of image analysis and data base systems have been developed to allow intercomparison of gels, and for spot quantitation. The TYCHO system allows images to be rapidly acquired, streaks and other imperfections removed, and images matched using non-linear stretching algorithms (10). The reliability of this process is demonstrated by systematically removing one at a time each spot from one of the two patterns being intercompared, regenerating the removed spot from the remaining data, and then comparing the regenerated spot position with the original true position. The average error between the original and the regenerated spots was a small fraction of a millimeter (9). Using Coomassie Blue staining, more than 100 proteins may be measured in liver samples with coefficients of variation of less than 15% (11). Using specific antiserum staining individual proteins may be identified in entire 2-D patterns transferred to nitocellulose or other suitable binding supports (12).

## Analysis of Milk

The analysis of milk illustrates the power of 2-D methods to resolve interesting problems. While the major components of human milk and cows milk have been identified on 2- D gels (14), hundreds of minor ones which can be resolved on silver-stained gels remain to be characterized. The contribution of 2-D methods to the solution of a central problem in milk coagulation important to cheesemaking is not well known and, since it has only appeared in the literature of human clinical chemistry, it is briefly reviewed here. Caseins, including kappa casein, have isoelectric points around pH 4.5, and all may be isoelectrically precipitated. The terminal peptide region of Kappa casein contains acidic amino acids and is multiply post-translationally modified by the addition of negatively charged groups including phosphate and sialic acid, to give the characteristic acid isoelectric point. However, when the negatively charge-modified region (termed the macropeptide) is cleaved off by chymosin, the remaining larger molecule (para kappa casein) bears a net positive charge. Para kappa casein, which was discovered to be very basic using 2-D gel analysis, then precipitates with the remaining caseins at neutral pH. Experimentally other similarly basic proteins will also precipitate alpha and beta caseins (14).

This discovery of the mechanism of the first step in cheese making is based on 2-D analysis where the native kappa casein was found at the acid end of the 2-D gel (i.e., to the left), while the larger cleavage product, para kappa casein, was found to be so basic that is was not on a regular 2-D pattern, but was only found when non-equilibrium first dimension conditions, specifically designed for the analysis of very basic proteins, were employed. Once it was realized how very basic para kappa casein is, the mechanism of protein precipitation in cheesemaking became apparent and was confirmed using other similarly basic proteins. 2-D analysis is therefore of great use in the study of protein processing, and offers many interesting opportunities to study protein changes during the ripening of cheese, and in the alteration of proteins in foods during processing and storage.

## Muscle Proteins

Muscle proteins have been extensively studied (see review in ref. 15) and this work provides the basis for future research on various meats. While many muscle proteins have been identified in samples of human and rabbit origin, we have found no published systematic study of meat proteins of various food animals, nor of the changes which occur during aging, storage, and spoilage. Such studies would be useful for the identification of adulterants, for studies on curing and aging, and in the evaluation of storage conditions.

## Analysis of Wheat Endosperm Proteins

High resolution two-dimensional electrophoresis has been applied to the analysis of endosperm proteins (16-22) and used to determine the chromosomal location of a number of different proteins in aneuploid strains of wheat. The method has not previously been used to look for electrophoretic protein variants within one strain. In this paper we describe individual analyses of 100 grains of Newton wheat (*Triticum aestivum* L.) with the procedures previously described (22). Newton (CI 17715) is a hard red winter wheat and is an increase of a single F4 plant derived from six parent strains (23). One dimensional electrophoresis has been used to detect variants of endosperm proteins, commonly termed biotypes (24,25), coexisting in one strain of wheat. They have not been systematically studied by two-dimensional electrophoresis however. The questions raised here are: (1) Can high-resolution 2-D electrophoresis using non-equilibrium conditions in the first dimension resolve reproducibly endosperm proteins in a series of 100 samples, (2) are electrophoretic proteins variants observed in a widely planted strain with this method, (3) will the method prove useful in strain characterization, and (4) will molecular inventories provided by 2-D maps prove useful and possibly necessary for patenting or registering new varieties.

Methods. Individual wheat kernels were ground in a small stainless steel mortar and extracted with a mixture containing urea, a nonionic detergent, and a reducing agent as previously described (22). Using a modification of our ISO-DALT system (3,4) we cast and ran both the first dimension non-equilibrium pH gradient gels and second-dimensional slab gels in batteries of 20. All gels were fixed and stained in Coomassie Brilliant Blue as described, and analyses were done on 8x10 inch high contrast prints. A Bausch and Lomb zoom transfer scope, model ZT4-H, was used to compare prints and to prepare maps of patterns.

Results. A map of Newton wheat proteins is shown in Fig. 1. Of the 100 grains analyzed, 47 maps contain the landmark area IV electrophoretic variants shown in Fig. 2, while two maps exhibited all three of the landmark area II electrophoretic variants shown in Fig. 3.

Discussion. The results indicate that electrophoretic variants exist in one sample of an inbred strain of wheat, that these variants are easily detected by high resolution 2-D electrophoresis, and that reproducible results may be obtained during an extended series of analyses. One major electrophoretic variant was found in 47% of the grains, while three minor electrophoretic variants occurred together in 2% of the grains analyzed. It appears that the genes for the three minor variants may be tightly linked, or that all may be the products of one gene, with the lower-molecular-mass variants produced by successive cleavage of two peptides off a larger variant.

It is important to distinguish between electrophoretic variants, posttranslational modification, physiological variants, and genetic variants. The electrophoretic variants described here could be true genetic variants produced either by mutations of existing genes in Newton, or by the introduction into the strain of new genes allelic

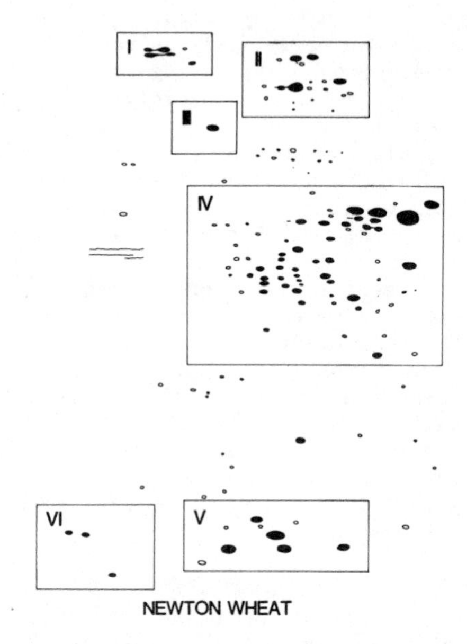


Figure 1. Map of major endosperm proteins of Newton wheat. The landmark areas indicated by rectangles are those previously described (9). The electrophoretic variants in this study occurred in landmark areas II and IV.

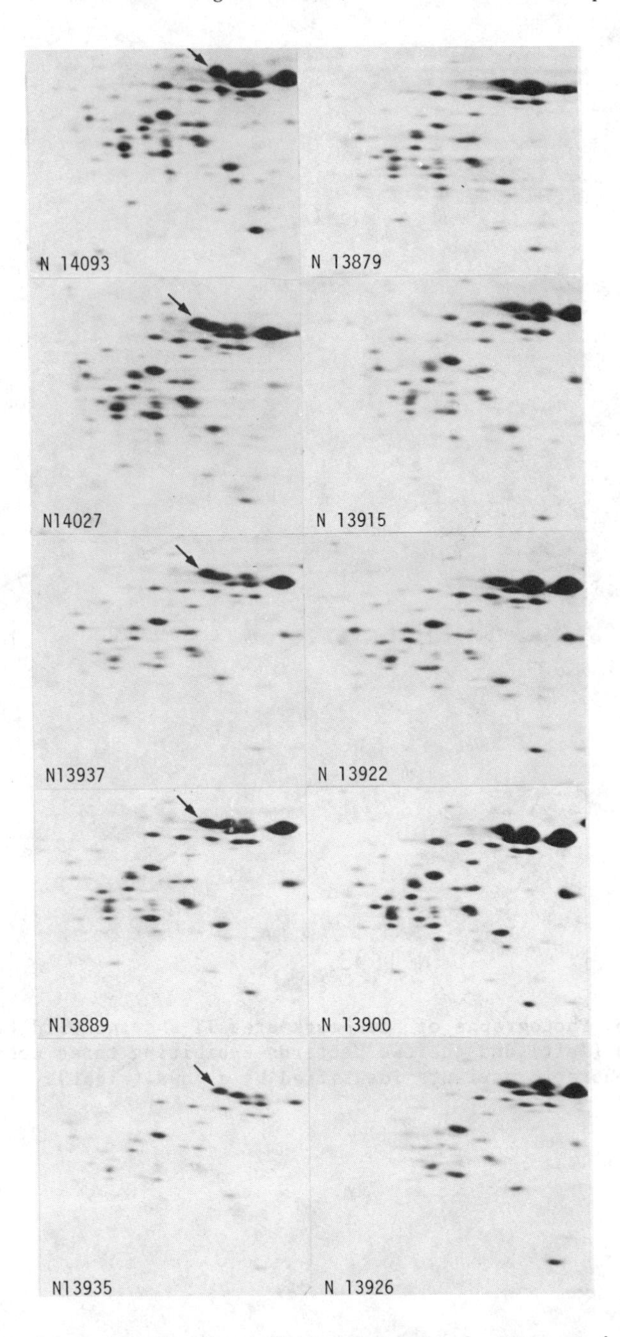


Figure 2. Photographs of portions of Coomassie blue stained gels showing landmark area IV. Representative wild type patterns are shown on the right, and patterns containing electrophoretic variants are shown on the left. Arrows identify variant protein.

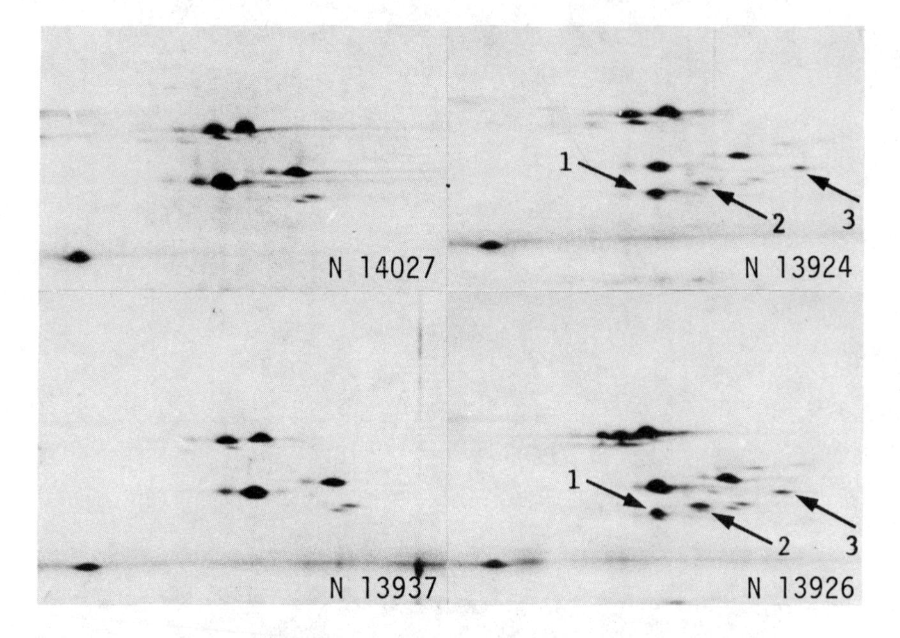


Figure 3. Photographs of landmark area II showing wild type patterns (left) and the two patterns exhibiting three coexpressed electrophoretic variants identified by arrows (right).

to those already present. In either case, breeding studies would show whether or not the gene products (proteins) behaved in a Mendelian fashions, whether a variant and a candidate wild type protein are allelic, and whether any of the genes are tightly linked. If an electrophoretic variant is a physiological variant, then the gene for that protein may behave in a Mendelian manner, but the appearance of the protein gene product would be controlled by a physiological or environmental variation as is the case, for example, with temperature mutants. New posttranslational modifications that are produced by a new or altered protein-modifying enzyme may also yield new spots that are not the direct product of altered genes. Sequencing of the gene responsible for a "wild type" and for a candidate genetic variant provides the definitive answer to the question of whether a given electrophoretic variant is indeed a direct genetic one. In the practice of clinical chemistry, electrophoretic variants that are not direct genetic variants are so rare that it is nearly always assumed that an electrophoretic variant is a genetic variant. A similar assumption has often been made in the interpretation of 1-D electrophoretic patterns of endosperm proteins. However, with the high resolution made possible by 2-D electrophoresis it is important to reexamine all possible sources of electrophoretic variants and to demonstrate experimentally the molecular and genetic basis of observed variants. In experimental studies of wheat endosperm proteins previously reported, altered electrophoretic patterns have not been seen in response to different environmental conditions (26-30), but have been noted in cases of extreme sulfur deficiency in wheat (31).

Since only a fraction of a grain is actually required for analysis, the seed embryo and part of the endosperm may be retained and germinated. Thus, 2-D electrophoresis may be used to screen several hundred individual kernels and the results used to select embryos that will produce uniform strains which may be crossed to demonstrate that candidate variants and wild-type pairs are indeed allelic. Whether the variants described here contribute in any important way to flour and baking quality can best be discovered by producing substrains homozygous for each.

If 2-D maps are to be used for varietal identification and for inclusion in patent and registration descriptions as we have suggested (22), then either residual variability should be eliminated from the strain, or variants should be included in varietal maps and descriptions. This suggestion arises because entire organisms and not genes or gene products are patented or registered under present interpretations of the relevant law in the United States (32).

When an electrophoretic variant is observed, it is important to discover which other protein in a pattern is the product of the allelic wild type gene. The positive identification of variant and wild-type pairs will require partial sequencing of the proteins involved and/or breeding studies in which plants homozygous for each gene of a pair are obtained. The most likely candidate for the wild-type protein for the polymorphism shown in Fig. 2 is the protein immediately to its right, which appears to be diminished in quantity when the polymorphism is present. For the three variants in Fig. 3, the protein immediately above variant number 1 may be its wild-type allele; however for variants 2 and 3 it is not possible to assign candidate wild-type spots. Methods for identifying posttranslational modification have been previously discussed (22).

## Genetic Purity

In any analytical series such as the one described, the questions of strain definition and of cultivar purity arise. For strains to be used in the Great Plains area of the United States, many breeders consider genetic heterogeneity necessary for best agronomic results. In contrast, homogeneous types are usually preferred in Europe and in Canada. In either case a description of the original strain as developed by the breeder is required if purity during extended cultivation is to be assessed. Any wheat grown on a large scale will inevitably become genetically contaminated with other germ plasm, or a batch of seed may be physically contaminated with a few kernels from another variety. Therefore the variants observed may have been present in the original plant and, if unrelated to any undesirable property of the cultivar, may never have been bred out of the strain. Alternatively the minor variant may have appeared later and may be present in a small fraction of existing stores of Newton wheat. The possibility also exists that one or both of the electrophoretically variant spots could be physiological variants what are expressed under some environmental conditions and not others. While only extreme growing conditions have been found to alter 1-D electrophoretic patterns of wheat endosperm proteins (31), the unlikely possibility of physiological variability can only be ruled out experimentally.

The basic questions relating to the design of experiments to settle these problems are whether the electrophoretic variants described were present in the original single parent plant from which the strain was derived, whether mew mutations have occurred during bulking and in production, whether new genetic material has been introduced by accidental crossing during the years since the variety was first produced, or whether the sample we have analyzed is unique and contaminated. Obviously a more extended series of studies is required to answer these questions concerning the past history of the variety Newton..

Prospectively, 2-D mapping of the parent strains and the cultivar registered would define the variety at the outset and would suggest the parental origin of each electrophoretic variant seen. Should electrophoretic variants be observed, they can be bred in or out at the very outset, or heterozygosity may be deliberately maintained using 2-D electrophoretic techniques to guide breeding. In addition, strains homozygous for each variant may be produced to determine whether any of the variant proteins are linked to a desired property, for example to baking quality. Once a cultivar is defined by 2-D mapping, then it is a simple matter to follow changes occurring with time by mapping small numbers of kernels obtained from many different locations. For retrospective studies, carefully randomized sampling of many different seed stocks during successive years will be required to answer some of the questions raised.

When a data base of 2-D maps of wheat proteins becomes available, it will be easier to identify contaminating kernels or their genes, to discover new proteins arising from adventitious crosses, and to identify the sources of such accidentally introduced genes. To maintain a new variety, it may be necessary to continually monitor each stage in seed production from breeder's seed through foundation seed, registered seed, on to the certified seed supplied to farmers. The studies reported here therefore constitute only a first step in the application of high-resolution two-dimensional electrophoresis to wheat cultivar identification and to wheat breeding.

## Acknowledgments

Part of this work was supported by the U.S.Department of Energy under contract No. W-31-109-ENG-38 to the Argonne National Laboratory. We express our appreciation to Dr. Rollin G. Sears, Agronomy Department, Kansas State University, for samples of Newton wheat, to to Dr. Jean-Paul Hoffmann for advice and helpful discussions, and to Sandra Tollaksen for assistance in performing the 2-D analyses.

## Literature Cited

1. O'Farrell, P. H. J. Biol. Chem. 1975, 250, 4007-21.
2. O'Farrell, P. Z.; Goodman, H. M.; O'Farrell, P. H. Cell 1977, 12, 1133-42.
3. Anderson, N. G.; Anderson, N. L.; Anal. Biochem. 1978, 85, 331-40.
4. Anderson, N. L.; Anderson, N. G. Anal. Biochem. 1978, 85, 341-54.
5. Anderson, N. L.; Anderson, N. G. Clin. Chem. 1984, 30, 1898-05.
6. Anderson, N. L.; Hickman, B. J. Anal. Biochem. 1979, 93, 312-20.
7. Tollaksen, S. L.; Edwards, J. J.; Anderson, N. G. Electrophoresis 1981, 2, 155-60.
8. Giometti, C. S.; Anderson, N. G.; Tollaksen, S. L.; Edwards, J. J.; Anderson, N. L. Anal. Biochem. 1980, 102, 47-58.
9. Taylor, J.; Anderson, N. L.; Anderson, N. G. Electrophoresis 1983, 4, 338-46.
10. Anderson, N. L.; Taylor, J.; Scandora, A. E.; Coulter, B. P.; Anderson, N. G. Clin. Chem. 1981, 27, 1807-20, 1981.
11. Anderson, N. L.; Nance, S.L.; Tollaksen, S. L., Giere, F. A.; Anderson, N. G. Electrophoresis 1985, 6, 592-9.
12. Anderson, N. L.; Nance, S. L.; Pearson, T. W.; Anderson, N. G.; Electrophoresis 1982, 3, 135-42.
13. Taylor, J.; Anderson, N. L.; Scandor, A. E.; Willard, K. E.; and Anderson, N. G. Clin. Chem. 1982, 28, 861-6.
14. Anderson, N. G.; Powers, M. T., and Tollaksen, S. L. Clin. Chem. 1982, 28, 1045-55.
15. Giometti, C. S. CRC Reviews in Clinical Laboratory Sciences 1982, 18, 79-109.
16. Brown, J.W.S.; Flavell, R.B., Theor. Appl. Genet. 1981, 59,349-49.
17. Brown, J.W.S.; Law, C.N.; Worland, A.J.; Flavell,R.B. Theoret. Appl. Genet. 1981, 59, 361-71.
18. Holt, L.M.; Astin, R.; Payne, P.I., Theoret. Appl. Genet. 1981, 60, 237-43.
19. Gabriel, D.W.; Ellingboe, A.H. Physiol. Plant Path. 1982, 20, 349-57.
20. Jackson, E.A.; Holt, L.M.; and Payne, P.I. Theoret. Appl. Genet. 1983, 66, 29-37.

21. Dunbar, B.D.; Bundman, D.S.; Dunbar, B.S. Electrophoresis 1985, 6, 39-43.
22. Anderson, N.G.; Tollaksen, S.L.; Pascoe, F.H.; Anderson, L. Crop Science 1985, 25, 667-674.
23. Heyne, E.G.; Niblett, C.L. Crop Science 1978, 18, 696.
24. Ellis, J.R.S.; Beminster, C.H. J. natn. Inst. agric. Bot. 1977, 14, 221-31.
25. Ellis, J.R.S. Phil. Trans. R. Soc. Lond. 1984, B304, 395-407.
26. Lee, J.W.; Ronalds, J.A. Nature 1967, 213, 844-6.
27. Wrigley, C.W. Biochem. Genet. 1970, 4, 509-6.
28. Meecham, D.K.; Kasarda, D.D.; and Qualset, C.O. Biochem. Genet. 1978, 16, 831-53.
29. Zillman, R.R.; Bushuk, W. Can. J. Plant Sci. 1979, 59, 281-6.
30. Lookhart, G.L.; Finney, K.F.; Bruinsma, B.L. Cereal Chem. 1984, 61, 496-9.
31. Wrigley, C.W.; duCros, D.L.; Fullington, J.G.; Kasarda, D.D. J. Cereal Sci. 1984, 2, 15-24.
32. Williams, S. B.,Jr. Science 1984, 225, 18-23.

RECEIVED November 25, 1986

# Chapter 10

# Applications of Isoelectric Focusing in Forensic Serology

**Bruce Budowle[1] and Randall S. Murch[2]**

[1]Laboratory Division, Forensic Science Research and Training Center, Federal Bureau of Investigation, Quantico, VA 22135
[2]Serology Unit, Laboratory Division, Federal Bureau of Investigation, Washington, DC 20535

The typing of certain genetically controlled polymorphic proteins present in human body fluids is an important aspect of the analysis of serological evidence. Until recently, the routine electrophoretic analysis of genetic markers of interest relied entirely upon conventional electrophoretic techniques. Isoelectric focusing (IEF) offers an effective alternative to conventional electrophoresis for genetic marker typing. This methodology resolves allelic products based upon their isoelectric point rather than charge-to-mass ratio. The advantages of employing IEF (particularly when utilizing ultrathin-layer gels) include: reduction of time of analysis, increased resolution, the possibility of subtyping existing phenotypes, increased sensitivity of detection, the counteraction of diffusion effects, and reduced cost per sample analysis. The IEF approach has been applied to the analysis of a number of genetic markers in bloodstains, semen stains and hair. The genetic markers of particular interest in this review are group-specific component, transferrin, and hemoglobin. Forensic serologists are interested in utilizing analytical methods for typing genetic markers in body fluids that are rapid, sensitive, highly informative, economic, reproducible and allow for conservation of sample. The more useful genetic markers are those which exhibit a large number of commonly occurring, alternate forms, which are relatively stable, and can be readily resolved and assayed. The analyses of the majority of genetically-controlled polymorphic systems of forensic interest have utilized the technique of electrophoresis. Historically, the development of electrophoretic systems for forensic analyses has centered upon conventional electrophoretic techniques (1, 2). Such systems have inherent drawbacks: diffusion of sample constituents during analysis, poor resolution, difficulty in manipulation of the gel, electroendoosmotic effects, extended analytical periods, low sensitivity of detection, and suboptimal methods of sample application which cause gel disturbance and foster band diffusion. The limitations of the efficacy of conventional electrophoretic methods lead to an increased number of inconclusive and negative results on casework samples. The nature and extent of these problems promote efforts within the forensic community to seek more powerful analytical methods for the typing of polymorphic protein systems.

An alternative electrophoretic approach for protein separation is isoelectric focusing (IEF). The sensitivity, resolving power, speed, economy and ease of assay make IEF a desirable tool for the forensic serologist. Unlike conventional electrophoresis, IEF employs a pH gradient which counteracts diffusion effects and concentrates proteins into sharp stable pH zones at the isoelectric points of the proteins. Subtle differences in the overall net charges of commonly occurring protein variants (which were not routinely demonstrable using conventional electrophoresis) can be resolved by IEF. Genetic markers for which additional phenotypes (called subtypes) have been observed by IEF include phosphoglucomutase-1 (PGM) (3, 4), esterase D (EsD) (5-7), transferrin (Tf) (8, 9), group-specific component (Gc) (10, 11), haptoglobin (12), and alpha-1 antitrypsin (Pi) (13, 14). Additionally, these narrower, sharper protein bands result in more protein per unit gel volume (compared with conventional electrophoresis) which, in effect, presents more protein for the subsequent assay. Thus, the sensitivity of protein detection of IEF is generally superior to conventional electrophoresis. The effect of concentrating protein into narrow zones during IEF also permits sample application to be in broad zones. Therefore, larger volumes of dilute or weak sample extracts can be applied to the gel with the expectation of IEF concentrating the proteins of interest (15). Thus, some forensic samples not previously typeable can now be analyzed using IEF.

Recent advances in IEF have shown that proteins can be separated on ultrathin-layer polyacrylamide gels ($\leq$ 0.36 mm thick gels) instead of thin-layer slab gels (1-2 mm thick gels) (16). The advantages of ultrathin-layer polyacrylamide gel isoelectric focusing (ULPAGIF) are increased resolution and sensitivity of detection, shorter separation and staining times, and reduced cost.

Ultrathin-layer polyacrylamide gels have a greater surface area-to-volume ratio than thicker gels and, therefore, are more effective at heat dissipation. Thus, higher field strengths can be applied to the gels and an increased concentration of synthetic carrier ampholytes can be incorporated into the gel with the expectation of superior resolution [resolution is proportional to the square root of the voltage (17, 18)]. Furthermore, the use of higher voltage gradients produces narrower bands (19-21), which results in more protein per unit gel volume. Since this, in effect, presents more protein for the subsequent assay, the sensitivity of detection of the system is increased. The applied higher field strengths also result in faster separation times making it possible to type more samples in less time than previously possible (7, 19, 22-27). Due to shorter diffusion pathways, staining times are greatly reduced (16, 28). Finally, by reducing the gel thickness, the quantities of reagents (and thus the cost) are reduced without compromising the number of samples analyzed.

Since the quality of case samples is often uncertain and frequently only minute amounts of sample are available for analysis, linearity, resolution, reproducibility and sensitivity are extremely important factors to consider when developing an ULPAGIF method for analysis of genetic markers in serological evidence. The method should be expected to resolve only the most prevalent phenotypes of a genetic marker system with little concern given to the identification of exceedingly rare variants. From a human genetics standpoint, it would be beneficial to be able to resolve and identify all rare variants that exist for each particular marker of interest. However, this is not practical in a forensic laboratory setting. The forensic scientist must consider the probabilities of encountering these rare variants versus the information obtainable by concentrating one's efforts on the common phenotypes, the need to conserve sample, the cost and the speed of analysis. For example, in typing erythrocyte acid phosphatase (EAP) in forensic samples on a day-to-day basis, it seems most reasonable for the forensic scientist to utilize a rapid and sensitive analytical method which resolves the common allelic products (A, B, C, R) in a wide variety of forensic samples. It is not to the benefit of the forensic scientist to use a method or multiple methods designed primarily to resolve rare variants especially when these have an expected frequency of less than 1 in 10,000. The benefits of the latter approach do not outweigh the detractions. Also, the forensic scientist should consider that samples rarely can be considered ideal and typing for rare variants should be undertaken with extreme caution. In most instances, the precise environmental and genetic history of forensic biological samples are unknown. Further, family studies usually cannot be performed in order to confirm the existence of the rare variant.

Despite the advantages of using ultrathin-layer polyacrylamide gels, there has been a reluctance to routinely use this methodology for genetic marker typing. Part of this reluctance can be attributed to serious concerns regarding a lack of reproducibility of banding patterns from gel to gel and a lack of linearity across a gel. Problems with variation in resolution and wavy patterns can lead to possible mistypings, increased inconclusive determinations and/or time consuming retyping of samples. These problems are surmountable and the solutions are well described in a recent review by Budowle (29). Briefly the factors to consider are:

## Casting Techniques

Care should be exercised in the manner the gels are cast. There are two general methods for casting ultrathin-layer polyacrylamide gels - capillary and flap techniques [see Allen (19) and Radola (25) for the methodology]. After gels are poured using the flap technique, a weight is placed on the top glass plate to seal the gel. If this weight is not evenly distributed, the resulting gel will not have a uniform thickness. The results are distorted protein patterns and undesired gel drying. Therefore, several glass plates should be used to evenly distribute the weight (24).

The capillary method in comparison to the flap technique is less time-consuming and allows for easier gel preparation. The major problem with the capillary method is avoiding the production of bubbles in the gel during pouring. This problem can be solved by lightly tapping the plate at the gel solution front as it is migrating under the plate by capillary action. This will produce an even flow of the gel solution and eliminate bubble formation. In the event that a bubble is trapped under the glass plate, simply slide the glass plate to expose the bubble, the bubble will dissipate, and then slide the plate back to its original position. After the gel polymerizes remove it from the casting tray and wrap it in plastic wrap. Gels can be maintained in the refrigerator in this condition from one to three weeks. In contrast, gels cast by the flap technique can be maintained at room temperature from three to six months (unpublished data).

The methods developed by Budowle (7, 21-24, 28, 30-32) routinely used 200 um thick gels instead of thinner gels used by others (25-27, 34-36). Even with the best of care in casting gels, there are still some slight variations in thickness within a gel. A 10% decrease in a portion of a 200 um thick gel (20 um) is less deleterious than a 10% decrease in a portion of a 100 um thick gel (10 um). The latter gel, at times, yielded uneven migration of bands, resulted in protein migrating along the gel surface instead of entering the gel and/or dried out (unpublished data and personal communication, R. C. Allen).

## Effect of Electrode Distance

As the distance between the electrodes is reduced the minor depressions in the gel present less of a problem (due to increased conductivity). Kinzkofer and Radola (26) have had great success with 50-100 um thick gels with inter-electrode wick distances of 1-3 cm. However, for the purposes of genetic marker typing these inter-electrode wick distances are impractical. The pH ranges of commercially available synthetic carrier ampholytes are not narrow enough for such distances and thus impose limitations on gel dimensions. Allelic products with slightly different isoelectric points lie too close together to resolve in this manner, especially when functional enzyme overlay assays are employed (diffusion problems). When the inter-electrode wick distance is 5-10 cm, as is necessary for most genetic marking typing systems, 200 um thick or thicker gels should be employed. Use of a gel with this thickness will contribute favorably to the reproducibility of the system.

## Effects of Sample Protein and Salt Concentrations

Excess protein loading is tolerated more by thicker gels than thinner ones. For example, in PGM analysis, 6-8 ug of protein can be applied to 200 um thick gels with an 8 cm inter-electrode wick distance containing 4% (w/v) pH 5-7 ampholytes (LKB) and produce linear PGM subtype patterns (24). Less than half that protein concentration can be applied to a similar gel, 100 um thick, without producing bowed PGM subtype patterns. Similar trends occur for salt loading. However, in the case of salts, diluting the sample (and still applying the whole sample to the gel) can reduce deleterious salt effects. It appears the limiting factor is the amount of salt per unit sample volume and not the total amount of salt. The effect of salts led Pflug (37) to develop a wedge-shaped gel for PGM subtyping of semen samples. The gels were 300 um thick at the anode decreasing in thickness to 50 um at the cathode. Pflug applied semen stain extracts at the thicker portion of the gel. There was less distortion of PGM due to the salt effects. The wedge-shaped gel combined the higher resolving capacity of ultrathin-layer gels with the loading capacity of thicker gels. In contrast, Budowle (24), utilizing a straight 200 um thick gel, a reduced inter-electrode wick distance of 8 cm and a much higher final voltage gradient (290 V/cm vs. 180 V/cm), observed no deleterious salt effects on PGM from semen stains.

It should be noted that protein loading tolerances are also dependent upon the pH range and the particular manufacturer of the synthetic carrier ampholytes. For example, more than twice the protein that can be applied to a pH 5-7 ampholyte (LKB) gel can be applied to a pH 4.5-5.5 Servalyte gel without any protein band distortion. While a gel containing pH 4.5-5.4 Pharmalytes can only handle 70% of the protein of the Servalyte gel (unpublished data).

## Effect of Field Strength

The primary advantage of ultrathin-layer polyacrylamide gels is the ability to use higher field strengths. Giddings and Dahlgren (17), Rilbe (18), and Allen (19, 20) have shown that resolution is proportional to the square root of the voltage gradient. Since these gels can more effectively dissipate Joule heat, higher field strengths can be utilized with the expectation of increased resolution. For gels with an inter-electrode distance of 5-10 cm, voltage gradients as high as 300-700 V/cm have been used (7, 19, 20, 21-26, 28-31). Furthermore, the higher voltage gradient can produce narrower protein bands (20, 22). The concentrating of protein bands into more narrow zones results in more protein-per-unit gel volume which, in effect, presents more protein for the subsequent assay. Allen and Arnaud (20) using the Rohament P enzyme test demonstrated that a ten-fold voltage gradient increase (50-500 V/cm) yielded a 2.3-fold increase in the number of bands resolved and a band-width decrease by a factor of 2.3 to 2.5. Budowle and Murch (22) observed that by increasing the voltage gradient from 340 V/cm to 460 V/cm a one-third decrease in the width of Pi bands could be realized. Many investigators (27, 33-36) have utilized ULPAGIF, but do not take obvious advantage of the application of higher field strengths. Usually these investigators used voltage gradients less than 200 V/cm. Genetic variants with very close isoelectric points, such as the C1 and C3 allelic products of Tf (38-40), the M3 and M2 allelic products of Pi (41), and rare variants of Hb, can be easily resolved when higher field strengths are employed (300 V/cm - 700 V/cm).

A problem that manifests itself with the use of certain ampholyte ranges and high field strengths is the appearance of "hot spots" (42). These "hot spots" are conductivity gaps in the gradient and serve as a limiting factor for the field strength that may be applied to the gel. With sufficiently high field strengths the "hot spots" become burn spots on the gel (19, 31). This is an undesirable effect, especially when the protein(s) of interest has yet to be resolved. The problem can be solved by adjusting one of the electrodes (after a prescribed time into the run) to bypass the area of the gel where the conductivity gap exists. There then will be a shorter inter-electrode wick distance, a more uniform conductance across the gel, no gel burning, and the run can be completed to obtain the desired resolution. This approach has been successfully used for subtyping Gc in bloodstains (31).

## Reproducing the Voltage Gradient

Not only is resolution dependent upon the voltage gradient, but for ultrathin-layer polyacrylamide gels, the voltage gradient is also important for gel-to-gel reproducibility and linearity across a gel. Budowle (21, 24) demonstrated that once the ideal voltage gradient conditions were empirically determined, the voltage gradient had to be reproduced at intervals for every ULPAGIF run. Although the gels were still focused with constant power, the power mode was adjusted at regular intervals depending upon the voltage parameter. By following this approach, the same results were obtained from gel to gel. To produce band linearity across a gel, it was found that the initial voltage (regardless of the inter-electrode wick distance) applied to the gel during prefocusing could not exceed 250 volts. As long as the conditions were reproduced, well-resolved, highly reproducible patterns were obtained Further, this approach appears to compensate for ambient temperature and humidity effects. Current can also be a limiting factor. However, the previously reported methods (21, 24) have maintained the current at low enough levels so as not to have an impact on gel reproducibility.

Reagents

Fresh stock solutions of recrystallized acrylamide should be considered for ULPAGIF. Chrambach, et al. (43) have shown that impurities, such as acrylic acid, will confer ion exchange properties to the gel. This can cause irreproducibility as well as artifacts between runs. Budowle (7) observed that lower grades of acrylamide appear to inhibit EsD activity. This is presumably analogous to the effects of impure acrylamide on mouse plasma esterases reported by Allen, et al. (44). Fresh stock solutions of acrylamide (less than one month old) should be used to avoid distortions originating in the anodal portion of the ultrathin-layer polyacrylamide gel. Older stock solutions cause distortion in the gel resulting in waviness across the gel so that similar proteins in different sample lanes can not be compared effectively. In addition, when a functional assay is utilized to visualize the polymorphic enzymes, the gels should sit overnight prior to use. Freshly poured gels tend to inhibit enzymatic activity. This phenomenon was readily observed for EsD and in particular for EAP (7, 23). The activity of the C band of EAP was completely inhibited in polyacrylamide gels that polymerized for only one to three hours. Thus, using fresh recrystallized acrylamide stock solution and permitting polyacrylamide gels to sit overnight prior to use are imperative for minimizing mistypings of genetic marker systems.

If the waviness which originates at the anode still persists, it is suggested to employ alternative anolytes. The anolyte for pH 5-7 ampholyte (LKB) gels for successful subtyping of PGM is phosphoric acid. However, when phosphoric acid is used as the anolyte with narrow-range Pharmalytes, an anodal distortion occurs. By using saturated L-aspartic acid instead of phosphoric acid the distortion is greatly reduced (22, 30, 32). Also, some narrow-range ampholytes which yield good separation of genetic variants, (such as Pharmalyte pH 4.2-4.9 for Pi (22), pH 4.5-5.4 for Gc (30, 31), and pH 6.7-7.7 for Hb (32)) may still exhibit an inherent instability in the gradient. To alleviate this waviness enhancing these narrow range ampholytes with small amounts (1:10) of wider range ampholytes (pH 4-6, pH 4-6.5 or pH 3-10) is required (30, 32). In addition to yielding more linear patterns across a gel, enhancing the pH gradient with wider pH range ampholytes does not appear to compromise the separation distances between allelic variant bands.

There is one constraint on this methodology which is out of the control of the laboratory. This is the batch-to-batch variation of ampholytes. To subtype Tf (21) pH 5-7 ampholytes (LKB) are used. While lots 48 and 50 produced the desired patterns demonstrated in the literature, it was impossible to resolve the C1C3 phenotype when lot 49 was used (unpublished data). These problems have been observed for Servalyte and Pharmalyte as well. To avoid these problems in the future ampholyte manufacturers will have to take responsibility for quality control.

Following the approaches described above, reliable, reproducible, and linear band patterns can be achieved using ULPAGIF for genetic marker typing as well as for most isoelectric focusing protein analyses. This is of particular concern for laboratories with heavy case loads. Time and expenses taken to rerun samples may be prohibitive. Further, when sample size is a limiting factor, as can be the case for forensic samples of supplies of extremely rare variants, multiple tests to obtain an analysis may prove to be impossible. If care is taken, seldom will a gel have to be rerun.

In a previous review on the applications of isoelectric focusing in forensic serology by Murch and Budowle (45), the use of ULPAGIF for typing the erythrocyte-borne genetic markers PGM, EsD and EAP in forensic samples was discussed. It was clearly shown that superior resolution, increased sensitivity of detection, reproducibility and rapid separation times were all possible with ULPAGIF for genetic marker typing of serological evidence. In fact, ULPAGIF of PGM, EsD and EAP is already being used on a routine basis in some forensic laboratories. In the remainder of this paper, the use of ULPAGIF for the forensic analyses of Gc, Tf and hemoglobin (Hb) will be presented.

## Group-Specific Component

Group-specific component is a serum protein involved with Vitamin-D transport (50). By conventional electrophoresis three common phenotypes -1, 2-1, and 2- can be observed. Constans and Viau (10) using polyacrylamide gel IEF demonstrated that the Gc 1 allele can be subtyped into the 1F and 1S alleles. Thus, six common Gc phenotypes could be observed in most populations where only three phenotypes could be detected using conventional electrophoretic methods (46). Budowle (30, 31) developed a ULPAGIF method for resolving the common Gc subtypes as well as the 1A1 allelic product. The technique combined the use of narrow range pH 4.5-5.4 Pharmalytes with high field strengths (final voltage gradient of 390 V/cm). Since resolution is proportional to the square root of the voltage (17-19) superior resolution was expected and observed compared with other IEF methods for Gc subtyping (10, 47, 48).

Due to the increase in discriminating probability [0.57 to 0.75 in the White population and 0.33 to 0.66 in the Black population (31)] achieved by subtyping Gc, there has been recent interest in applying IEF for the analysis of Gc derived from bloodstains. Kido et al (49) demonstrated that Gc in bloodstains maintained at room temperature could be subtyped up to four months. Budowle (31) investigated 86 laboratory-prepared bloodstains and confirmed the findings of Kido et al (49). Budowle showed that even after six months 81.4% of the bloodstains could still be typed for Gc. There was no preferential loss of the 1F, 1S or 2 allelic products although a differential loss of the anodal and cathodal Gc 1 bands was observed. The anodal Gc 1 bands tended to fade before the cathodal Gc 1 bands resulting in inconclusive calls.

While the data suggested that subtyping of Gc derived from bloodstains was feasible, one problem remained to be solved. Actin readily complexes with Gc causing an anodal shift of the Gc bands (44-53). The actin is released from platelet membranes when platelets lyse during bloodstain formation (49-53). The Gc-actin complex can be disrupted in the presence of urea (54, 55). Budowle (30) reported that urea extraction of bloodstains was necessary or mistypings of Gc could occur. Further, as Gc is dissociated from actin, there is more free Gc for the subsequent assay. With the high resolution ULPAGIF procedure described by Budowle (30, 31), the primary Gc bands and the secondary Gc bands (Gc-actin complex) easily can be resolved (see ref. 33 for diagram). Careful study of the patterns will permit successful subtyping of Gc derived from bloodstains even with the existence of these secondary bands.

A preliminary validation study (unpublished data) demonstrated that subtyping of Gc by the method described by Budowle (31) markedly improved the rate of conclusive phenotype determinations for casework specimens. Two hundred and sixty-six known liquid blood specimens obtained from cases submitted to the FBI Laboratory were analyzed for Gc by both ULPAGIF and conventional electrophoresis (33). The data revealed that 92.1% were conclusively typed by ULPAGIF compared with 63.2% by conventional electrophoresis. There were no phenotype discrepancies between the two procedures within their respective analytical capabilities.

Transferrin (Tf)

Transferrin, a serum protein marker involved with iron transport, is used as a forensic genetic marker for bloodstain analysis. It is often run contemporaneously with Gc by immunofixation conventional agarose gel electrophoresis (56-58). The most common Tf allele, C, occurs at a frequency greater than 98% for most populations (59). Using conventional electrophoretic methodology the discriminating probability is 0.02 and 0.15 for Whites and Blacks, respectively (60). Thus, the statistical value of Tf for individualization is limited. Kuhnl and Spielmann (8, 9) using IEF determined that the greatest degree of polymorphism for Tf, in fact, exists within the commonly occurring TfC allele. The TfC allele can be subdivided into three alleles - C1, C2 and C3 - producing six possible phenotypes. Thus, the discriminating probability of Tf is increased to 0.56 and 0.43 for Whites and Blacks, respectively.

Budowle and Scott (28), Dykes, et al. (61), and Carracedo, et al. (62) have described IEF techniques followed by silver staining which produced sensitive assay systems for subtyping Tf derived from bloodstains. In fact, the method of Budowle and Scott (28) permitted successful typing of Tf in six-month-old bloodstains maintained at room temperature and three-month-old bloodstains maintained at 37 C. The sensitivity of detection of this method is attributable to the stability of Tf, the concentrating effect of ULPAGIF and silver staining. Silver staining has been shown to be 100 - 500 times more sensitive than staining with coomassie blue for protein detection (19, 63-65). Also, due to the shorter diffusion pathways with ultrathin-layer gels the silver staining procedure takes less than 15 minutes compared with hours for thicker gels (63).

Unlike Tf derived from serum, Tf from bloodstains present a secondary band for each allelic product. Transferrin has two iron binding sites and the monoferric form of Tf is the most stable. It appears that in bloodstains either binding site is available for iron. Depending upon the site occupied by iron the monoferric forms will have slightly different isoelectric points, and both monoferric forms will exist within a bloodstain. Thus, the monoferric form of Tf will produce two protein bands for each allelic product observed after ULPAGIF. This hypothesis is supported by the fact that these secondary bands do not appear with the apo- or diferric forms of Tf (unpublished data). The knowledge of the existence of these secondary monoferric Tf bands will enable the forensic serologist to interpret the Tf patterns without erroneously typing samples.

## Hemoglobin

Hemoglobin (Hb), the oxygen-binding protein in erythrocytes, has long been used in the forensic analysis of bloodstains and liquid blood samples from human donors (1). Five common variants (A, F, S, C, A2) and numerous less common variants can be resolved by conventional electrophoretic or isoelectric focusing methods. Adult hemoglobin (Hb A) is the prevailing form in humans above approximately 6 months of age. Fetal hemoglobin (Hb F) is prevalent in fetuses after 10 weeks of gestation and increases in concentration until birth at which time it begins to rapidly disappear. Its identification in criminalistics is useful in instances relating to death investigations regarding fetuses and investigations relating to self-induced abortions or child abandonment immediately after birth. Hemoglobins S and C are also useful markers for clinical diagnoses and can imply racial origin of a donor. The S and C variants are carried by approximately 9% and 2% of the U.S. Black population, respectively (1, 32).

Hemoglobin variants have been typed by conventional electrophoresis on cellulose acetate or agarose and by IEF (1, 32, 66-70). Budowle and Eberhardt (32) recently developed an ULPAGIF method for typing the A, F, S, C and a number of rare variants, which is presently used for the analysis of bloodstained evidence submitted to the FBI Laboratory. The method employs pH 6.7-7.7 Pharmalytes in an ultrathin-layer gel with an inter-electrode wick distance of only 5.0 cm. The result is a rapid screening method for Hb that takes only 25-30 minutes-comparable to cellulose acetate. Further, the distances between the A-F, F-S and S-C were 4 mm, 7 mm and 12 mm, respectively, compared with 3 mm, 5.5 mm, and 7.5 mm, respectively, of other IEF methods (69) which use a 10 cm inter-electrode wick distance. Once again, the use of increased field strengths (700 V/cm) and narrow range ampholytes permit high resolution, rapid screening of genetic marker variants. Further, using an approach described by Altland (71), a large number of samples can be screened for Hb in a relatively short period of time. A center anode strip was used with a cathode strip at both the top and bottom of a single gel. Thus, two gels sharing a common electrode (anode) were run on one plate. As many as 150 samples can be typed at one time (32).

## Conclusion

Isoelectric focusing, particularly ULPAGIF, offers advantages over previous electrophoretic methods used in forensic serology. These include increased resolution of protein bands, increased sensitivity of detection, the possibility of subtyping existing phenotypes, the counteraction of diffusion effects, reduction of time of analysis, and reduced cost. Thus, the forensic serologist has the ability to increase the number of conclusive determinations for EAP, PGM, EsD, Hb, Gc and Tf, as well as type weak or dilute samples where it was previously not possible with conventional electrophoretic methods.

## Disclaimer

This is publication number 86-6 of the Laboratory Division of the Federal Bureau of Investigation. Names of commercial manufacturers are provided for identification only and inclusion does not imply endorsement by the Federal Bureau of Investigation.

## Literature Cited

1) Gaensslen, R. E. "Sourcebook in Forensic Serology, Immunology and Biochemistry"; U.S. Department of Justice, National Institute of Justice: Washington, D.C., 1983, pp. 59-62, pp. 421-520.
2) Divall, G. B. Electrophoresis 1985, 6, 249-258.
3) Bark, J. E.; Harris, M. J.; Firth, M. J. For. Sci. Soc. 1976, 16, 115-120.
4) Kuhnl, P.; Schmidtmann, U.; Spielmann, W. Hum. Genet. 1977, 35, 219-223.
5) Olaisen, B.; Siverts, A., Jonassen, R.; Mevag, B.; Gedde-Dahl, T. Hum. Genet. 1981, 57, 351-353.
6) Divall, G. B. For. Sci. Int. 1984, 26, 255-267.
7) Budowle, B. Electrophoresis 1984, 5, 314-316.
8) Kuhnl, P.; Spielmann, W. Hum. Genet. 1978, 43, 91-95.
9) Kuhnl, P.; Spielmann, W. Hum. Genet. 1979, 50, 193-198.
10) Constans, J.; Viau, M. Science 1977, 198, 1070-1071.
11) Constans, J.; Viau, M.; Cleve, H.; Jaeger, G., Quilici, J.C.; Palisson, M.J. Hum. Genet. 1978, 41, 53-60.
12) Teige, B.; Mevag, B.; Olaisen, B.; Pedersen, L. Proc. 10th Int. Cong. Soc. For. Haemagen., 1983, pp. 373-379.
13) Allen, R. C.; Harley, R. A.; Talamo, R. C. Amer. J. Clin. Path. 1974, 62, 732-739.
14) Frants, R. R.; Erickson, A. W. Hum. Hered. 1976, 26, 435-440.
15) Budowle, B. For. Sci. Int. 1984, 24, 273-277.
16) Gorg, A.; Postel, W.; Westermeier, R. Anal Biochem. 1978, 89, 60-70.

17) Giddings, J. C.; Dahlgren, H. Sep. Sci. 1971, 6, 345-356.
18) Rilbe, H. Ann. N.Y. Acad. Sci. 1973, 209, 11-22.
19) Allen, R. C. Electrophoresis 1980, 1, 32-37.
20) Allen, R. C.; Arnaud, P. Electrophoresis 1983, 4, 205-211.
21) Budowle, B. Electrophoresis, 1985, 6, 97-99.
22) Budowle, B.; Murch, R. S. Electrophoresis, 1985, 6, 523-525.
23) Budowle, B. Electrophoresis, 1984, 5, 254-255.
24) Budowle, B. Electrophoresis, 1984, 5, 165-167.
25) Radola, B. J. Electrophoresis, 1980, 1, 43-56.
26) Kinzkofer, A.; Radola, B. J. Electrophoresis, 1981, 2, 174-183.
27) Goedde, H. W.; Brenkmann, H. G.; Hirth, L. Hum. Genet., 1981, 57, 434-436.
28) Budowle, B.; Scott, E. For. Sci. Int. 1985, 28, 269-275.
29) Budowle, B., In "Proc. Electrophoresis Soc. Amer.," Reeder, D. J.; ed.; U.S. Department of Commerce, Gaithersubrg, MD, 1986, pp. 1-12.
30) Budowle, B. Electrophoresis 1986, 7, 141-144.
31) Budowle, B. For Sci. Int. 1986, (in press).
32) Budowle, B.; Eberhardt, P. Hemoglobin 1986, 10, 161-172.
33) Pflug, W.; Vigne, U.; Bruder, W. Electrophoresis 1981, 2, 327-330.
34) Divall, G. B.; Ismail, M. For. Sci. Int. 1983, 22, 253-263.
35) Gill, P.; Sutton, J. G. Electrophoresis 1985, 6, 23-26.
36) Gill, P. and Sutton, J. G., Electrophoresis 1985, 6, 274-279.
37) Pflug, W. Electrophoresis 1985, 6, 19-22.
38) Kuhnl, P.; Spielmann, W. Hum. Genet. 1978, 43, 91-95.
39) Kuhnl, P.; Spielmann, W. Hum. Genet. 1979, 50, 193-198.
40) Gorg, A.; Weser, J.; Westermeier, R.; Postel W.; Weidinger, S.; Patutschnick, W.; Cleve, H. Hum. Genet. 1983, 64, 222-226.
41) Gorg, A.; Postel, W.; Weser, J.; Weidinger, S.; Patutschnick, W.; Cleve, H. Electrophoresis 1983, 4, 153-157.
42) Righetti, P. G.; Drysdale, J. W. In "Isoelectric Focusing," North-Holland, Amsterdame 1976, p. 463.
43) Chrambach, A.; Jovin, T. M.; Svendsen, P. J. and Rodbard, D., In: "Methods of Protein Separation" Castimpoolas, N.; ed.; Plenum Press, New York, 1976, pp. 27-144.
44) Allen, R. C.; Propp, R. A.; Moore, D. J. Histochem. Cytochem. 1965, 13, 249-254.
45) Murch, R. S.; Budowle, B. J. For. Sci. 1986, 31, 869-880.
46) Cleve, H. Israel J. Med. Sci. 1973, 9, 1133-1146.

47) Hoste, B. Hum. Genet. 1979, 50, 75-79.
48) Kueppers, F.; Harpel, B. Hum. Hered. 1979, 29, 242-249.
49) Kido, A.; Oya, M.; Komatsu, N.; Shibata, R. For. Sci. Int. 1984, 26, 39-43.
50) Van Baelen, H.; Bouillon, R.; DeMoor, P. J. Biol. Chem. 1980, 255, 2270-2272.
51) Emerson, D.; Arnaud, P.; Galbraith, R. M. Amer. J. Rep. Immunol. 1983, 4, 185-189.
52) Kimura, H.; Shinomiya, K.; Yoshida, K.; Shinomiya, T. For. Sci. Int. 1983, 22, 49-55.
53) Shinomiya, K. For. Sci. Int. 1984, 25, 255-263.
54) Westwood, S. Electrophoresis 1984, 5, 316-318.
55) Goldschmidt-Clemont, P.; Galbraith, R. M.; Emerson, D.; Nel, A.; Werner, P.; Lee, W. Electrophoresis 1985, 6, 155-161.
56) Wraxall, B.; Bordeaux, J.; Harmor, G. "Bloodstain Analysis System. Final Report"; Aerospace Corporation, El Segundo, CA 1978, Aerospace Corporation Sub-Contract No. 67854.
57) Alper, C.; Johnson, M. A. Vox Sang. 1969, 17, 445-452.
58) Budowle, B.; Gambel, A. M. Crime Lab. Dig. 1984, 11, 12-13.
59) Dykes, D. D.; DeFurio, C.; Polesky, H. Electrophoresis 1982, 3, 162-164.
60) Sensabaugh, G. In "Forensic Science Handbook"; Saferstein, R., Ed.; Prentice-Hall, Inc.: New Jersey, 1982; pp. 338-415.
61) Dykes, D. D.; Miller, S.; Polesky, H. Electrophoresis 1985, 6, 90-93.
62) Carracedo, A.; Concheiro, L.; Requena, I.; Lopez-Rivadula, M. For. Sci. Int. 1983, 23, 241-248.
63) Merril, C.; Goldman, D.; Sedman, S.; Ebert, M. Science 1981, 211, 1437-1438.
64) Oakley, B.; Kirsch, D.; Morris, R. Anal. Biochem. 1980, 105, 361-363.
65) Sammons, D.; Adams, L.; Nishizawa, E. Electrophoresis 1981, 2, 135-141.
66) Brown, I. R.; Grech, J. L. Life Sciences 1971, 10, 191-194.
67) Jeppson, J. O.; Berglund, S. Clin. Chim. Acta 1972, 40, 153-158.
68) Cossu, G.; Manca, M.; Pirastru, M. G.; Bullita, R.; Bosisio, A. B.; Gianazza, E., Righetti, P. G. Amer.J. Haem. 1982, 149-157.
69) Basset, P.; Braconnier, F.; Rosa, J. J. Chromatog. 1982, 227, 267-304.
70) Black, J. Hemoglobin 1984, 8, 117-127.
71) Altland, K. In "Electrofocusing and Isotachophoresis"; Radola, B. J.; Graesslin, D., Eds.; Walter de Gruyter: Berlin, 1977, p. 295.

RECEIVED August 4, 1986

# Chapter 11

# Biophysical Characterization by Agarose Gel Electrophoresis

**Philip Serwer**

**Department of Biochemistry, The University of Texas Health Science Center, San Antonio, TX 78284-7760**

Electrophoresis in agarose gels is being developed for fractionating mixtures of multimolecular particles by the particles' size, shape, flexibility and solid support free electrophoretic mobility ($\mu o$). Advances in the use of sieving to determine size, shape, flexibility and $\mu o$ are described. A procedure of two-dimensional agarose gel electrophoresis can be used to determine distributions of size and $\mu o$ for mixtures of heterogeneous particles.

The size and shape of macromolecules and multimolecular aggregates have been determined by electron microscopy, centrifugation, light scattering, x-ray scattering, measurement of viscosity and measurement of diffusion (reviewed in refs. 1-2). Empirically, the motion of such particles through gels is retarded by the gel and at least some of this retardation increases as the particle's size increases (sieving). Comparison of a spherical particle's sieving with the sieving of spherical particles of known size has been used to determine the size of a particle (3-5). Sieving can be measured by either: (a) dividing the gel into beads and measuring the rate at which particles migrate through a column of these beads, driven by flow of liquid through the column (gel filtration); because penetration of the beads decreases as the size of the particle increases, the rate of passage through the column increases as the particle's size increases (4,6), or (b) measuring the rate of an electrically charged particle's motion through an undivided gel, driven by an electrical field (gel electrophoresis); this rate decreases as the particle's size increases.

Procedures of sieving for measuring a particle's size are less expensive (in time and cost per sample analysed) than the other procedures indicated above. In addition, procedures of sieving usually require samples that are less purified and less concentrated than the samples required for the other procedures above. Procedures of sieving are performed with comparatively simple equipment and do not require more than two days of training for most investigators. The disadvantage of using sieving is that the characteristics of gels used are sufficiently uncertain so that an empirical calibration of each gel must be performed with particles of known size and shape

0097-6156/87/0335-0158$06.00/0

(markers). Below, progress in the use of sieving during agarose gel electrophoresis to determine the size and shape of a particle is described.

## Electrophoretic Mobility As a Function of Gel Concentration

Spheres. Semilogarithmic plots of the electrophoretic mobility, $\mu$ ( = velocity/voltage gradient), as a function of gel concentration have been found to be linear for monomolecular proteins in starch (7) and polyacrylamide (3,4) gels. Because Ferguson (7) was the first to make such plots, these plots are sometimes called Ferguson plots. The slope ($K_R$) of a Ferguson plot increases as the size of a spherical particle increases. Thus, $K_R$ is a measure of the particle's size. The size-dependence of $K_R$ distinguishes sieving from viscous retardation, the latter independent of size.

It is known that agarose gels of any given concentration are stronger than and have larger pores than starch and all but the most crosslinked polyacrylamide gels at the same concentration (reviewed in ref. 5). The more dilute agarose gels used for electrophoresis (0.04-0.10%; see ref. 5) permit entry of particles 0.3-0.4 µM in radius (P. Serwer and S.J. Hayes, unpublished observations). Agarose gels, therefore, can be used for the electrophoresis of particles big enough so that their shape is known from either electron microscopy or one of the other techniques indicated above. The dimensions of such particles have, in some cases (see below), also been determined and can be used to calibrate the sieving of agarose gels. Whenever dimensions have been determined by a technique that, unlike electron microscopy, does not require drying of the specimen, such dimensions obtained with hydrated particles are used to calibrate the sieving of agarose gels. This is done to avoid drying-induced alterations in size (see ref. 8).

Semilogarithmic plots of $\mu$ vs. $A$ for spherical viruses 13-42 nm in outer radius (R) are, like the comparable plots for monomeric proteins (<13 nm in R) in starch and polyacrylamide gels, linear when the percentage of agarose ($A$) is between 0.05 and 0.9. However, for $A$ values above 0.9 these plots decrease in magnitude progressively more rapidly than a linear plot (convex curvature) (reviewed in ref. 5).

The data in the region of linearity of semilogarithmic $\mu$ vs. $A$ plots for agarose gels are described by:

$$\mu = \mu o' \cdot e^{-K_R \cdot A} \qquad (1)$$

$\mu o'$ is $\mu$ extrapolated to an $A$ of 0 on a semilogarithmic $\mu$ vs. $A$ plot. The value of $\mu o'$ is determined by the $\mu$ of a particle in the absence of a gel ($\mu o$), and the flow rate of buffer through the gel (electro-osmosis or EEO). EEO is independent of the $\mu o$ of particles subjected to electrophoresis and has been described by $\mu_E$, the velocity of buffer flow divided by the voltage gradient (5):

$$\mu o' = \mu o + \mu_E \qquad (2)$$

By comparing a particle's $K_R$ to the $K_R$ of particles of known R, R can be determined ± 8% (5).

To improve accuracy in measuring R, sieving in gels more concentrated than 0.9% (i.e., in the region of convex curvature in Ferguson plots for particles 13-42 nm in R) was measured by using plots of μ vs. A, instead of log |μ| vs. A. These plots were surprisingly linear for 6.5% hydroxyethylated agarose and the A obtained by extrapolation to a μ of 0 (Ao) is a monotonically decreasing function of R (9). From Ao, R is determined ± 4%.

Rods. To develop sieving for the determination of a particle's shape, the sieving of rods and spheres is compared. Unlike spheres, rods can orient to minimize their sieving. That is, rods can migrate end-first through a gel (reptation). X-ray diffraction patterns of tobacco mosaic virus (TMV), a rod-shaped (radius = 9 nm; length = 300 nm) plant virus (10), after electrophoretic migration into polyacrylamide gels reveals preferential orientation in the direction of migration (11). Dependence of the sieving of TMV on flow rate during molecular sieve chromatography in agar gels is observed (12). This latter observation also suggests end-first orientation of TMV during migration through gels.

Semilogarithmic plots of μ vs. A for TMV, and also for the rod-shaped (radius = 4.5 nm; length = 895 nm) bacteriophage, fd (13), and length variants of these viruses are found to be linear for A below a value that decreases as the length of the rod increases (14). In contrast to the results with spheres, as A increases these plots develop concave curvature (14), suggesting the occurrence of reptation. Thus, sieving can be used to distinguish a rod-shaped from a spherical virus. Among the questions to be answered in the future are: How accurately can the shape and dimensions of a 10-1000 nm rigid particle of completely unknown shape be determined by sieving? Can sieving be used to determine the shape of monomolecular proteins, normally below 10 nm in their greatest dimension?

Random Coils. In an agarose gel, a random coil should have a spherical envelope in the absence of the electrical field. However, during electrophoresis, the random coil may deform to a coil with a rod-shaped envelope. Evidence that such deformation occurs during the agarose gel electrophoresis of double-stranded DNA is: (a) concave semilogarithmic μ vs. A plots (15) and (b) an increase in resolution by molecular weight that is achieved when the direction of the electrical field is orthogonally and periodically alternated (16,17; see also the theory in ref. 18). The concavity in semilogarithmic μ vs. A plots increases as the voltage gradient and DNA molecular weight increase. Among the questions to be answered in the future are: Can reptation caused by deformation be distinguished from the reptation of a naturally rod-shaped particle? How accurately can the flexibility of a particle be measured by use of sieving?

## Two-dimensional Electrophoresis: Concept

For determining R and μo, the procedures described above have two limitations: (a) If particles vary in R and μo, there is no way for determining the distribution of one of these two characteristics, independent of the other. (b) Because measuring $K_R$ requires the comparing of a particle's μ in one gel with its μ in another gel and because the occurrence of curvature imposes an upper limit to usable A in μ vs. A plots (see above), experimental error in measurement of R has not been less than ± 4%. Both of these limitations are overcome by the use of two-dimensional agarose gel electrophoresis, as described below (details of experimentation are in refs. 19,20).

To determine R by two-dimensional agarose gel electrophoresis, initially a comparatively dilute agarose gel (first-dimension gel) is embedded within a more concentrated agarose slab (second-dimension gel), as shown in Figure 1. Samples are layered in a well (indicated in Figure 1) at the origin of the first-dimension gel and are subjected to electrophoresis through the first-dimension gel. Comparatively little sieving is experienced in the first-dimension gel and ideally there would be no detectable sieving. After the first electrophoresis, a second electrophoresis is conducted with the electrical field orthogonal to its direction for the first electrophoresis. During the second electrophoresis, the particles migrate into the more concentrated second-dimension gel. The *A* used for the second-dimension gel is the highest *A* possible without either preventing a particle from entering the gel or requiring excessive time of electrophoresis (20). As the *A* of the second-dimension gel increases, the separation by R in the second-dimension gel increases.

If sieving is a function of a spherical particle's R, and not either $\mu o$ or internal contents (see the data and discussion in ref. 20), the R of a spherical particle determines the angle, $\theta$, between the direction of the first-dimension gel and the direction of a line (size line) connecting the effective origin of electrophoresis (0 in Figure 1) with the position of the particle in the gel (Figure 1). Values of $\theta$ increase as R decreases, and the magnitude of $\mu o$ increases as the distance from the origin along a size line increases.

To determine R as a function of $\theta$, size standards are mixed with the particles to be analysed. Accuracy of $\pm$ 0.3 nm in measuring R from $\theta$ for particles with R's of ~30 nm has been achieved using such a calibration (20). Achieving this level of accuracy requires: (a) size standards with R known to a comparable accuracy, and (b) appropriate choice of *A* for the second-dimension gel. To avoid trial and error in finding *A* for the second-dimension gel, a systematic procedure based on knowledge of agarose pore size as a function of *A* should be developed.

Because R is a unique function of $\theta$ in Figure 1, the procedure of two-dimensional electrophoresis described can be used to determine the distribution of R for particles heterogeneous in either R or $\mu o$ (or both). Assuming homogeneity of $\mu o$ and no detectable sieving in the first-dimension gel, particles with variable R would all be found on a straight line (dotted line for $\mu o_1$ in Figure 1), perpendicular to the first-dimension gel. However, in reality, as R increases eventually detectable sieving occurs in the first dimension gel, causing the dotted line to become curved (curved solid line #1 tangent to the dotted line in Figure 1). If particles (all spheres) in the gel have one of two possible values of $\mu o$, particles will all fall on two curved lines (1 and 2 in Figure 1). This analysis can be extended to include additional $\mu o$'s that differ sufficiently so that any two particles of the same R and different $\mu o$ form two separate bands.

Mixtures of solid spheres variable in R and homogeneous in $\mu o$ have not yet been analysed by the procedure in Figure 1. However, a mixture of linear double-stranded DNA's, variable in molecular weight and homogeneous in $\mu o$, has been thus analysed and all such

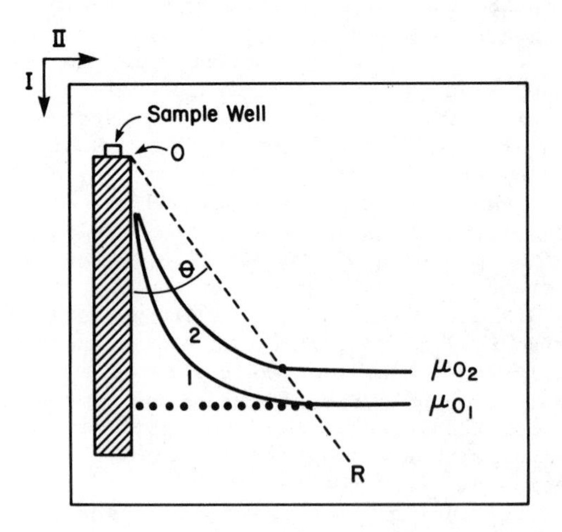


Figure 1. The gel for two-dimensional agarose gel electrophoresis. The first-dimension gel is shaded; the second dimension gel is not. The directions of the first (I) and second (II) electrophoresis are indicated by arrows.

DNA's are found on a curve similar to either curve 1 or curve 2 in Figure 1 (19,21).

The $\mu o'$ for all particles that fall on a curve such as curve 1 in Figure 1 is the $\mu$ in the first-dimension gel determined for particles that form the portion of curve 1 that is parallel to the dotted line (i.e., the smaller particles). Accuracy as high as ± 2% in measuring $\mu o'$s can be achieved with previously-described (22) procedures of control for voltage gradient and temperature. In case R is constant and $\mu o$ varies, all bands are found on one size line. If discrete bands are not resolved, a single ellipsoidal band with its longer axis coincident with the size line is formed (see refs. 19, 23).

## Two-dimensional Electrophoresis: Some Applications

Latex Spheres. Though the above procedure of two-dimensional agarose gel electrophoresis was developed for analysis of bacteriophage assembly pathways, it also has other applications. For instance, the homogeneity of chemically-made spheres, such as latex spheres, can be tested. After two-dimensional electrophoresis, a band formed by carboxylated latex spheres 30 nm in R has its center on the same size line as a band formed by the related spherical bacteriophages, T3 and $\phi$II, both 30.1 nm in R (as expected) (Figure 2). However, the band formed by the carboxylated latex is more than three times wider than the band formed by T3 in a direction not parallel to the 30 nm size line. This latter observation indicates heterogeneity in R for the latex spheres. Possibly two-dimensional agarose gel electrophoresis can also be used in the analysis of commercial mixtures with latex spheres (paints, for instance).

Double-stranded DNA. It has previously been shown (19,21) that a mixture of linear, double-stranded DNA's of variable length forms an arc like one of the arcs in Figure 1, after fractionation by two-dimensional agarose gel electrophoresis. Because the sieving of open circular DNA increases more rapidly than the sieving of linear DNA as A increases (15,19), open circles variable in molecular weight form an arc closer to the first-dimension gel than the arc formed by linear DNA (19). However, unlike the arcs in Figure 1, the arcs for linear and circular double-stranded DNA must eventually become coincident as $\theta$ increases, because these double-stranded DNA's all have the same $\mu o$ (15).

The results of the previous paragraph indicate that a mixture of linear and open circular DNA's can be analysed to determine the amount and the mass distribution of each by use of a single two-dimensional agarose gel electrophoresis. However, if forms other than linear and open circular double-stranded DNA's are present, additional data are needed to complete such an analysis. The behavior of some branched DNA's has been determined (21). Additional studies of the behavior of branched, closed circular and single-stranded DNA's are needed to increase the capabilities of two-dimensional agarose gel electrophoresis for analysing mixtures of heterogeneous DNA's.

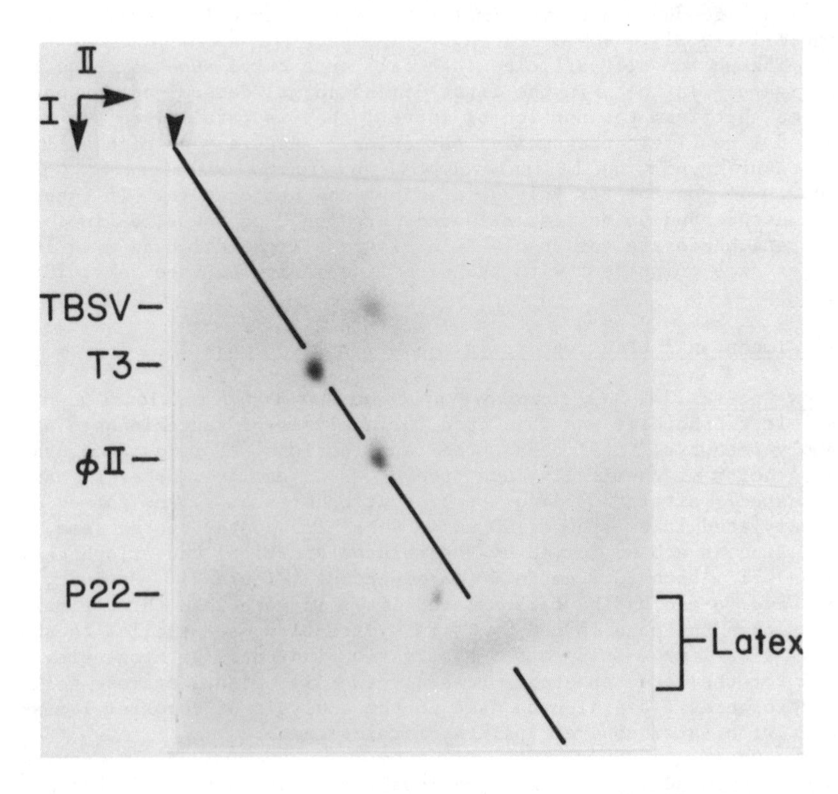


Figure 2. Two-dimensional agarose gel electrophoresis of carboxylated latex spheres. The following mixture was made: 24 µl of 2.5% carboxylated latex spheres with an R of 30 nm (purchased from Polysciences, Inc.); 6.0 µl containing 4 µg each of bacteriophages T3 and φII (R = 30.1 nm), 2 µg of bacteriophage P22 (R = 31.4 nm); 1.0 µl containing 3 µg of tomato bushy stunt virus (TBSV; R = 17.0 nm) (20); and 28 µl of 6% sucrose, 400 µg/ml bromphenol blue in 0.025 M sodium phosphate, pH 7.4, 0.001 M $MgCl_2$ with 1.1% Triton X-100. Of this mixture, 50 µl was subjected to two-dimensional agarose gel electrophoresis, as previously described (20), in a buffer that contained 0.025 M sodium phosphate, pH 7.4, 0.001 M $MgCl_2$, 0.5% Triton X-100. The concentrations of the first and second-dimension gels were 0.15 and 2.4% agarose (Seakem LE agarose from the Marine Colloids Division of the FMC Corporation), respectively. Electrophoresis in the first direction was performed at 2.0 V/cm for 9.0 hr., and in the second direction was performed at 1.8 V/cm for 25.0 hr. The gel was stained with ethidium bromide (20). The directions of the first (I) and second (II) electrophoresis are indicated by the arrows. All particles migrated toward the anode. The size line for R = 30.1 nm is drawn. The effective origin of electrophoresis (see Figure 1) is at the intersection of the size line and the arrowhead.

## Conclusion

The studies presented above suggest the use of sieving during agarose gel electrophoresis for the determination of several biophysical properties of particles with dimensions between 10 and 400 nm. For most problems, the accuracy now achievable in the determination of R and $\mu o$ appears to be sufficient and, for R, could not be improved without improvements in the accurancy of R for size markers. The areas ripe for advance appear to be: (a) improved procedures for determination of shape and flexibility, (b) characterization of additional random coils (branched DNA, for instance) and (c) development of automated scanning densitometry for obtaining distributions of R and $\mu o$ in patterns of heterogeneous particles after two-dimensional agarose gel electrophoresis.

## Acknowledgments

I thank Shirley J. Hayes for performing the experiment in Figure 2 and Anna Uriegas for typing this manuscript. Work done in the author's laboratory was supported by grants from the National Institutes of Health (GM24365 and AI22568) and the Robert A. Welch Foundation (AQ-764).

## Legend of Symbols

$\mu$, electrophoretic mobility; A, percentage of agarose in a gel; $\mu o'$, $\mu$ extrapolated to an A of 0; $\mu o$, $\mu$ in the absence of a gel; $\mu_E$, term that corrects $\mu o'$ for electro-osmosis; $K_R$, negative of the slope of semilogarithmic plot of $\mu$ vs. gel concentration; R, radius of a sphere; TMV, tobacco mosaic virus.

## Literature Cited

1. Bull, H.B. "An Introduction to Physical Biochemistry"; F.A. Davis: Philadelphia, 1971; Chap. 8,10,11,14.
2. Cantor, C.R.; Schimmel, P.R. "Biophysical Chemistry"; W.H. Freeman: San Francisco, 1980; Chap. 10.
3. Rodbard, D.; Chrambach, A. Anal. Biochem. 1971, 40, 95-134.
4. Morris, C.J.O.R.; Morris, P. Biochem. J. 1971, 124, 517-528.
5. Serwer, P. Electrophoresis 1983, 4, 375-382.
6. Laurent, T.C.; Killander, J. J. Chromatog. 1964, 14, 317-330.
7. Ferguson, K.A. Metabolism 1964, 13, 985-1002.
8. Serwer, P. J. Ultrastruct. Res. 1977, 58, 235-243.
9. Serwer, P.; Allen, J.L. Electrophoresis 1983, 4, 273-276.
10. Klug, A.; Caspar, D.L.D. Adv. Virus Res. 1960, 7, 225-325.
11. Barkas, B.V.; Kornev, A.N.; Mesianzhikov, V.V.; Poglazov, B.F.; Turkin, A.I. Doklady Akademii Nauk. SSSR. 1979, 245, 736-739.
12. Steere, R.L. Science 1963, 140, 1089-1090.
13. Newman, J.; Swinney, H.L.; Day, L.A. J. Mol. Biol. 1977, 116, 593-606.
14. Serwer, P.; Merrill, G.; Moreno, E.T.; Herrmann, R. In "Protides of the Biological Fluids"; Peeters, H., Ed.; Pergamon: Oxford, 1985; Vol. 33, pp. 517-520.
15. Serwer, P.; Allen, J.L. Biochemistry 1984, 23, 922-927.
16. Schwartz, D.C.; Cantor,C.R. Cell 1984, 37, 67-75.

17. Carle, G.F.; Olson, M.V. Nucleic Acids Res. 1984, 12, 5647-5664.
18. Lumpkin, O.J.; Déjardin, P.; Zimm, B.H. Bioploymers 1985, 24, 1573-1593.
19. Serwer, P. Anal. Biochem. 1985, 144, 172-178.
20. Serwer, P.; Hayes, S.J.; Griess, G.A. Anal. Biochem. 1986, 152, 339-345.
21. Bell, L.; Byers, B. Anal. Biochem. 1983, 130, 527-535.
22. Serwer, P. Electrophoresis 1983, 4, 227-231.
23. Wieme, R.J. "Agar Gel Electrophoresis"; Elsevier: Amsterdam, 1965.

RECEIVED August 11, 1986

# Chapter 12

# Pulsed Electrophoresis

**David C. Schwartz**

**Department of Embryology, Carnegie Institution of Washington, 115 West University Parkway, Baltimore, MD 21210**

Pulsed gel electrophoresis uses newly discovered physical phenomena to size separate giga-dalton molecules in agarose gel. Large DNA molecules are easily deformable random coils in solution, and suitably pulsed electrical fields (applied across a gel) dynamically modulate the DNA coil conformation immersed in a gel matrix. Although not fully understood, conformational changes affect electrophoretic mobilities which are strongly dependent on pulse frequency. Experimentally, one or more sets of electrodes are used, and field gradients appear to enhance simultaneous size resolution over a broad range of molecular weights. Ancillary methodology was developed using DNA embedded in agarose to reliably prepare intact giga-dalton sized DNA molecules suitable for biochemical and pulsed electrophoresis analysis. This combination of new methodologies has generated novel approaches for solving difficult problems concerning genome organization.

This article is written to serve as a succinct introduction and review of pulsed gel electrophoresis, which is an entirely new electrophoretic method. Briefly, the method utilizes pulsed electrical fields of varying frequencies (depending on molecular weight) to selectively modulate electrophoretic mobilities. Presently, pulsed electrophoresis has not yet been applied to a broad range of separation problems and is still in need of an encompassing theoretical foundation. Despite this, pulsed electrophoresis has been instrumental in solving a diversity of biological problems. Furthermore, we can confidently predict that as more researchers gain familiarity with pulsed electrophoresis, its diversity of application will increase.

Large DNA molecules, defined in this article as greater than

0097-6156/87/0335-0167$06.00/0

$50 \times 10^6$ daltons, exist in solution as random coils. Since the hydrodynamic radii of large DNA molecules are clearly bigger than typical agarose pore sizes, the molecules must somehow deform in order to enter the gel matrix (1, 2, 3). The nature of this distortion is not known with any certainty, although there is evidence that coils may compress to some degree (4) or undergo reptation (1). Due in part to this molecular deformation, simple quantitative relationships based on mass and measured electrophoretic mobilities are difficult to calculate. Reliable quantitative data can be obtained only with strict regard to experimental variables such as field strength, and, moreover, quantitative relationships drawn from a given data set are often limited to specific size ranges.

The size resolution of conventional agarose gel electrophoresis has a practical upper limit of about $50 \times 10^6$ daltons (5). Above this molecular weight size resolution is lost: large molecules possess the same electrophoretic mobilities as much smaller molecules. Contrary to common lab knowledge, large DNA molecules will enter agarose gels, although they will not be resolved. The resolution size limit has been pushed as high as $500 \times 10^6$ by drastically lowering the gel concentration and using weak electrical fields (5). Unfortunately, these conditions impose serious experimental difficulties such as limiting the amount of sample that can be successfully loaded and drastically increasing running times. Despite these drawbacks, conventional agarose gel electrophoresis is a common tool for analysis of small DNA molecules in molecular biology. Intact DNA molecules can be incredibly massive: the largest human chromosome contains a DNA molecule 7 inches long. Organisms maintain biological function of DNA by condensation with proteins to produce compact structures called chromosomes. These can be visualized by light microscopy in their most condensed phase as the familiar "H" shaped structures. This condensation also protects the DNA molecule from shear forces. When isolated from the protective protein complex, intact DNA molecules become extraordinarily prone to breakage. Simple laboratory procedures such as pipetting or even stirring will produce enough shear to break large DNA molecules (6).

Some major biological problems are uniquely solvable by analysis of large DNA molecules (examples will be illustrated later in this review). With this in mind, the "insert" methodology was developed to allow convenient isolation and chemical manipulation of intact DNA molecules (2, 7). Insert methodology produces large amounts of intact, usable DNA by lysing cells within an agarose gel matrix. Since large DNA molecules do not readily diffuse, reagents and enzymes can be diffused into the insert (which is actually a small agarose block) while the DNA remains embedded. This methodology, coupled with pulsed electrophoresis, permits biologists to incorporate direct analysis of high molecular weight DNA into their particular biological system and experimental repertoire. Simultaneously, it has provided a series of new physical phenomena to be investigated.

## Pulsed Electrophoresis

Introduction. For simplicity and convenience, variants of pulsed electrophoresis will be divided into one of two groups, depending upon the number of fields employed: one- and two-dimensional pulsed electrophoresis. Regardless of the number of fields utilized, pulsed electrophoresis probably exploits dynamic polymer conformation induced by pulsing the applied electrical field (2, 8, 9). Conformational changes of a polymer during electrophoresis can differentially modulate electrophoretic mobilities according to variables such as molecular weight or secondary structure (1). In the case of high molecular weight DNA molecules, this dynamic behavior occurs on a very macroscopic time scale (2, 7, 8). This phenomenon is amply illustrated by DNA viscoelastic behavior as determined by Zimm and co-workers. They measured the longest relaxation time of a bacterial chromosome (E. coli, $2.5 \times 10^9$ daltons) to be about 60 seconds (10).

Two-dimensional pulsed electrophoresis (TDP) is presently the most commonly used variant of pulsed electrophoresis. It was first introduced in 1983 as pulsed field gradient gel electrophoresis (PFG) (8); with the equipment modified by other workers, it sometimes appears in the literature as orthogonal electrical field alternating gel electrophoresis (OFAGE) (11). TDP uses two electrode pairs perpendicularly oriented to each other with the applied field alternately pulsed. PFG (a methodological subset of TDP) utilizes at least one field gradient to simultaneously resolve a broad size range of DNA molecules. By varying the frequency of field alternation, defined as the pulse time, different molecular size ranges are optimally resolved (2, 7, 8). In essence, this tunable separation constitutes an electrophoretic spectroscopy method.

One-dimensional pulsed electrophoresis (ODP) can size fractionate large DNA molecules. One experimental arrangement applies a relatively short electrical pulse followed by a much longer period of zero field strength (2,12). Presumably, during the field-on condition, molecular conformation is perturbed and relaxation then takes place during the field-off condition. As with TDP electrophoresis, mobility can be modulated with different pulse times.

Conventional DNA Gel Electrophoresis: Theoretical Background. Only recently has a comprehensive molecular theory for DNA gel electrophoresis been available. Lumpkin et al. (9) have used the polymer reptational model to quantitatively account for electrophoretic mobility dependence on field strength (first introduced by Lerman and Frisch (3)). Briefly, the reptational model for DNA agarose gel electrophoresis assumes that the polymer moves through the gel in a manner analogous to a worm moving through a burrow; that is, the polymer "tail" follows in the same path tube as the "head". Lumpkin et al. further propose that random coil dimensions show field strength dependence through a mechanism whereby the "head" statistical segments of the polymer are preferentially aligned relative to the field (9). Lumpkin et al. showed theoretically that such an alignment is possible even at low field strengths (1 volt/cm). Their derived expression equating polymer velocity and dimensions is:

$$\langle v_{cm} \rangle = \langle h_x \rangle QE/L_t f, \quad (1)$$

where $v_{cm}$ is the velocity of the polymer center of mass parallel to the field, $h_x$ is the component of the polymer's end-to-end vector parallel to the field, Q is the total polymer effective charge, $L_t$ is the tube contour length, and f is the polymer translational frictional coefficient. As field strength (E) is increased, resolution drops, since $h_x$ approaches the polymer contour length, which is proportional to the tube length, and Q/f is constant. In other words the polymer becomes more aligned with the applied electrical field. This alignment is analogous to that observed for the Kerr effect in birefrigence experiments except that much lower field strengths can be used.

Stellwagen measured birefringence of DNA smaller than $1x10^6$ daltons in agarose gel using the customarily high field strengths (1000v/cm) common to this technique and found field induced polymer alignment (13). However, Lumpkin et al. have predicted a measurable birefringence effect with larger DNA molecules using much lower field strengths than are commonly used for DNA electrophoresis (9).

By extrapolating to zero field strength it is possible to find an inverse relationship between electrophoretic mobility and molecular weight (14). This relationship contrasts with that obtained with a Ferguson plot (15) normally used for proteins, but also frequently and often erroneously used for DNA. However, the inverse relationship is fully predicted by reptation theory (3, 9). In summary, determining precise DNA molecular weights using gel electrophoresis requires careful attention to experimental detail and judicious choice of data workup.

Two-Dimensional Pulsed Field Electrophoresis: Theoretical Background. Applying the polymer reptation model to conventional DNA electrophoresis is proving to be quite successful, especially in the case of field strength dependent electrophoretic mobilities. Many researchers have applied the reptational model to pulsed electrophoresis. Lumpkin et al. (9) have derived an expression for the time needed by a DNA molecule to reach a steady state orientation with an applied electrical field. This time can be related to the pulse time dependent mobilities seen in TDP by postulating a conformational transition from one orientation to another. As discussed by Lumpkin et al., minimum resolution should occur when molecules are all aligned.

Slater and Noolandi have followed Lumpkin et al.'s approach to the same problem with a similar, semi-quantitative derivation (16). They stress that "non-ideal" electrophoretic behavior stems from the non-gaussian conformation that DNA coils adopt during electrophoresis, as characterized by stretching. As defined in the Lumpkin and Zimm (1) as well as the Lumpkin et al. (9) derivation, the polymer remains in the path tube without any segmental leakage. In terms of a worm in a burrow, this means that none of the worm is able to move in a tube not already defined by the head. For TDP electrophoresis there is no direct experimental evidence for this postulated behavior. On the contrary there may indeed be considerable tube leakage, which may contribute to the mechanism of separation in TDP electrophoresis.

At first glance, postulating no tube leakage during DNA

electrophoresis is reasonable since the DNA persistence length is nearly the same as most estimates of agarose pore size (as pointed out by Lumpkin et al. (9)). Intuitively, since persistence length is a measure of polymer stiffness (DNA is fairly stiff with a persistence length of about 500Å), energetically it is unfavorable for a coil segment to leak out of the path tube to any appreciable extent. Countering this intuition is the fact that a DNA coil is quite deformable and that the total electrical forces present on a giga-dalton DNA molecule are considerable. Schwartz has taken tube leakage into account in describing the ODP and TDP electrophoretic spectrum (a plot of mobility versus pulse frequency) as well as other associated phenomena, including the inability of DNA larger than 2 giga-daltons to enter agarose gel (2). The Lumpkin et al. theory (9) based on no tube leakage erroneously predicts an unlimited molecular size capacity for agarose (or even acrylamide) gels despite contrary experimental evidence (2).

Field gradients play an important role in enhancing size resolution by electrophoresis. A field gradient is merely a variation in field strength over a given area. Precisely how field gradients affect size resolution in pulsed field gradient gel electrophoresis is not known; however, the method utilizes a minimum of one field gradient to generate a series of effective pulse times. The pulse time required to produce a minimum electrophoretic mobility is roughly inversely related to the field strength (2). As a DNA band moves through a gel slab during PFG, it not only encounters field strength variation but also a variation in the angle between the two electrical fields. The electrodes may be perpendicularly oriented to each other, but the field gradient will produce a series of angles between it and the other field. The optimum angle between the two electrical fields for obtaining maximum size resolution should be $90^{o}$, as predicted from the work of Lumpkin et al. (9).

One Dimensional Pulsed Electrophoresis Separation Mechanism: Background. Jamil and Lerman (12) have investigated the effect of a single pulsed electrical field on the electrophoretic mobility of fairly small DNA molecules in agarose and polyacrylamide gels. They changed both the field on- and off-times and found that both variables can significantly retard electrophoretic mobility. The degree of retardation varied with molecular weight, and an empirically derived equation was presented to describe the data. Jamil and Lerman's results can be due to modulation of electrophoretic mobilities by some type of molecular relaxation process. Presumably, during the field off-time molecular relaxation can occur and randomize any molecular orientation and conformational deformation achieved prior to the off-time. Such perturbations could conceivably alter a measured mobility relative to a relaxed state mobility.

Schwartz conducted ODP experiments using giga-dalton sized molecules (derived from yeast chromosomes) and obtained similar results (2) to those of Jamil and Lerman. However, the resolution mechanisms operating in the large and small molecular size ranges could be quite different. This could be partially due to size dependence of molecular relaxation mechanisms in a gel.

Briefly, the separation mechanism outlined by Schwartz for large DNA molecules involves the transition of a randomly coiled DNA molecule immersed in a gel matrix (as conceived by Lumpkin et al. (9)) to a minimally gaussian, oriented chain followed by diffusional relaxation back to a gaussian conformation (2). Unlike TDP, ODP relies totally on diffusive forces to effect conformational relaxation. To estimate the time need for each of these processes, Schwartz first calculated a gel frictional coefficient for a persistence length sized DNA molecule and used this value to estimate a Zimm relaxation time in a gel. The time needed to orient a large DNA coil immersed in a gel was estimated by Schwartz by first modeling the initial electrophoretic behavior as a multi-reptational process: that is, a large, gaussian coil (immersed in a gel matrix) may simultaneously enter many tubes. For successful transport through the gel, the coil must eventually pick only one tube. Initially, the mobility should be retarded until the coil moves through only one tube. An estimation of the the degree of mobility retardation can be obtained from deGennes's derivation of the branched chain frictional coefficient in a polymer melt (17) or more appropriately a gel matrix (as applied by Lerman and Frisch (3)). Here we are drawing a hydrodynamic analogy between multi-tube reptation and polymer branching. deGennes's work indicates that a branched chain frictional coefficient should increase exponentially with the size or number of branches.

Pulsed Electrophoresis: Experimental Methodology

TDP presents many experimental difficulties involving the presence of more than one set of electrodes in a conductive media as well as the task of preparing and loading giga-dalton sized DNA molecules. We will examine some equipment designs as well as the insert method for preparing giga-dalton sized DNA molecules. This section is intended to provide the researcher with enough information to apply this methodology in the laboratory.

The Problem and Solution of Electrode Interaction. Placing two electrodes pairs in a conductive media such as electrophoresis buffer and powering one set will not generate a uniform, homogeneous field. The unpowered electrode pair would preferentially conduct the field, thereby causing distortion (2). In fact, a single unpowered electrode, depending upon the orientation in the electrical field, can generate both hydrogen and oxygen. To produce uniform homogeneous fields, electrically invisible, spectator electrodes must be utilized.

The problem was solved (2, 7) by splitting a straight horizontal platinum electrode into a vertical array with each member attached to a diode and then onto a common bus (hence named diode isolated electrodes). A diode can be thought of as an electrical ratchet because it confines electrical current to unidirectional flow. This new electrode design permits more than one electrode pair to coexist simultaneously in the same conductive media without any appreciable interaction.

Two-Dimensional Pulsed Electrophoresis: Equipment Design.

Presently, there are several published instrumentation designs for generating multiple electrical fields as needed for TDP. The first type is depicted (2, 7) in Figure 1. It shows simple use of the diode isolated electrodes as previously described, plus a provision for cooling the running buffer and gel. The gel is run in a submarine manner so that circulation and cooling is simultaneously accomplished. By utilizing electrodes present in the arrays, any combination of uniform or gradient electrical fields can be produced. (When one or more field gradients are used, it is commonly named PFG.) For example, a single field gradient can be generated by utilizing a single North electrode and an entire South array, while a uniform field results from powering the entire East and West arrays.

Another design uses ordinary electrodes (18) (Figure 2) to produce two field gradients. Unfortunately, this design cannot produce uniform fields due in part to electrode interactions. However, the apparatus is easy to construct and provides excellent resolution of yeast chromosomal DNAs. As with the previous design, there are provisions for cooling and buffer recirculation.

Sample Preparation. Traditional DNA preparation procedures (6) cannot produce intact, very high molecular weight DNA molecules. Zimm and his colleagues solved this problem for viscoelastic measurements (10) by using a cell lysis solution containing high EDTA concentrations to chelate divalent cations essential for the activity of many nucleases and a combination of proteinase K and detergent to denature and digest protein. In addition, all lysis is performed in situ to eliminate any shear mediated DNA breakage. Unfortunately, it is difficult to prepare high molecular weight DNA for electrophoresis using this protocol. Schwartz and Cantor (7) adopted this protocol for electrophoresis by embedding cells in a small agarose plug; lysis is accomplished by diffusion of the lytic solution into the plug, called an "insert". Since the deproteinized DNA molecules are enormous, diffusion out of the insert is virtually nil.

Practical Considerations for Running PFG Gels. As previously discussed, PFG uses a minimum of one field gradient. Much of the existing experimental data has been obtained using PFG with gradient and homogeneous fields. Therefore we will confine most of our discussion to practical considerations of this experimental configuration.

Sample inserts are prepared in a mold to snugly fit wells cast in agarose and are gently pressed into them prior to a run. The most commonly used buffer is a low ionic strength buffer consisting of Tris, boric acid and EDTA (TBE). Typical PFG potentials used with a 20 x 20 cm apparatus are 310 and 120 volts for the gradient and uniform fields, respectively (2, 7). Generally 1% agarose is used, and HGT (high gelling temperature) agarose made by FMC Corporation appears to give the best results as judged by band sharpness and running time (2). A 70 second pulse time provides good separation for molecular weights ranging from 100 - 2,000 megadaltons, as encountered with chromosomally sized yeast DNA molecules (2, 7). Likewise, using these conditions a 30 second pulse time will

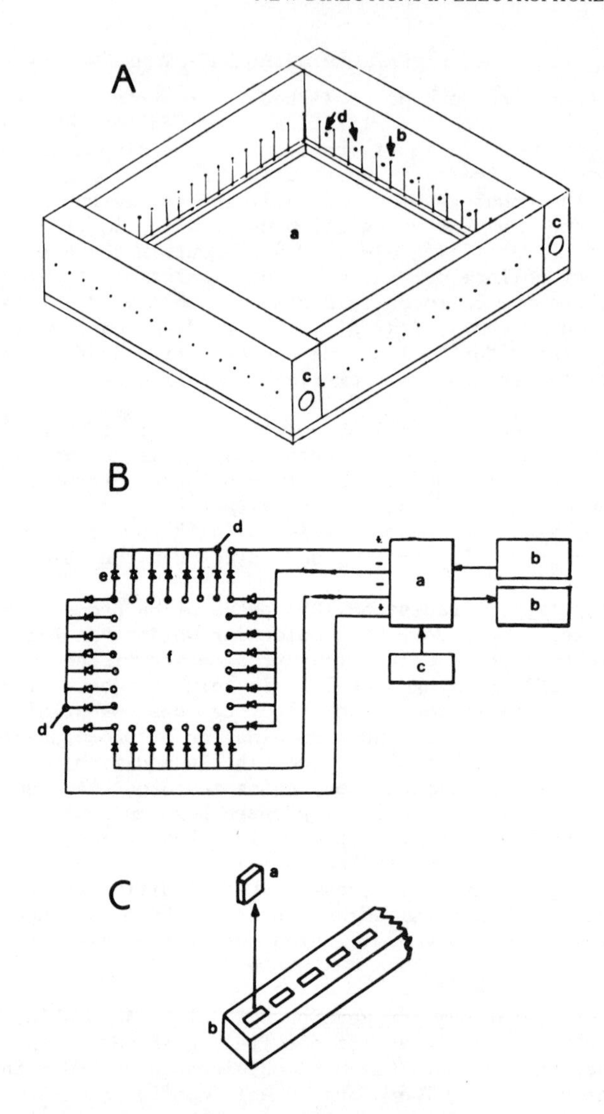


Figure 1. Pulsed Electrophoresis Equipment. Panel (A) depicts a PFG electrophoresis box with (a) being the space for the slab gel and (b) showing the vertical platinum electrodes which are connected to diodes as depicted in panel (B). (c) shows the buffer circulation input and output and (d) the buffer circulation ports.

Panel (B) shows the wiring block diagram with (a) being the switching relay, (b) being two power supplies, (c) being a timer controlling the switching, and (d) being SPST switches. (e) shows the diode-platinum electrode arrays flanking the electrophoresis chamber as shown in panel (A), and (f) the gel location within that chamber.

Panel (C) depicts an insert maker, with (b) being a mold machine made with plexiglas, and (a) the resulting insert.

optimally separate molecules in the 350 megadalton range. Examples of electrophoretic mobility as a function of pulse time (2) are shown in Figure 3. The figure demonstrates that as the pulse time is increased, degenerate bands are resolved into their components. At very long pulse times no appreciable resolution should be apparent (2, 7). It can further be noted that resolution takes place over about a 2-3 fold mobility range.

Figure 3 illustrates the effect of two different agarose concentrations on electrophoretic mobility (2). Aside from the mobility retardation effect, qualitatively there is no dramatic effect upon resolution in this concentration range. However, it is preferable to use a lower gel concentration since it shortens run times.

## Biological Applications of Pulsed Field Gradient Gel Electrophoresis

Organisms such as yeast and unicellular parasites (i.e., trypanosomes) have relatively small genomes containing chromosomal DNA molecules in the 100 - 2,000 megadalton range (2). Consequently, these genomes can be successfully resolved or, adapting cytological terminology, a molecular karyotype can be produced. In the early stages of pulsed electrophoresis research, prior extensive genetic analysis in yeast provided important information on the physical nature and sizes of very large DNA molecules. Later, knowledge obtained from PFG analysis of yeast was instrumental in establishing genetics in organisms such as unicellular parasites. In this section we will discuss some applications of PFG to the biology of these organisms.

Yeast. The yeast Saccharomyces cerevisiae has a genome size of approximately 14,000 kb (6) and 17 chromosomes ranging in size from 300 to more than 2000 kb (19). (In this section we will use kilobasepairs, 1 kb = 660,000 daltons, since daltons are now rarely used in biology to describe nucleic acid size.) Early evidence on its genome makeup was obtained from sedimentation, viscoelastic and genetic studies. Unfortunately these methods cannot definitively establish the number and absolute physical sizes of chromosomes. Unlike higher organisms, yeast chromosomes cannot be visualized in the light microscope. PFG provided the first molecular separation of yeast chromosomes, allowing their count to be firmly established (2, 7). Figure 4 shows a typical yeast molecular karyotype (2).

With a yeast molecular karyotype in hand, biologists can quickly and conveniently monitor genetic alterations and obtain a genomic overview previously unattainable. Some examples include mapping the location of specific genes to chromosomes, monitoring chromosome breakage, uncovering chromosomal size polymorphisms and analyzing the chromosomal distributions of multiple copy genes.

PFG can vastly simplify genetic analysis as demonstrated in its application to analysis of yeast chromosomal translocations. Briefly, a chromosomal translocation involves the transposition and attachment of a piece of one chromosome onto another. Translocations are not infrequent genetic events and occur in virtually every organism including man. However, studying these events in organisms whose chromosomes cannot be visualized by light microscopy used to

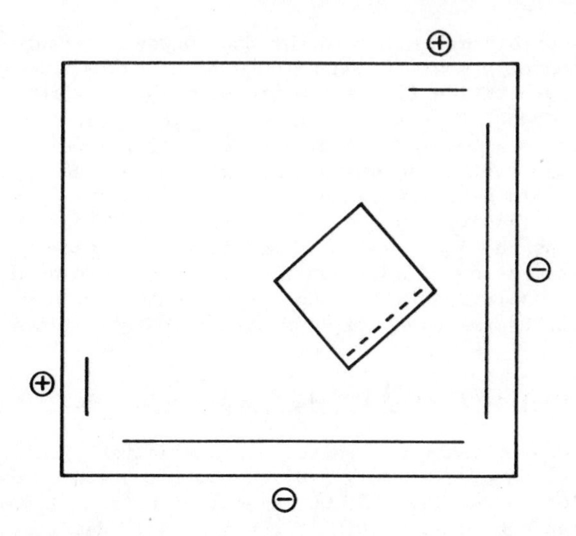

Figure 2. OFAGE Apparatus. Long and short lines depict platinum electrodes with polarities marked. Square in center depicts slab gel with illustrated wells.

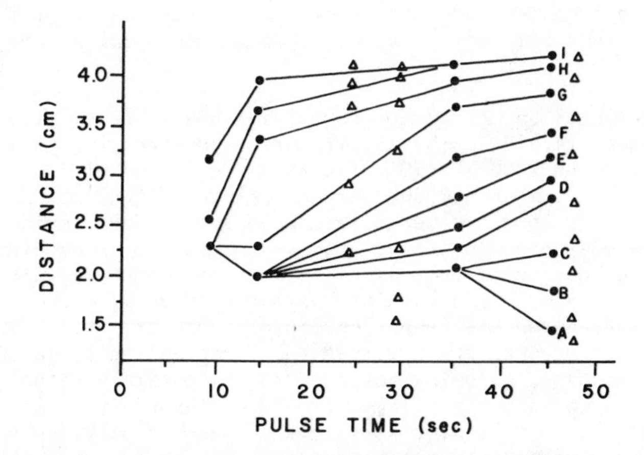


Figure 3. Change of Electrophoretic Mobility of Yeast Chromosomal DNAs as a Function of Pulse Time and Gel Concentration. The dots denote 1.0% high gelling temperature agarose gels and triangles denote 1.5% low endoosmosis agarose gels. All gels were loaded identically with yeast strains D273-10B/A1 and DBY782 and run with voltages of 200 (North-South; field gradient) and 84 volts (East-West; uniform field) in a 10 x 10 cm apparatus. Low endoosmosis agarose gels were run for 10 hours; high gelling temperature for 7 hours. Mobility measurements were made from ethidium stained gel photographs, measuring from the left hand corner of the third slot (containing yeast DBY782) to the center of a band. Relative discrepancies in mobility from gel to gel were calculated from the relative mitochondrial DNA band mobilities and found to be negligible within the accuracy of band measurements, +/-0.6 mm.

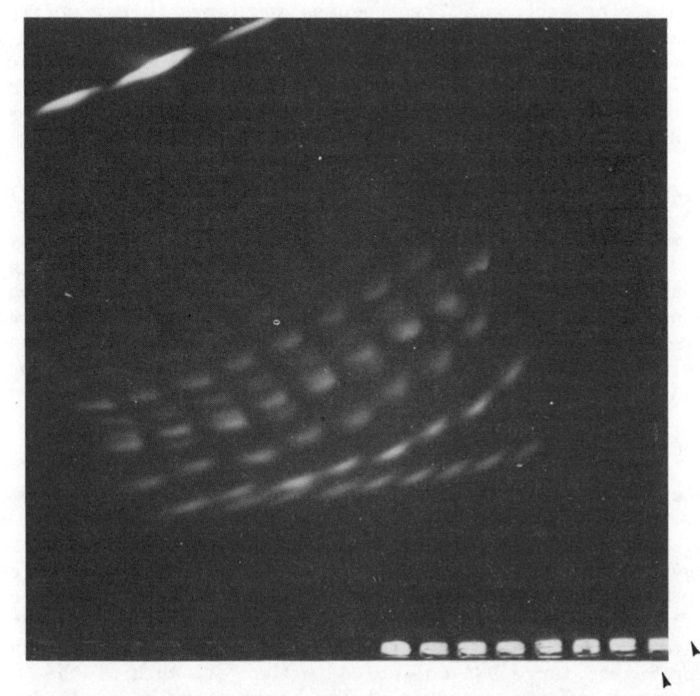


Figure 4. Yeast Molecular Karyotype. This illustrates an ethidium bromide-stained 1.0% high gelling temperature gel (20 x 20 cm) run with a 70 second pulse time and 325 volts (North-South; field gradient) and 130 volts (East-West; uniform field) for 30 hours. Lane 1: marker DNAs G phage (630 kb), T2 phage (180 kb) and T7 phage (40 kb). Other lanes, yeast strains D273-10B/A1 loaded alternatively with DBY782. Note variation of chromosome sizes between strains.

require a fair amount of genetic manipulation and could be impossible to detect without prior knowledge of the identity of participating chromosomes. Figure 5 shows the PFG analysis of a translocation between a small and large chromosome(2, 7). The resulting products are two medium-sized chromosomes with visibly shifted gel band positions. The analysis of a chromosomal translocation can thus be reduced to studying a photograph.

Trypanosomes. *Trypanosoma brucei* is a vicious unicellular parasite carried by the tsetse fly that is responsible for causing an often fatal sleeping sickness. Trypanosomes sometimes infest entire regions in Africa, making these regions inhospitable to both man and cattle. Its success as a parasite is due in part to antigenic variation: it can change its surface proteins and thus continually elude the host's immune system. Until recently virtually no genetics existed for trypanosomes, and the molecular details of their antigenic switching was unknown (20).

Trypanosome chromosomes, like those of yeast, are not visible with a light microscope. PFG electrophoresis was able to provide the first molecular karyotype for trypanosomes (21). This karyotype allowed the first direct study of the parasite's genome and quick elucidation of many molecular features of antigenic variation that would not have been possible using traditional methodology. Besides aiding the search for medical treatments of parasitic diseases, new biological paradigms gleaned from this analysis will help in studying the molecular biology of higher organisms.

## Biochemistry Using Giga-Dalton DNA Molecules

Inserts contain intact DNA molecules in a shear free environment that is also quite permeable -- permeable enough for enzymes to freely diffuse throughout the insert and react with DNA. Molecular biology utilizes a powerful battery of enzymes to cut, ligate and modify DNA molecules. Normally, these reactions are carried out in solution with relatively small DNA molecules using experimental conditions oblivious to any shear considerations. However, by leaving intact large DNA molecules in the gel, much of the experimental repertoire of molecular biology can be applied to these molecules, rendering them experimentally useful and accessible (2). A prime example is the use of restriction endonucleases (enzymes that cleave DNA molecules at specific sequences) to cleave intact DNA molecules. The restriction maps generated from cleavage positions reveal genomic organization. Preparative fractionation of restriction enzyme digests can be used to isolate specific fragments. One application would be isolation of the gene responsible for Huntington's disease, a fatal, dominantly inheritable mental disorder often manifesting itself in midlife, and whose biochemical basis is unknown. Researchers have identified an anonymous DNA segment marking a chromosomal region containing the putative disease gene (22). This marker permits the identification of Huntington's disease gene carriers, both pre- and postnatally, yielding important information for parents, prospective parents, or unexpressed victims. Attempts to isolate this chromosomal region from the human genome by using infrequently cutting restriction

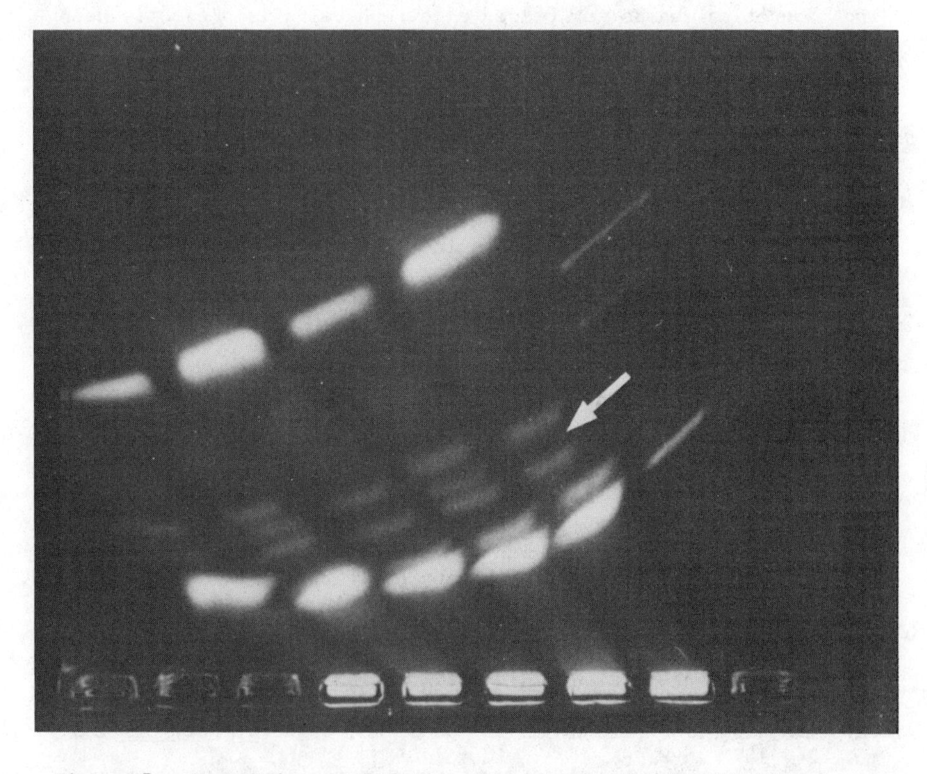

Figure 5. Yeast Chromosomal Translocation Analysis. Ethidium bromide-stained gel. The first lane was loaded with size markers as in Figure 4. Other lanes were alternately loaded with wild-type yeast (no translocations) and a strain containing a translocation between chromosomes 3 and 12. Experimental conditions were as in Figure 3, using 1.5% low endoosmosis agarose gel. The arrow indicates the position of the wild-type chromosome and the absence of the translocated chromosome.

enzymes followed by PFG fractionation are now in progress (2). The hope is to close in on the disease gene using ever closer markers. If successful this approach would become the method of choice for human genetic analysis and perhaps even for providing a detailed physical map of the human genome.

## Summary

The theoretical basis for pulsed electrophoresis is presently being constructed, and new associated phenomena are constantly being discovered. The lack of a firm physical understanding for this new electrophoretic phenomenon has not, however, prevented its successful application to many important biological problems.

Some future developments should include applying pulsed electrophoresis to studies of very small DNA molecules, proteins, macromolecular complexes, cells and synthetic polymers. Fully understanding the physical mechanisms responsible for pulsed electrophoretic fractionation should enable researchers to extend resolution to even larger DNA molecules and fully generalize use of this new electrophoretic effect.

## Literature Cited


1. Lumpkin, O. J. and Zimm, B. H. Biopolymers 1982, 21, 2315-2316.
2. Schwartz, D. C. Ph.D. Thesis, Department of Chemistry, Columbia University, NY, NY, 1985.
3. Lerman, L. S. and Frisch, H. L. Biopolymers 1982, 21, 955.
4. Serwer, P. and Allen, J. L. Biochem. 1984, 23, 922-927.
5. Fangman, W. L. Nucl. Acids Res. 1978, 5, 653-664.
6. Roberts, T. M., Lauer, G. D. and Klotz, L. C. 1974, CRC Critical Review of Biochem. 1974, 3, 349-449.
7. Schwartz, D. C. and Cantor, C. R. Cell 1984, 37, 67-75.
8. Schwartz, D. C., Saffran, W., Welsh, J., Haas, R., Goldenberg, M., and Cantor, C. R. Cold Spring Harbor Symp. on Quant. Biology 1983, XLVII, 189-195.
9. Lumpkin, O. J., Dejardin, P. and Zimm, B. H. Biopolymers 1985, 24, 1573-1593.
10. Klotz, L. C. and Zimm, B. H. Macromolecules 1972, 5, 471-481.
11. Carle, G. F. and Olson, M. V. Proc. Natl. Acad. Sci. U.S.A. 1984, 82, 3756-3760.
12. Jamil, T. and Lerman, L. S. J. of Biomolec. Struct. and Dynamics 1985, 2, 963-964.
13. Stellwagen, N. C. J. of Biomolec. Struct. and Dynamics 1985, 3, 299-314.
14. Stellwagen, N. C. Biochem. 1983, 22, 6180-6185.
15. Ferguson, K. A. Metabolism 1964, 13, 985-1002.
16. Slater, G. W. and Noolandi, J. Biopolymers 1986,
17. deGennes, P. G. "Scaling Concepts in Polymer Physics", Cornell University Press, New York, N.Y.,1974, pp. 219-244.
18. Carle, G. F. and Olson, M. V. Nucl. Acids Res. 1984, 5, 653-665.
19. Mortimer, R. D. and Schild, D.. In Appendix to "The Molecular Biology of Yeast Saccharomyces: Life Cycle and Inheritance"

(updated as of November 1984); J. N. Strathern, E. W. Jones and J. R.Broach, Eds., Cold Spring Harbor Laboratory, Cold Spring Harbor, N.Y., 1984.

20. Borst, P. and Cross, G. A. M. Cell 1982, 29, 291-303.
21. Van der Ploeg, L. H. T., Schwartz, D. C., Cantor, C. R. and Borst, P. Cell 1984, 37, 77-84.
22. Gusella, J. F., Wexler, N. S., Conneally, P. M., Naylor, S. L., Anderson, M. A., Tanzi, R. E., Watkins, P. C., Ottina, K., Wallace, M. R., Sakaguchi, A. Y., Young, A. B., Shoulson, I., Bonilla, E. and Martin, J. V. Nature 1983, 306, 234-238.

RECEIVED September 19, 1986

# Chapter 13

# Capillary Zone Electrophoresis

**James W. Jorgenson**

**Department of Chemistry, University of North Carolina at Chapel Hill, Chapel Hill, NC 27514**

Capillary zone electrophoresis is a technique which permits rapid and efficient separations of charged substances in an instrumental format. Buffer-filled capillaries with typical dimensions of 50 microns i.d. and 100 cm length are used as the separation chambers. Applied potentials as high as 30 KVolts are used to drive the electrophoretic process. Such high potentials promote rapid migration of zones, while minimizing zone spreading. A simple theory of zone spreading and resolution is presented. A physical description of the system and some of its operating characteristics is provided. Separating performance of the system is described and example separations shown. Limitations of the system, particularly with regard to the separation of proteins, are discussed and future areas for research are suggested.

## Background

The various techniques of modern electrophoresis are a powerful and versatile approach to separation and analysis of substances, especially proteins and polynucleotides. Separation modes such as isoelectric focusing and SDS-gel sieving electrophoresis are quite effective in their own right, and when combined in a two-dimensional format, form a technique of unrivalled resolving power. But modern electrophoresis, as practiced, is a rather labor intensive approach to analysis. Making gels, sample application, staining and destaining gels are time-consuming tasks. Furthermore, the techniques for band detection, including staining and use of a densitometer for quantitation, are characterized by limited dynamic range and linearity. Indeed much of the interest in applying HPLC to separation and analysis of biopolymers stems from the fact that HPLC is a highly instrumental technique with autosamplers and on-line detectors connected to computers for data acquisition and analysis.

0097-6156/87/0335-0182$06.00/0

A fully instrumental version of electrophoresis would seem to be a desirable goal and several systems have been described. Hjerten(1) employed zone electrophoresis in a gel-free tube of three mm inner diameter. In this system thermal convection was minimized by rotation of the tube around its longitudinal axis. Zone detection was accomplished with an UV absorption detector which scanned the length of the tube. Kolin(2) described several free zone electrophoresis systems which used flow in serpentine and helical paths to combat thermally driven convection. Catsimpoolas(3) described an instrumental system with scanning detection for following the course of isoelectric focussing in gel-filled tubes. These techniques have not come into widespread use presumably due to their complexity.

An alternative approach to instrumental electrophoresis is to carry out the separation in a capillary tube of sub-millimeter diameter. Mikkers et al.(4) demonstrated free zone electrophoresis in a capillary tube on equipment designed for separations by isotachophoresis. Thormann et al. carried out electrophoresis in gel-free chambers of rectangular cross section (28,29). Tubes of small diameter offer many advantages as a format in which to do electrophoresis. Joule heat is generated uniformly throughout the tube cross section but is dissipated through the tube walls. This results in a temperature gradient across the tube with the fluid in the center being warmer than the fluid at the wall. This temperature gradient results in a density gradient (leading to convective flow) as well as viscosity and pH gradients across the tube. Convective flow may lead to spreading of electrophoretic zones, while viscosity and pH gradients may also act as a cause of zone dispersion. The use of smaller diameter tubes reduces the magnitude of temperature differences within the tube. The temperature difference between the fluid in the center and at the wall is roughly proportional to the square of the tube's i.d.(5,6,7)

Decreasing tube diameter has other beneficial effects as concerns zone spreading. Convective flow is damped by the drag of the stationary tube wall acting on the viscous fluid. This effect, called the "wall effect" by Mikkers et al.(4), becomes increasingly effective in preventing convective flow as the tube diameter is decreased. Also, radial pH and viscosity gradients are only undesirable to the extent that individual analyte molecules spend an inordinate amount of time in either the tube center or near the wall(8). Diffusion of analyte molecules tends to randomize their radial occupancy allowing a molecule to "sample" all portions of the tube cross section. This diffusional averaging leads to narrower zones. The effectiveness of diffusional averaging is increased as the tube diameter is reduced. Thus use of tubes of decreased diameter leads to smaller temperature differences and furthermore acts to minimize the zone spreading effects of any residual temperature gradients. These compounded effects argue strongly for the use of small diameter capillaries. One obvious difficulty in this conclusion is that it necessitates small sample volumes, and thus places great demands on detection sensitivity.

Another important group of mechanisms of zone spreading may be collected under the heading of sample overloading. The sample, if sufficiently concentrated, may perturb the physical and chemical properties of the electrophoresis medium.(4,8,9) For instance, polymeric analytes may increase the viscosity of the medium, and thus decrease electrophoretic mobility within the zone. Most analytes contain acidic and/or basic groups and thus can shift the pH within the zone, again affecting mobility. And probably most important, the analytes are themselves ions, and may alter the electrical conductivity of the medium, leading to distortion of an otherwise homogenous electric field and resulting in zone spreading. All of these difficulties argue for a very low ratio of analyte concentration to concentration of supporting electrolyte (buffer) in order to minimize these sample overloading effects. Again the obvious difficulty with this conclusion is that it will place great demands on detection sensitivity, since sample concentration must be kept low.

In the ultimate limiting case, if all other mechanisms of zone spreading can be rendered insignificant, zone broadening in zone electrophoresis will be dominated by a seemingly unpreventable mechanism, longitudinal diffusion. As diffusion in liquids is rather slow, this may result in rather narrow zones. It is possible to describe the separating power of a zone electrophoresis system in terms of theoretical plates in analogy with chromatography(10). Since an instrumental zone electrophoresis system will produce data in the form of an electropherogram, closely analogous to a chromatogram, the use of theoretical plates to describe performance has a similar merit (and limitations) in electrophoresis as it does in chromatography. Jorgenson and Lukacs(8) predicted that if longitudinal diffusion was the only significant source of zone broadening, the number of theroretical plates, N, is given by:

$$N = \frac{\mu V}{2D} \quad (1)$$

where $\mu$ and D are the analyte's electrophoretic mobility and diffusion coefficient, and V is the voltage applied to the system to drive the electrophoretic separation. Since mobility and diffusion coefficient are not easily altered in a way to increase N, high applied voltages are the most direct way to high separation efficiencies. It must be remembered that this prediction is based strictly on the assumption that longitudinal diffusion is the dominant mechanism of zone broadening, a condition which may be difficult to realize in practice. It must also be borne in mind that an infinitely narrow band of injected sample is assumed. It is interesting to note that tube length does not enter into this equation. Thus in principle, very short tubes may be used to promote rapid analyses, while not jepardizing separation efficiency. However, in practice this is not entirely true, as will be shown.

## Description of System and its Basic Operating Characteristcs

A schematic of the capillary zone electrophoresis (CZE) system is shown in figure 1. A capillary tube is filled with buffer and suspended between two reservoirs filled with the same buffer. The end at which high voltage is applied is surrounded by a plexiglass safety interlock box. This box must be opened to gain access to the electrode and reservoir, and opening the box automatically shuts off the high voltage power supply while simultaneously shorting its output to ground through the high voltage relay. This relay is built in-house, using a solenoid connected to a switch mechanism with an 8 cm gap when open. Attention to safety is extremely important, as a power supply which can deliver as much as 30 KVolts is used in this system. The other end of the capillary is dipped into a buffer reservoir in which the electrode is connected through an ammeter to ground. Electrical currents in CZE depend on capillary dimensions, buffer conductivity and applied potential, but typical operating currents are between 10 and 100 microamps.

Sample is most easily introduced into the capillary using an electromigration technique(8). The buffer reservoir at the high voltage end is replaced with a reservoir containing sample. High voltage is applied for a specific amount of time (usually a few seconds) and then turned off. This brief application of voltage migrates a narrow band of sample into the capillary. Now the sample reservoir is removed, and the buffer reservoir is replaced. High voltage is again applied, and the electrophoretic run begins. This sample introduction technique is simple and effective, introducing sample with minimal band broadening. However it does act to discriminate against the various components of the sample, based upon their mobilities.

Detection is usually accomplished by "on-column" fluorescence(11) and UV absorption(12). The electric field, conductivity, and thermal detecters which are effective in capillary isotachophoresis are of insufficient sensitivity to be generally useful in CZE(13). Where feasible, fluorescence detection is attractive due to its great sensitivity. However, many samples, such as proteins, do not appear to be good candidates for fluorescence detection. The intrinsic fluorescence of proteins is quite variable from protein to protein, and is usually rather weak. Fluorescence labelling of proteins is also difficult, as there is a tendency to produce a complex assortment of multiple (partially labelled) products. Proteins have been detected most effectively by UV absorption detection. In both fluorescence and UV absorption, detection is carried out "on-column", shining the incident light beam onto the capillary and measuring either fluorescence or transmitted light. In this regard, capillaries fabricated from fused silica are vastly superior to those from glass, due to their excellent UV transparency and extremly low background luminescence.

Figure 2 shows the effect of tube i.d. on the measured separation efficiency of dansyl-labelled isoleucine(14). The tubes were all 100 cm long Pyrex type 7740 borosilicate glass capillaries, filled with pH 6.86, 0.05 M phosphate buffer, and operated at a potential of 15 KVolts. Only the tube diameters were varied. Clearly, below 80 microns, a performance of approximately 250,000 theoretical plates is obtained, and no significant improvement is seen at smaller diameters. Tube diameters larger than 80 microns result in a precipitous decrease in plates. This is in general agreement with the zone broadening considerations described earlier. Figure 3 shows the effect of tube length on the separation efficiency of the same analyte(14). All tubes were 75 micron i.d., filled with the same phosphate buffer as before, and operated at an applied potential of 15 KVolts. Tubes of approximately 100 cm and longer yield consistently high plate numbers approaching 230,000. Tubes longer than 100 cm do not improve separation efficiency, but do result in significantly longer analyses. Tubes shorter than 100 cm show a dramatic loss in plates. This is presumably due to "thermal overloading" of the system. Shorter capillaries offer lower electrical resistance, and thus at constant voltage, carry higher currents. Power dissipated (the product of voltage and current) by the capillary increases while available surface area over which to dissipate this heat decreases. Below a length of 100 cm, even a capillary of 75 microns i.d. cannot dissipate heat efficiently enough to prevent significant temperature gradients and their attendant zone broadening phenomena.

Figure 4 shows the effect of applied potential on the separation efficiency of fluorescamine-labelled hexylamine(8). The analyte was run in a pH 6.86, 0.05 M phosphate buffer. At the lower applied potentials, general agreement with equation 1 is seen, with plates being proportional to applied potential. Beginning at approximately 20 KVolts, this relationship begins to fail. This is again likely to be the result of thermal overloading. Figure 5 shows the results of an indirect measurement of the average temperature of the buffer inside of three different capillaries as a function of applied potential(15). All three capillaries were 75 μm i.d. x 100 cm long. The buffer filling each capillary was again a pH 6.86, 0.05 M phosphate buffer. Temperature was measured by first measuring the conductivity of this buffer as a function of temperature. The conductance of the buffer filled capillary was then measured as a function of applied potential, and from this the temperature was inferred. With all three capillary materials a significant increase in temperature is seen as potential is increased. These elevated temperatures are not only important from the point of view of zone broadening mechanisms, but also in regard to the stability of thermally labile analytes, such as many proteins.

Figure 6 shows the effect of analyte concentration on separation efficiencey(14). The analyte is dansyl-labelled isoleucine, run in a pH 6.86, 0.05 M phosphate buffer. The capillary was 75 microns i.d., 100 cm long, and the applied potential 30 KVolts. Achieving the highest separation efficiencies requires the use of low analyte concentrations, due to the effects of sample overloading, as described earlier. As analyte concentration approaches $1x10^{-5}$ M, there appears to be some plateau in performance being approached. This is expected, as at some point sample overloading should become insignificant. An alternative way to minimize sample overloading is to use higher concentrations of buffer salts. This approach has its limits, as increased salt concentrations lead to increased electrical currents, power dissipation, and thus thermal overloading.

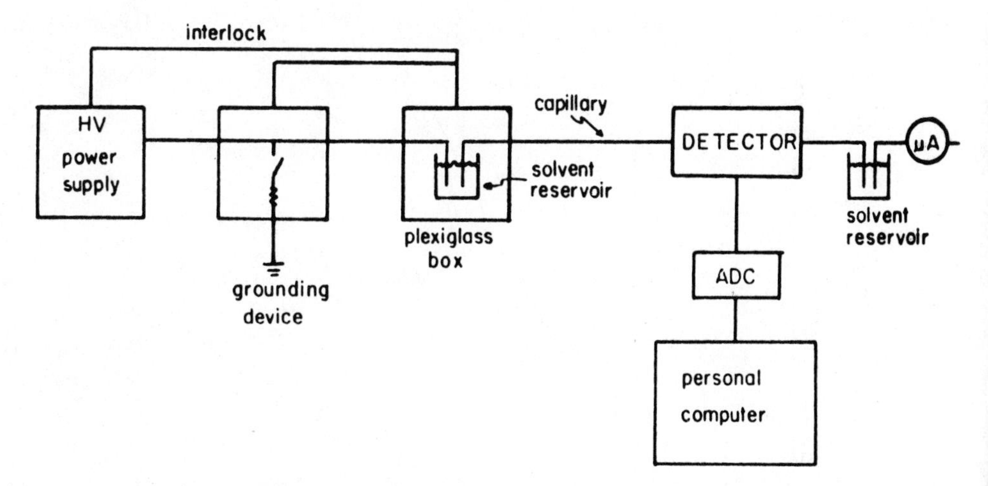


Figure 1. Schematic of CZE system. Reproduced with permission from Ref. 27. Copyright 1986 Walbroehl, Y.

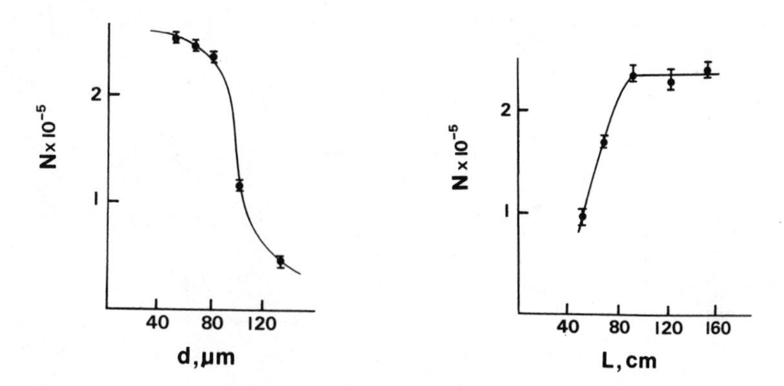


Figure 2 (left). Separation efficiency as a function of tube inner diameter. Reproduced with permisssion from Ref. 14. Copyright 1985 J. High Res. Chromatogr. Chromatogr. Commun.

Figure 3 (right). Separation efficiency as a function of tube length. Reproduced with permission from Ref. 14. Copyright 1985 J. High Res. Chromatogr. Chromatogr. Commun.

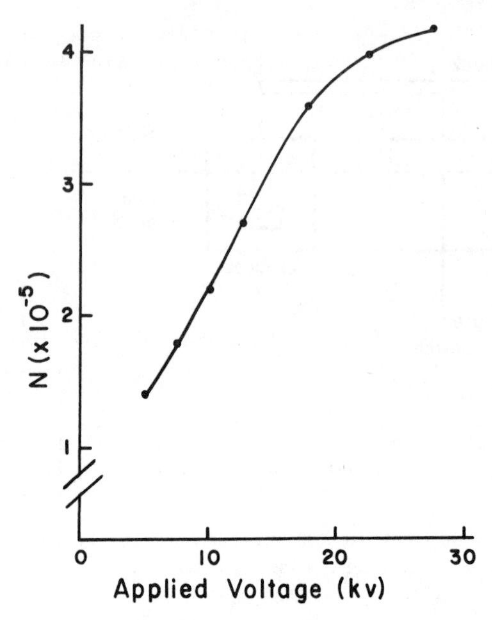


Figure 4. Separation efficiency as a function of applied potential. Reproduced with permission from Ref. 15. Copyright 1981 *Anal. Chem*.

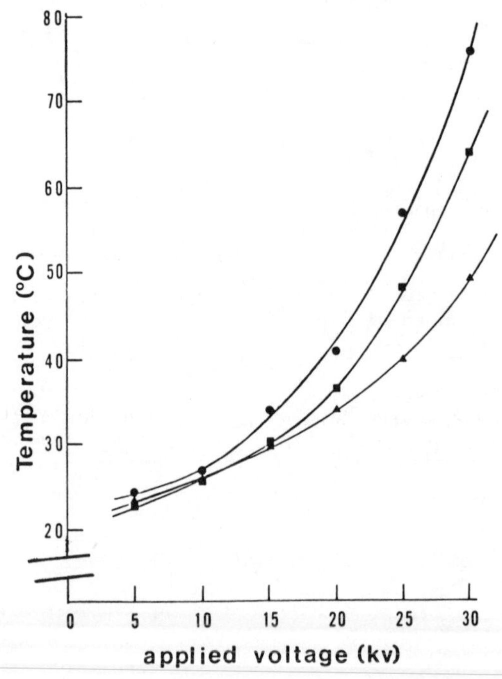


Figure 5. Mean buffer temperature as a function of applied potential. All capillaries 75 micron i.d. ● = teflon; ■ = fused silica; ▲ = pyrex. Reproduced with permission from reference 15. Copyright 1983 Lukacs, K. D.

It is evident that the operating performance, in terms of theoretical plates, involves a complex interplay of capillary dimensions, buffer concentration, applied potential, and analyte concentration. Although general trends are clearly apparent, a successful and accurate detailed quantitative theory of zone broadening as a function of these and other parameters does not yet exist, and may prove to be so complex as to be beyond realization. None the less, effective operating parameters can be found relatively easily by experiment.

## System Performance

An electropherogram of a group of dansyl-labelled amino acids, detected by fluorescence, is shown in figure 7(16) . Good separation efficiency and a relatively rapid analysis time are evident. The analytes migrate in order of charge, with the more positively charged basic amino acids being detected first, and the negatively charged acidic amino acids coming out last. The direction of migration of all analytes, regardless of sign of net charge, is from positive to negative. Even negatively charged analytes migrate to the negative electrode. This is due to a strong electroosmotic flow of buffer toward the negative electrode. This flow is strong enough at pH 7 to sweep most ions, regardless of charge, toward the negative electrode. Fortunately electroosmotic flow exhibits a virtually perfectly flat flow profile and thus is insignificant as a cause of zone spreading(17). Different analyte ions are still separated in the presence of electroosmotic flow, as their electrophoretic mobilities are simply superimposed upon the electroosmotic flow. Electroosmosis can actually affect resolution of zones in CZE(8). If resolution, Rs, is defined in a manner analogous to that in chromatography(10), then the resolution of two zones in CZE is given by:

$$Rs = 0.177\ (\mu_1-\mu_2)\left[\frac{E}{D(\bar{\mu}+\mu_{osm})}\right]^{\frac{1}{2}} \qquad (2)$$

where $\mu_1$, and $\mu_2$ are the electrophoretic mobilities of the two analytes, D is the average of their diffusion coefficient, $\bar{\mu}$ is the average of their mobilities, and $\mu_{osm}$ is the electroosmotic flow coefficient (electroosmotic flow velocity in an electric field of unit strength). From this equation it may be seen that the greatest resolution may be obtained when the electroosmotic flow is roughly equal in magnitude but opposite in sign (direction) to the analyte's mobilities. This will yield higher resolution, but at the expense of longer analysis times. Figure 8 shows the effect of electroosmotic flow on the resolution of some dansylated amino acids(8). In the upper electropherogram the analytes were run in an untreated glass capillary which exhibits relatively rapid electroosmotic flow. In the lower electropherogram, this same capillary was silylated with trimethylchlorosilane (TMCS) and then filled with buffer and the same set of analytes run again. TMCS, by reacting with many of the surface silanols, eliminates some of the surface charge and thus reduces the electroosmotic flow. The result of improved resolution and longer analysis time is obvious.

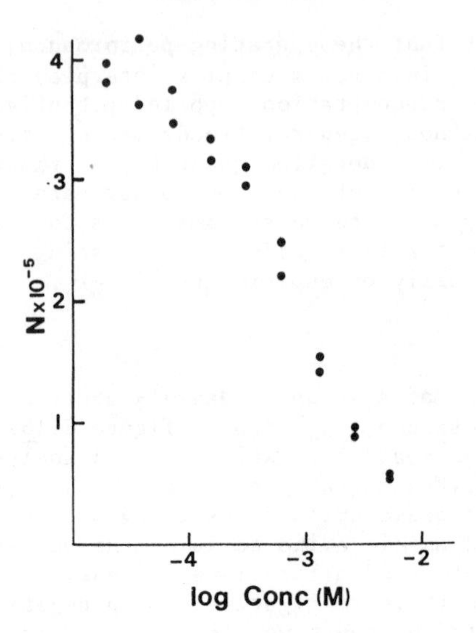


Figure 6. Separation efficiency as a function of analyte concentration. Reproduced with permission from Ref. 14. Copyright 1985 J. High Res. Chromatogr. Chromatogr. Commun.

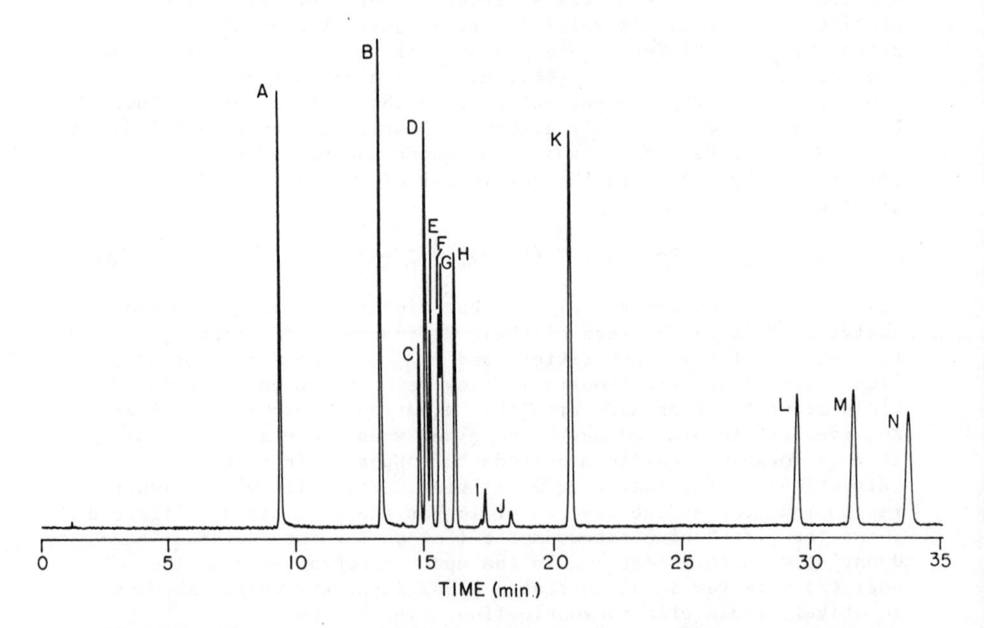


Figure 7. Electropherogram of dansyl amino acids. A = ε-labelled lysine; B = dilabelled lysine; C = isoleucine; D = methionine; E = asparagine; F = serine; G = alanine; H = glycine; I and J = unknown impurities; K = dilabelled crystine; L = glutamic acid; M = aspartic acid; N = cystic acid. Reproduced with permission from Ref. 16. Copyright 1984 J. High Res. Chromatogr. Chromatogr. Commun.

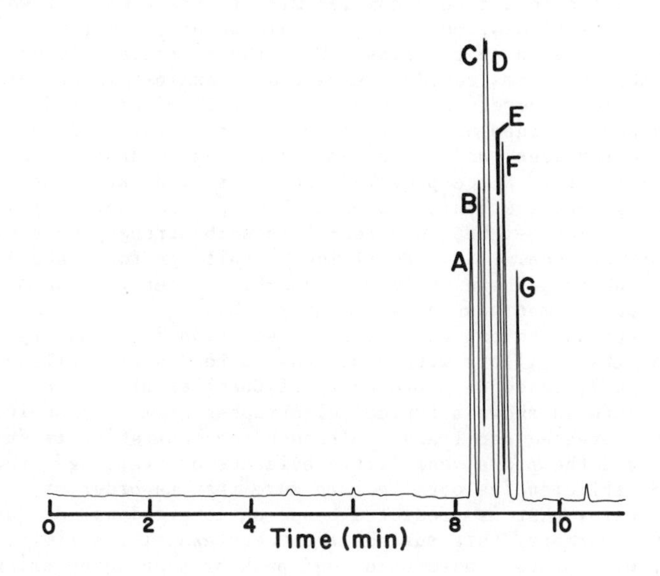


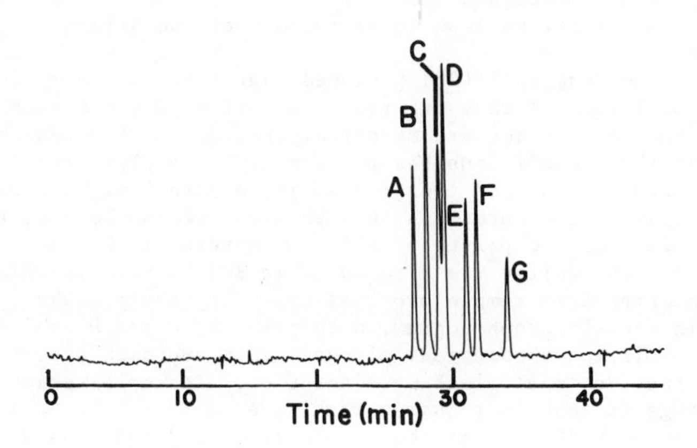


Figure 8: Electropherograms of dansyl amino acids. Upper: in untreated glass capillary. Lower: in glass capillary treated with trimethylchlorosilane. A = asparagine; B = isoleucine; C = threonine; D = methionine; E = serine; F = alanine; G = glycine. Reproduced with permission from Ref. 8. Copyright 1981 Anal. Chem.

Figure 9 shows the separation of fluorescamine labelled peptides obtained from a tryptic digest of reduced and carboxymethylated egg white lysozyme(18). Very high sepration efficiencies for this rather complex mixture are seen. It was hoped that this kind of performance would be obtained for yet larger analytes such as proteins. With their smaller diffusion coefficients, proteins could be expected to exhibit in excess of several million theoretical lates in CZE if longitudinal diffusion dominates zone broadening. Unfortunately, proteins tend to adsorb strongly to surfaces, and any adsorption leads to dramatic zone broadening in CZE. Since proteins contain so many kinds of functional groups (cationic, anionic, hydrophobic, polar) and are of high molecular weight, they tend to adsorb strongly to a wide variety of surfaces. It is a difficult challenge to create a surface to which proteins will not adsorb, but realization of this goal will be an important development in CZE.

One approach to prevent protein adsorption is to modify the surface of the capillary with a silane. A bonded diol silane ("glycophase"), based on a procedure of Chang et al.(19) was tried. Figure 10 shows a typical electropherogram of proteins from such a treated capillary. Although the separation is fairly efficient and the peaks show little evidence of "tailing", the separation efficiencies obtained are more than an order of magnitude below what is predicted based on longitudinal diffusion alone. Furthermore, this surface treatment exhibits a limited lifetime, with protein adsorption and peak broadening becoming more noticeable after only a few days of use. In addition, any such silylation treatments are only stable in a pH range of 2 to 7. Basic pH conditions lead to rapid loss of the silane by hydrolysis.

Lauer and McManigill(20) proposed that protein adsorption could be minimized if they are run in a buffer pH where both proteins and the surface are negatively charged. They reasoned that under these conditions the protein might be electrostatically repelled from the surface thus preventing adsorption. Figure 11 shows a separation of proteins in a pH 8.24 tricine buffer, a pH above the isoelectric points of all the proteins in the sample(21). The buffer was also 40 mM in KCl to help minimize zone-broadening from sample overloading. Very sharp peaks are evident in the electropherogram, with peaks B, C and D exhibiting nearly one million theoretical plates. This approach appears highly effective in eliminating adsorption. Its only serious disadvantage is that it requires working at a pH on the basic side of the isoelectric point of the sample proteins, and thus does not give a great deal of flexibility in operating conditions.

This discovery by Lauer and McManigill suggested to me that much of the protein adsorption might be due to ion exchange interactions between cationic sites in the protein and cation exchange sites (silanoate groups) on the fused silica surface. As in ordinary ion exchange chromatography, this interaction could be weakened by raising the concentration of competing ions in the buffer. Figure 12 shows the results of CZE of proteins in a pH 9 Ches buffer with 0.25 M $K_2SO_4$ added in an effort to decrease ion exchange interactions(22). This approach apparently works since

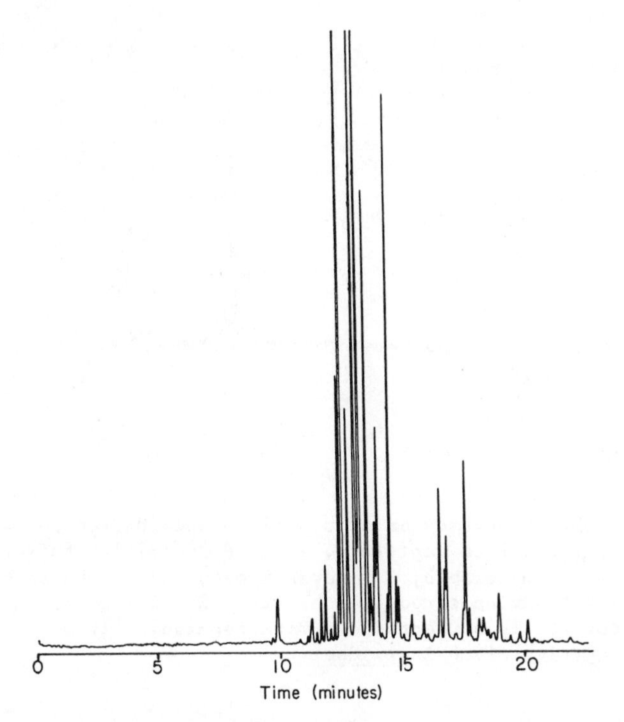


Figure 9. Electropherogram of fluorescamine-labelled peptides obtained from a tryptic digest of reduced and carboxymethylated egg white lysozyme. Reproduced with permission from Ref. 18. Copyright 1981 J. High Res. Chromatogr. Chromatogr. Commun.

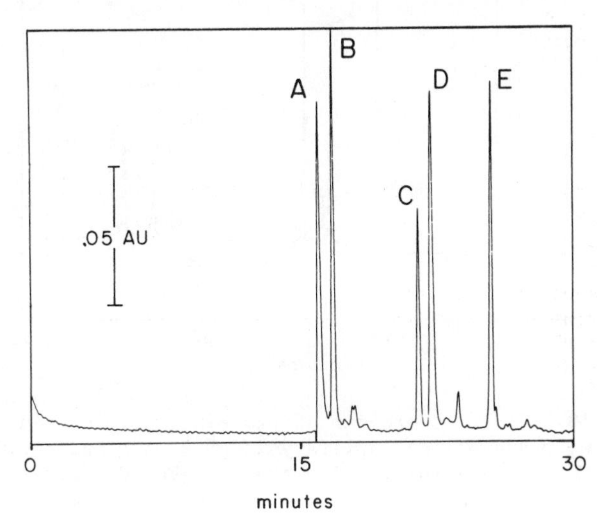


Figure 10. Electropherogram of protein standards run in a glycophase-treated fused silica capillary, in pH 7.0 phosphate buffer. A = lysozyme; B = cytochrome c; C = ribonuclease; D = chymotrypsinogen; E = horse myoglobin.

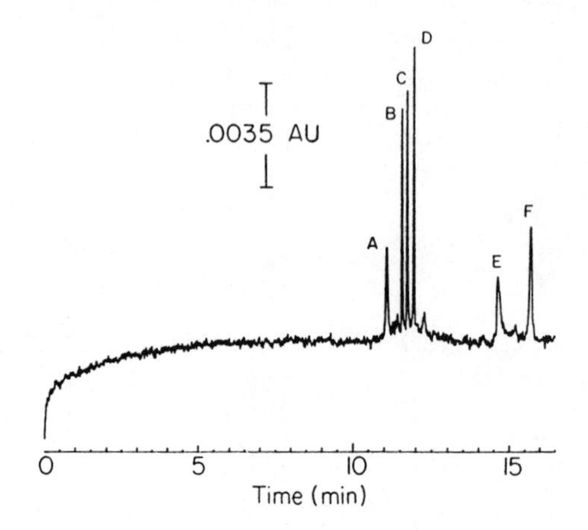


Figure 11. Electropherogram of protein standards run in an untreated fused silica capillary, in pH 8.24 tricine buffer. A = sperm whale myoglobin; B = horse myoglobin; C = human carbonic anhydrase; D = bovine carbonic anhydrase; E = ß-lactoglobulin B; F = ß-lactoglobulin A. Reproduced with permission from Ref. 21. Copyright 1986 Anal. Chem.

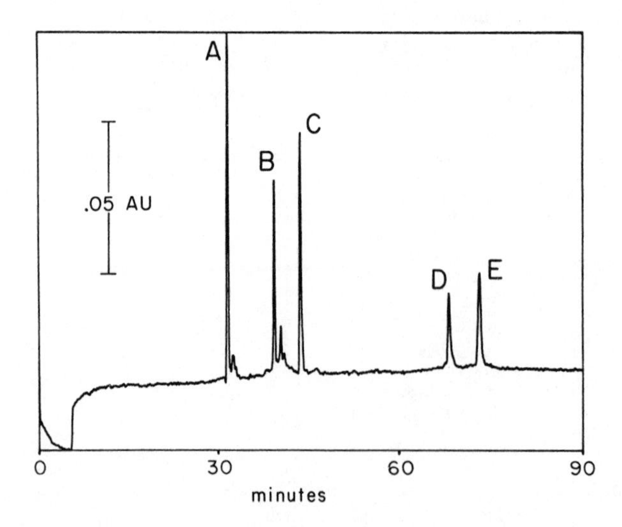


Figure 12. Electropherogram of protein standards run in an untreated fused silica capillary in pH 9 Ches buffer, with 0.25 M KCl. A = lysozyme; B = trypsinogen; C = myoglobin; D = ß-lactoglobulin B; E = ß-lactoglobulin A. Reproduced with permission from Ref. 22. Copyright 1986 Trends Anal. Chem.

lysozyme, with an isoelectric point of 11, is still migrated as a relatively sharp zone, even though it is 2 pH units below its isoelectric point. In general the peaks in this electropherogram are quite sharp. However, the time scale of the analysis is longer than usual. This is due to the fact that the run was done in a 25 micron i.d. capillary, 50 cm long, with only 5 KVolts of potential instead of the more usual 20 KVolts. The high salt concentrations required the use of lower voltages and smaller i.d. capillaries in order to avoid serious thermal overloading.

These electropherograms of proteins also serve to illustrate another significant problem with proteins. The signal to noise ratio in these separations is not extremely high. All three separations were monitored by on-column UV absorption detection. For CZE to be truly successful, detection limits for proteins must be improved by roughly two orders of magnitude. This is a formidable and yet important goal.

CZE is versatile in permitting unusual separation chemistries to be investigated with relative ease. An example is shown in figure 13, which is an electropherogram of dansyl-labelled d,l-amino acids separated in a buffer containing a copper complex of l-histidine(23). Separation of the d and l isomers is made possible by their differential association with the copper-l-histidine complex. Another unusual electropherogram is shown in figure 14(24). This is the separation of neutral organic molecules by their hydrophobic interaction with tetrahexylammonium ion. In this case, the more hydrophobic the analyte, the more it "binds" to the hydrophobic cation, and the faster it migrates through the system. This "hydrophobic interaction electrophoresis" gives a new tool to aid separation by electrophoresis. Although this electropherogram shows a separation of relatively small neutral molecules, a similar effect might be used to aid in separation of proteins based in part on their relative hydrophobicities.

Some of the necessary future developments in CZE are clear. Capillaries with surfaces non-adsorptive toward proteins are important. Perhaps more important are detection schemes for proteins which are vastly more sensitive than present detectors. In my lab we are constructing and testing autosamplers and micro-fraction collectors for CZE, both of which function under microcomputer control. Hjerten(25,26) has begun to explore the possibilities of gel-filled capillaries and of isoelectric focusing in capillaries. All of this work has as its goal a versatile and powerful instrumental version of electrophoresis complementary to modern HPLC.

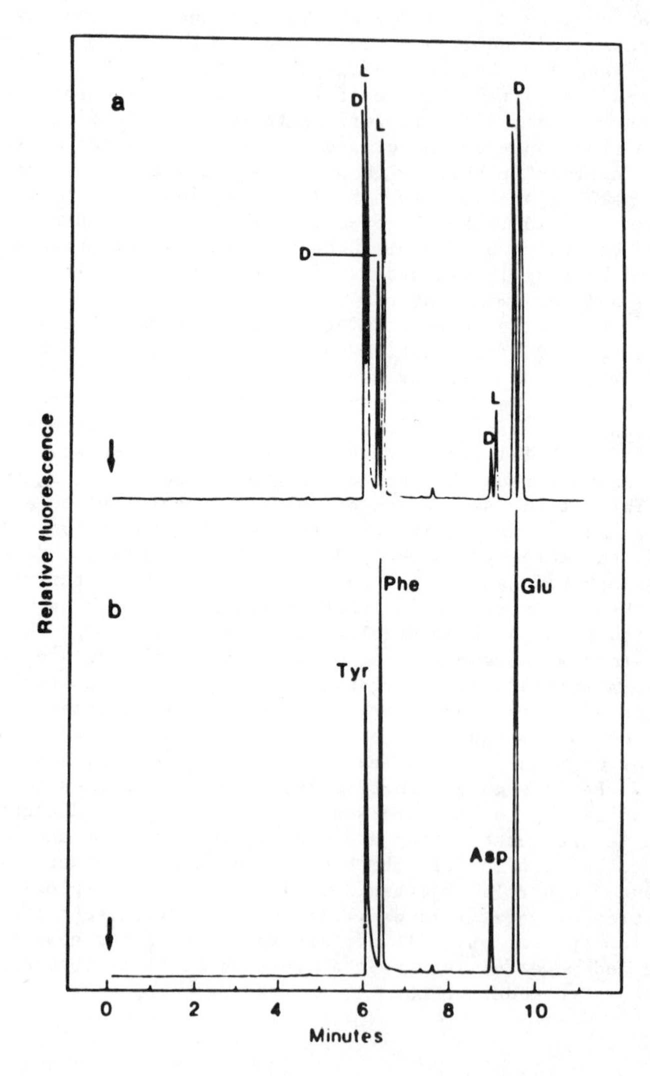


Figure 13. Electropherograms of D, L-dansyl amino acids with upper: Cu(II)-L-histidine electrolyte at pH 7. lower: Cu(II)-D, L-histidine electrolyte at pH 7. Reproduced with permission from Ref. 23. Copyright 1985 Science.

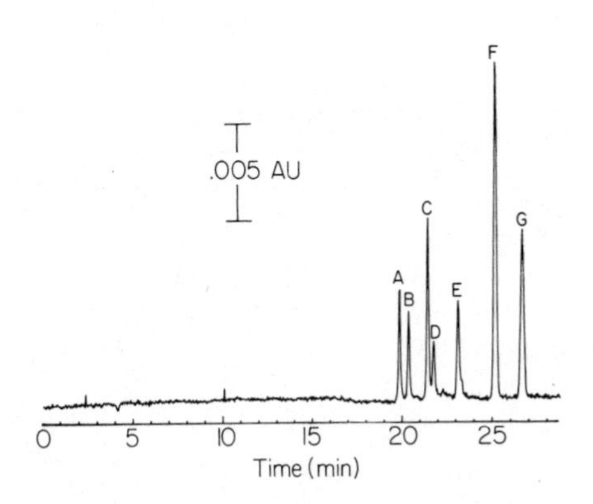


Figure 14. Electropherogram of neutral organic compounds in 50/50 acetonitrile/water with 0.025 M tetrahexylammonium perchlorate. A = benzo-(GHI) perylene; B = perylene; C = pyrene; D = 9-methyl-anthracene; E = naphthalene; F = mesityl oxide; G = formamide. Reproduced with permission from Ref. 24. Copyright 1986 Anal. Chem.

Acknowledgment

Support for this work was provided by a grant from the Alfred P. Sloan Foundation, the Hewlett-Packard Corporation, and the National Science Foundation under Grant CHE-8213771.

## Literature Cited

1. Hjerten, S., Chromatogr. Rev. 1967, 9, 122.
2. Kolin, A.; In "Electrophoresis - A Survey of Techniques and Applications, Part A: Techniques"; Deyl, Z., Ed.; Elsevier: Amersterdam, 1979; Chap. 12.
3. Catsimpoolas, N., ibid., Chap. 9.
4. Mikkers, F. E. P.; Everaerts, F. M.; Verheggen, Th. P.E.M.; J. Chromatogr. 1979, 169, 11.
5. Wieme, R. J.; In "Chromatography: A Laboratory Handbook of Chromatographic and Electrophoretic Methods 3rd Ed."; Heftmann, E., Ed.; Van Nostrand Reinhold: New York, 1975; Chap. 10.
6. Hinkley, J.O.N.; J. Chromatogr. 1975, 109, 209.
7. Brown, J. F.; Hinkley, J. O. N.; ibid, 218.
8. Jorgenson, J. W.; Lukacs, K. D.; Anal. Chem. 1981, 53, 1298.
9. Bier, M.; Palusinski, O. A.; Mosher, R. A.; Saville, D. A.; Science 1983, 219, 1281.
10. Giddings, J. C.; Sep. Sci. 1969, 4, 181.
11. Green, J. S.; Jorgenson, J. W.; J. Chromatogr. 1986, 352, 337.
12. Walbroehl, Y.; Jorgenson, J. W.; J. Chromatogr. 1984, 315, 135.
13. Everaerts, F. M.; Beckers, J. L.; Verheggen, Th. P.E.M.; "Isotachophoresis: Theory, Instrumentation and Applications"; Elsevier, Amsterdam, 1976.
14. Lukacs, K. D.; Jorgenson, J. W.; J. High Res. Chromatogr. Chromatogr. Commun. 1985, 8, 407.
15. Lukacs, K. D.; Ph.D. Thesis, University of North Carolina, Chapel Hill, N.C. 1983.
16. Green, J. S.; Jorgenson, J. W.; J. High Res. Chromatogr. Chromatogr. Commun. 1984, 7, 529.
17. Rice, C. L.; Whitehead, R.; J. Phys. Chem. 1965, 69, 4017.
18. Jorgenson, J. W., Lukacs, K. D.; J. High Res. Chromatogr. Chromatogr. Commun. 1981, 4, 230.
19. Chang, S. H.; Gooding, K. M.; Regnier, F.E.; J. Chromatogr. 1976, 120, 321.
20. Lauer, H. H.; McManigill, D.; Anal. Chem. 1986, 58, 166.
21. Jorgenson, J.W.; Anal. Chem. 1986, 58, 743A.
22, Lauer, H. H.; McManigill, D.; Trends Anal. Chem. 1986, 5, 11.
23. Gassmann, E.; Kuo, J. E.; Zare, R. N.; Science 1985, 230, 813.
24. Walbroehl, Y.; Jorgenson, J. W.; Anal. Chem. 1986, 58, 479.
25. Hjerten, S.; J. Chromatogr. 1983, 270, 1.
26. Hjerten, S.; Zhu, M.-D.; J. Chromatogr. 1985, 346, 265.
27. Walbroehl, Y.; Ph.D. Thesis, University of North Carolina, Chapel Hill, N.C. 1986.
28. Thormann, W., Arn, D., Schumacher, E., Electrophoresis 1984, 5, 323.
29. Thormann, W., Arn, D., Schumacher, E., Sep. Sci. Technol. 1985, 19, 995.

RECEIVED November 25, 1986

Chapter 14

# Capillary Isotachophoresis

F. M. Everaerts and Th. P. E. M. Verheggen

Laboratory of Instrumental Analysis, Eindhoven University of Technology, P.O. Box 513, 5600 MB Eindhoven, the Netherlands

Isotachophoresis (displacement electrophoresis) belongs to one of the four basic electrophoretic separation techniques: Moving boundary electrophoresis, zone electrophoresis and isoelectric focusing.
In the article presented capillary isotachophoresis is described. Basic principles are outlined and attention is given to qualitative and quantitative evaluation of isotachophoretic analysis in which conductimetric, potential gradient, photometric (UV-absorption and fluorimetric) detectors are used.
Separation of anions and cations is shown and a way for automatic signal processing is given.

The development of isotachophoresis in principle started in 1896 with the theoretical treatment of Kohlrausch [1] and his colleagues. He described [1] the basis of all electrophoretic separation techniques. In 1942 Martin [2] first tried an isotachophoretic experiment in a tube filled with gelatin (i.d. 0.25 inch). He detected a front between potassium and sodium with a ring of 32 thermocouples and registered this transition with a 1 mV thread recorder. He continued his experiments in 1946, using a ca 1 m capillary tube with an inner diameter of ca 0.5 mm. In these days he separated chloride (leading ion), acetate, aspartate and glutamate (terminating ion). For detection he used a differential copper-constantan thermocouple and a 1 mV thread recorder.
From 1964-1974 A.J.P. Martin has been nominated as extraordinary professor at the Eindhoven University on the chair entitled "The theory of Analogues". In these days he started, together with F.M. Everaerts, a project called "Displacement electrophoresis", later-on called Isotachophoresis [3].
In the early days, thermocouples (Cu-constantan; 15 µm) were used as detector. The introduction of the UV-detector and the con-

0097-6156/87/0335-0199$06.50/0

ductivity (potential gradient) detector, widened the scope of especially capillary isotachophoresis in the (bio) chemical field. For more detailed information is referred to the references [3,4 and 5].

Isotachophoresis is applicable in many fields from low to high molecular weight substances. Due to its high accuracy, high resolution, short time for analysis and tremendous flexibility (the choice of operational conditions to perform analysis is simple because no special columns or conditioning is needed), this technique desires more attention than obtained, this holds especially for the biochemical field.

## THEORY

Although capillary isotachophoresis is described in this chapter, the theory, in principle, holds for all kind of separations (on paper, on cellulose-acetate, in gels and in preparative modes). In isotachophoresis, a steady-state configuration is obtained as the result of a separation process that proceeds according to the moving boundary principle. Although this separation process is a transient state, it is governed by the same regulation function concept [1] as the steady state.

Quantitative and qualitative descriptions of the transient state provide information on the time needed for an isotachophoretic separation [3,4].

Moreover, such a description requires the definition of resolution in isotachophoresis and shows the result that can be expected from optimalization procedures. In electrophoresis the migration velocity, v, of a constituent, i, is given by the product of effective mobility, $\bar{m}_i$, and the local field strength, E.

$$v_i = \bar{m}_i E \tag{1}$$

The electrical field strength is vectoral so the effective mobilities can be taken as charged quantities, positive for constituents that migrate in a cathodic direction and negative for those migrating anodically. As a constituent may consist of several forms of subspecies in rapid equilibrium, the effective mobility represents an ensemble average. Ignoring constituents consisting of both positively and negatively charged subspecies in equilibrium, we can take concentrations with a sign corresponding to the charge of the subspecies. Thus, the total constituent concentration, $\bar{c}_i$, is given by the summation of all the subspecies concentrations, $c_n$.

$$\bar{c}_i = \sum_n c_n \tag{2}$$

Following the mobility concept, as described in [3], the effective mobility is given by:

$$\bar{m}_i = \sum_n \frac{c_n m_n}{\bar{c}_i} \tag{3}$$

where $m_n$ is the ionic mobility of the subspecies. In dissociation equilibria, the effective mobility can be evaluated using the degree of dissociation, a.

$$\bar{m}_i = \sum_n a_n m_n \tag{4}$$

The degree of dissociation can be calculated once the equilibrium constant, K, for the subspecies and the pH of the solution are known. For a restricted pH range, a very useful relationship has been given by the Henderson-Hasselbalch equation

$$pH = pK_a \pm \log \{ \frac{1}{a} - 1 \} \tag{5}$$

where $pK_a$ is the negative logarithm of the protolysis constant; the positive sign holds for cationic subspecies, the negative sign for anionic species.

All electrophoretic processes are essentially charge-transport processes that obey Ohm's law. In electrophoresis, this law is most conveniently expressed in terms of electrical current density, J, specific conductance, k, and electrical field strength, E.

$$J = kE \tag{6}$$

The specific conductance is given by the individual constituent contributions

$$k = F \Sigma_i \bar{c}_i \bar{m}_i \tag{7}$$

where F is the Faraday constant.

The equation of continuity for any electrophoretic process states

$$\frac{\partial}{\partial t} \cdot \bar{c}_i = - \frac{\partial}{\partial t} \{ - \frac{\partial}{\partial t} D_i \bar{c}_i + v_i \bar{c}_i \} \tag{8}$$

where t and x are time and place coordinates, respectively, and D is the diffusion coefficient. Neglecting diffusional dispersion, we can apply equation (8) for each constituent and the overall summation of the constituents gives

$$\frac{\partial}{\partial t} \sum_i \bar{c}_i = - \frac{\partial}{\partial t} \cdot E \sum_i \bar{m}_i \bar{c}_i \tag{9}$$

In combination with specific conductance (equation 7) and modified Ohm's law (equation 6), it follows that

$$\frac{\partial}{\partial t} \sum_i \bar{c}_i = 0 \qquad \text{or} \qquad \sum_i \bar{c}_i = \text{constant} \tag{10}$$

For monovalent weakly ionic constituents, equation (8) can be written as

$$\frac{\partial}{\partial t} \cdot \bar{c}_i = - \frac{\partial}{\partial t} \cdot E m_i c_i \tag{11}$$

where $m_i$ and $c_i$ are the mobility and the concentration of the charged species, i. Division by $m_i$ and application of the resulting relationship for each constituent, gives

$$\frac{\partial}{\partial t} \sum_i \frac{\bar{c}_i}{m_i} = - \frac{\partial}{\partial t} \cdot E \sum_i c_i \tag{12}$$

Electroneutrality, however, demands $\sum_i c_i = 0$, so

$$\frac{\partial}{\partial t} \sum_i \frac{\bar{c}_i}{m_i} = 0 \qquad \text{or} \qquad \sum_i \frac{\bar{c}_i}{m_i} = \text{constant} \tag{13}$$

Equation (13) is well known as the Kohlrausch regulating function [1]. In an electrophoretic system, different zones can be present, where a zone is defined as a homogeneous solution separated by moving and/or stationary boundaries. We can apply the continuity principle (equation 8) to a boundary and derive the general form of the moving boundary equation.

$$\bar{m}_i^K \bar{c}_i^K E^K - \bar{m}_i^{K+1} \bar{c}_i^{K+1} E^{K+1} = v^{K/K+1} (\bar{c}_i^K - \bar{c}_i^{K+1}) \tag{14}$$

where $v^{K/K+1}$ represents the drift velocity of the separating boundary between the zones K and K+1. In the case of a stationary boundary, the boundary velocity is zero and equation (14) reduces to

$$\frac{\bar{m}_i^{K+1} \bar{c}_i^{K+1}}{\bar{m}_i^K \bar{c}_i^K} = \frac{E^K}{E^{K+1}} = \text{constant} \tag{15}$$

From equation (15) it follows directly that for monovalent weak and strong electrolytes all ionic subspecies are diluted or concentrated over a stationary boundary to the same extent, because

$$\frac{c_i^{K+1}}{c_i^K} = \text{constant} \tag{16}$$

From the moving boundary equation (equation 9) it follows directly that, in a separation compartment of uniform dimensions at constant electrical driving current, all boundary velocities within the isotachophoretic framework are equal and constant. According to Joule's law, heat generation will occur resulting in different regions that are moving or stationary. In order to reduce the effects of temperature, relative mobilities, r, can be introduced. Obviously the leading ion constituent, L, provides the best reference mobility

$$r_i = \frac{m_i}{m_L} \tag{17}$$

Moreover, as in most isotachophoretic separations, for simplicity, only the counter-ion constituent, C, will be present, the reduced mobility, k, can be introduced

$$k_i = \frac{1 - r_C}{r_i - r_C} \tag{18}$$

Using the derived equations it is possible to calculate all dynamic parameters of analytical importance [3]. Moreover, the model considerations can be extended to moving boundary electrophoresis as well as zone electrophoresis. Due to the fact that diffusional effects play no role in the steady-state of isotachophoresis and it does e.g. in zone electrophoresis, the zone-characteristics will be different.

As in all differential migration methods, the criterion for separation in isotachophoresis depends simply on the fact that two ionegenic constituents will separate whenever their migration rates in the mixed state are different. For two constituents i and j, this means that according to equation (1) their effective mobilities in the mixed state must be different.

$$\frac{\bar{m}_i}{\bar{m}_j} \neq 1 \tag{19}$$

When the effective mobility of i is higher than that of j the latter constituent will migrate behind the former. Consequently, two monovalent weakly ionic constituents fail to separate, if the pH of the mixed state causes the effective mobilities of these constituents to be equal.
A unique feature of isotachophoresis is that after the separation process has been completed, all electrophoretic parameters remain constant with time. Assuming a uniform current density, all sample constituents between the leading-terminating electrolyte migrate at identical speeds. Moreover, at constant current density local migration rates will be constant. In this steady state, resolution values of stacked constituents will be either unity or zero. The basic features of the steady-state configuration have been discussed [3.4].

## THE SEPARATION PROCESS

The application of the above equations and definitions and the resulting implications are best illustrated by using a relatively simple two-component sample. We shall deal with the case where all constituents involved are monovalent weak electrolytes. Although essentially immaterial, we shall consider a separation compartment of uniform dimensions at a constant electrical driving current and at constant temperature. The separation process and some relevant information are given in Fig. 1. For more detailed information is referred to the references [5a and 5b].
It should be emphasized that, within the separator, three different regions are present and each has its own regulating behaviour. The regulating functions (equations 10 and 13) are the mathematical expression of this regulating behaviour and locally they cannot be overruled by the electrophoretic process. All changes in electrophoretic parameters, e.g. concentration (conductance), pH and temperature, will be in agreement with the local regulating function. It is obvious to use these parameters for a universal detection of the zones of the various constituents. Photometric and radiometric detectors can be used for specific zone detection.

As is shown is Fig. 1 the signals derived from a conductivity detector (potential gradient, thermometric) give general characteristics. From these signals qualitative and quantitative information can be deduced. Various parameters can be given to list the qualitative information, so that other laboratories working with different equipment can make use of it. For a correct qualitative evaluation it is always important to know whether the temperature difference between leading electrolyte and terminating electrolyte is large or small. The final temperature not only influences the effective mobility (2% per °C), but also the $pK_a$ values, especially of cationic species. For quantitative evaluation the temperature is of less influence, as long as calibration curves are used. If ionic species are present that are migrating faster than the leading ion this can be seen in the linear signal of the conductivity detector, which indicates a conductivity lower than that of the leading ion. The qualitative information of all other ions, migrating in between the leading zone and the terminating zone is not lost. If desired, a leading ion can be chosen with a higher effective mobility to have all constituent ions migrate between the leading and terminating zone.
If an ion is present in the sample with an effective mobility equal to that of the leading ion, the conductivity of the leading electrolyte is not changed. Qualitative information can still be obtained by measuring the retardation of the appearance of the first separation boundary. Of course in this case the conductivity and zone lengths of all other zones are not influenced by this ionic species.
For the qualitative aspects of an isotachopherogram, commonly, values of conductivity, potential gradient or temperature are used ($h_X$). Sometimes the reduced stepheight ($h_X-h_L$) or the stepheight-unit value (SU = 100 $h_X/h_L$) is used, where $h_X$ is the

value of a sample constituent and $h_L$ is the value of the leading constituent.
The reference-unit value [RU = $100(h_X-h_L)/(h_R-h_L)$] is also used, where $h_R$ is the value of a reference constituent. This has a greater accuracy than that of all other values.
For the quantitative aspects the zone-lengths are measured. There is a linear relationship between the zone length of an ionic species and the amount of that ionic species introduced as a sample, assuming the electric current is stabilized (Fig. 1). Calibration curves can be made or the information of time measurement is handled via a (micro) computer system, if necessary making use of a calibration constant [3].

## THE EQUIPMENT USED FOR CAPILLARY ISOTACHOPHORESIS

The equipment consists of two electrode compartments (Fig. 2: a,i) which are directly connected with the current stabilized, power supply, an injection block (Fig. 2: d) and a narrow bore tube. To prevent a hydrodynamic flow between the two electrode compartments, a semi-permeable membrane (Fig. 2: h) is mounted. The separation compartment is a narrow bore tube of Teflon[R] (PTFE) with an inside diameter of 0.2 mm and an outside diameter of 0.4 mm. This diameter was found to be optimal because the temperature difference between the various zones is small. Moreover, the convective disturbances are small and the zone profile is optimal. The sample can be introduced with a microlitre syringe into the injection block (Fig. 2: d).
In isotachophoretic analysis the sample ions are separated in consecutive zones according to their effective mobilities. All zones have their characteristic features: temperature, conductance, pH and potential gradient. Moreover, a zone may have an absorption or optical rotation; alternatively fluorescence or radioactive compounds may be present. A thermometric detector [3] (e.g. a thermocouple made of Cu-Constantan wires with a diameter of approximately 25 µm) was developed initially as the detection system. This detector is mounted on the outside of the narrow tube. The response of the thermometric detector is rather low and sensitivity is less than that of the high resolution detectors: conductometric, potential gradient and UV-absorption detector. From thermometric detection universal information can be derived. From the potential gradient detector and the conductivity (Fig. 2: p), with micro-sensing electrodes (10 µm Pt-Ir 10%) in direct contact with the electrolytes, universal information can also be derived. Generally the conductivity probe (housing) is made of acrylic resin, although for non-aqueous solutions Delrin (PTFE) or a well-chosen araldite can be used. The electrodes are mounted so that the electrolyte remains surrounded by an uninterrupted cylindrical wall. A contact cement has been used for the construction of the probe. The cell volume is approximately a few nl.
In our equipment [3] a UV-absorption is also mounted. The UV-source is a microwave mercury electrodeless lamp. The UV-light is guided by a quartz rod of optical quality into a slit with a diameter of 0.1 mm. The UV-light passes the narrow bore tube and

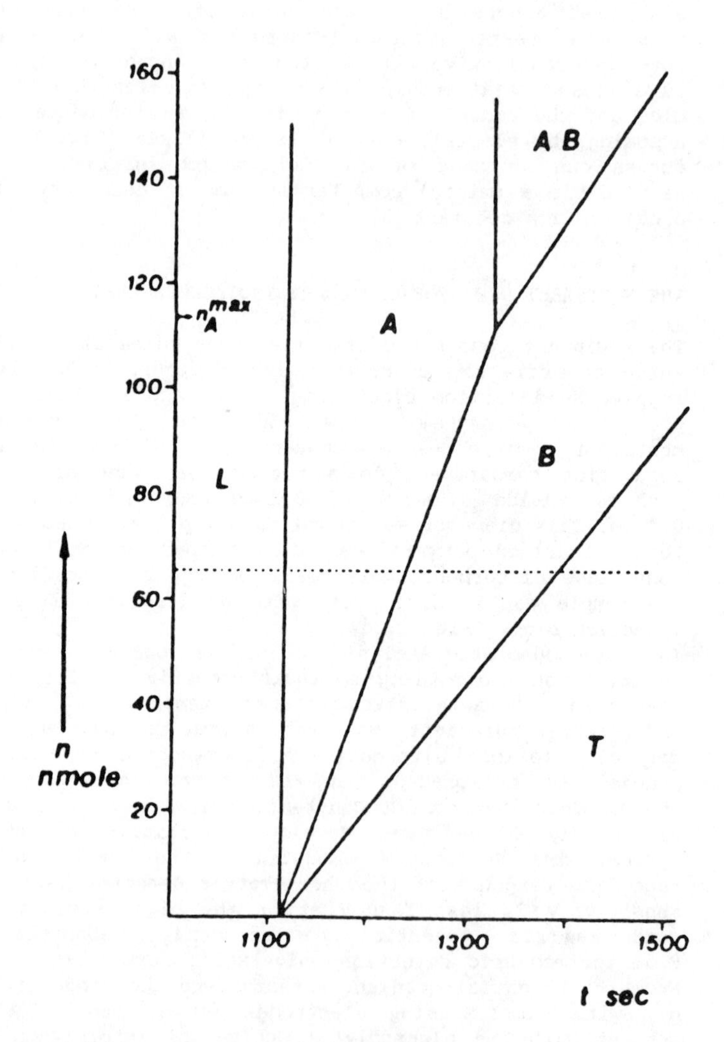


Figure 1. Resolution lines for a two-constituent mixture (operation conditions are given in Table I). L = chloride; A = formate; B = glycolate; T = propronate; AB = mixed-zone; $E^{L,A,B,T}$ = electrical field strength; n = amount sampled. Sample $c_{formate}$ = 0.05 M; $c_{glycolate}$ = 0.05 M; $pH_{sample}$ = 3.00. Reproduced with permission from Ref. 5. Copyright 1979 Elsevier.

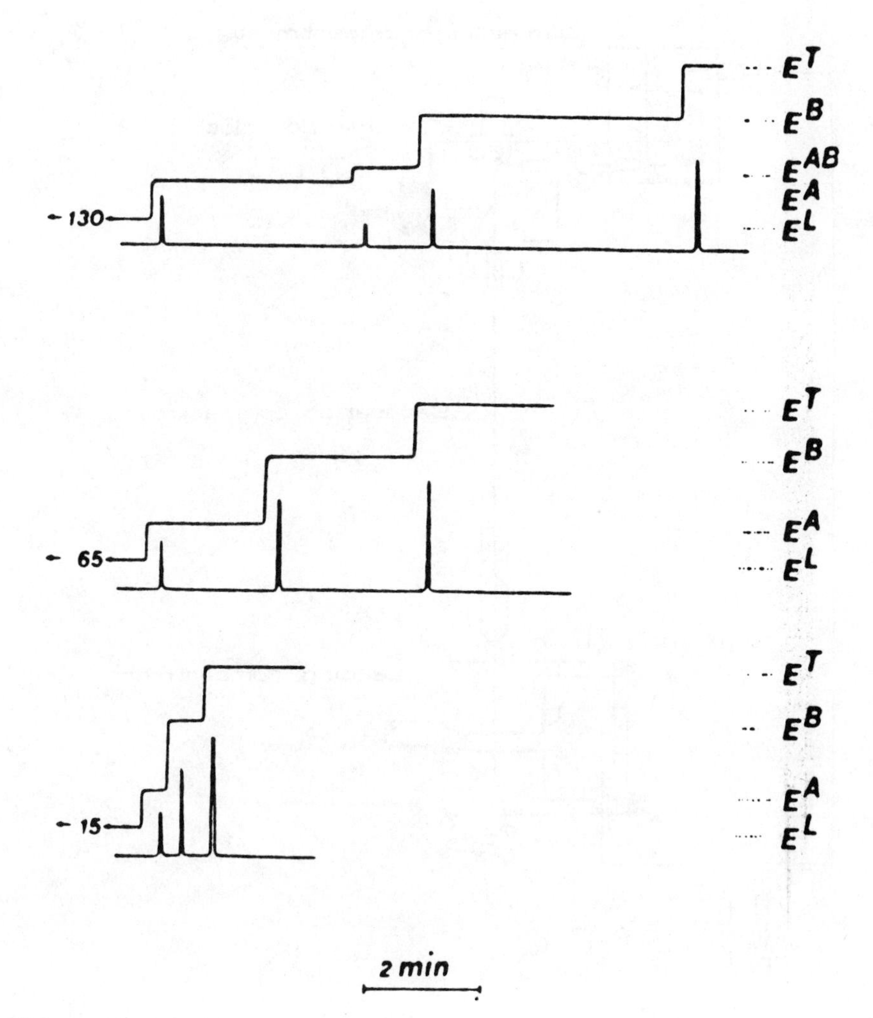


Figure 1. Continued.

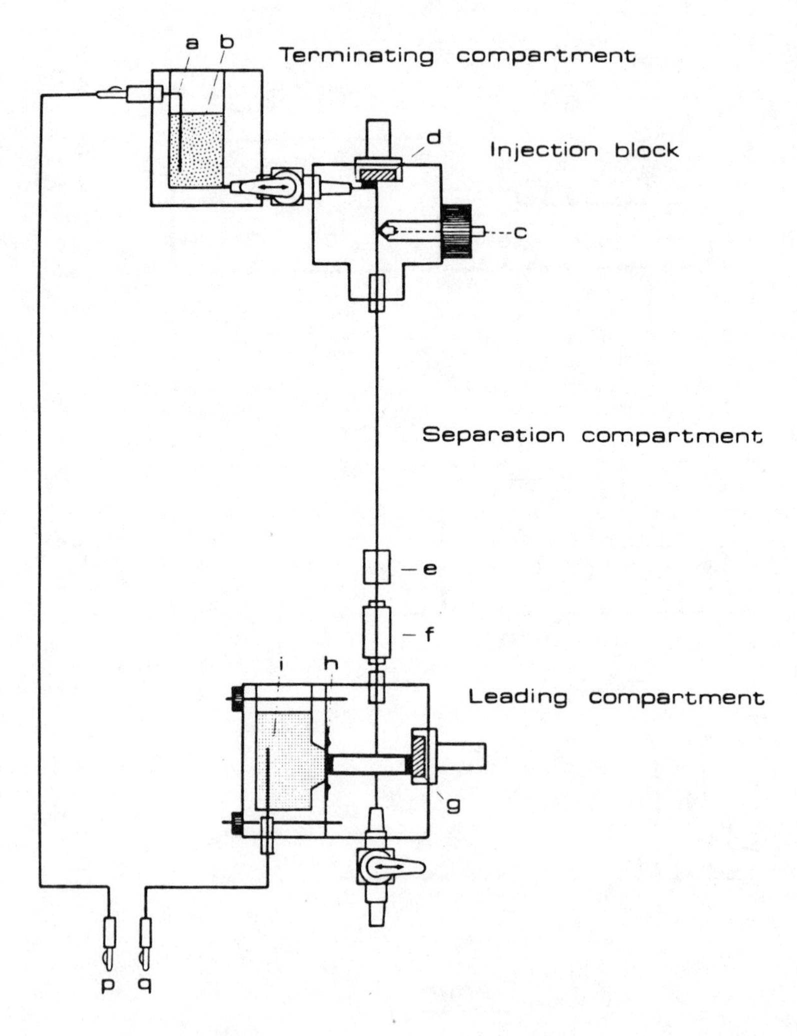


Figure 2. Schematic diagram of an ITP-apparatus. a = PT-electrode; b = terminating electrolyte; c = drain; d = silicone septum; e = UV-detector; f = conductivity (potential gradient) detector; g = silicone septum; h = semi-permeable membrane (e.g. cellulose acetate); i = Pt-electrode. p and q lead to a current-stabilized power supply (20 kV). The separation compartment is a PTFE-capillary (I.D. = 0.2 mm, O.D. = 0.35 mm, length is about 20 cm). Reproduced with permission from Ref. 25. Copyright 1983 Elsevier.

is guided by another quartz rod towards a UV-sensitive photodiode. The wavelength is selected by an interference filter. Teflon$^R$-lined valves are used at various places in the equipment, for the connection with the electrolyte reservoirs and the drain. Figure 2 shows clearly distinguishable parts in the electrophoretic equipment: the reservoir for the terminating electrolyte (Fig. 2: b); the place where the sample can be introduced (Fig. 2: d); the separation compartment; the places where the detectors are mounted (Fig. 2: e,f).

## APPLICATION

It is difficult to describe all possible fields of application of analytical isotachophoresis. To predict feasability for isotachophoresis as an analytical method, the ratio of molecular weight to effective charge can be used. Generally this ratio should not exceed 3000. Needless to say the compound must have a sufficient solubility in the solvent chosen. A brief survey of applications will be given and further information can be found in the references [3,5,6].

In the last seven years about nine hundred papers have been published on these subjects: proteins, nucleotides, inorganic anions, amino acids, drugs, enzyme reactions, metals, organic bases, organic acids, interaction studies, preparative applications (mainly in the field of proteins).

Detailed information can be found in the application notes [7,8] of LKB-Produkter AB (Bromma, Sweden) and of Shimadzu Co (Kyoto, Japan). From LKB-Produkter AB (Bromma, Sweden) a list of references is available: Acta Isotachophoretica [9]. A more recent review is given by P. Bocek et al. [9a]. Because it is difficult to describe all possible fields of application an analysis is selected that gives a survey of applicable on-line detectors used in our laboratory and certainly soon to be available in commercial ITP-equipment.

In Fig. 3 an analysis is shown of vitamin B constituents, analyzed as cations at pH= 5.0 with 0.01 M potassium/acetate as leading electrolyte and $H^+$ as terminating ion. The thermal detector, as used in the early days of isotachophoresis [10], is not given in this survey. Although accurate, the thermal detector provides insufficient resolution and almost all information collected with a thermometric detector can be obtained using the high-resolution conductometric (potential gradient) detector. Another characteristic isotachopherogram is shown in Fig. 4a. This isotachopherogram shows the steady state in the analysis of a wine sample. It was analyzed at low pH for organic and inorganic acids and an AC conductivity detector (3) was used. In standard isotachophoretic equipment minimal detectable amounts are of the order of 100 pmol. Specially adapted equipment and/or sample pretreatment can lower the minimal detectable concentration to c. 1 µM.

On-line and off-line combinations of ITP-MS, ITP-HPLC and ITP-HPLC-MS are under investigation by groups in Vienna and Bratislava [11] and Eindhoven [12]. On-line ITP-HPLC looks particularly promising and the results obtained are comparable

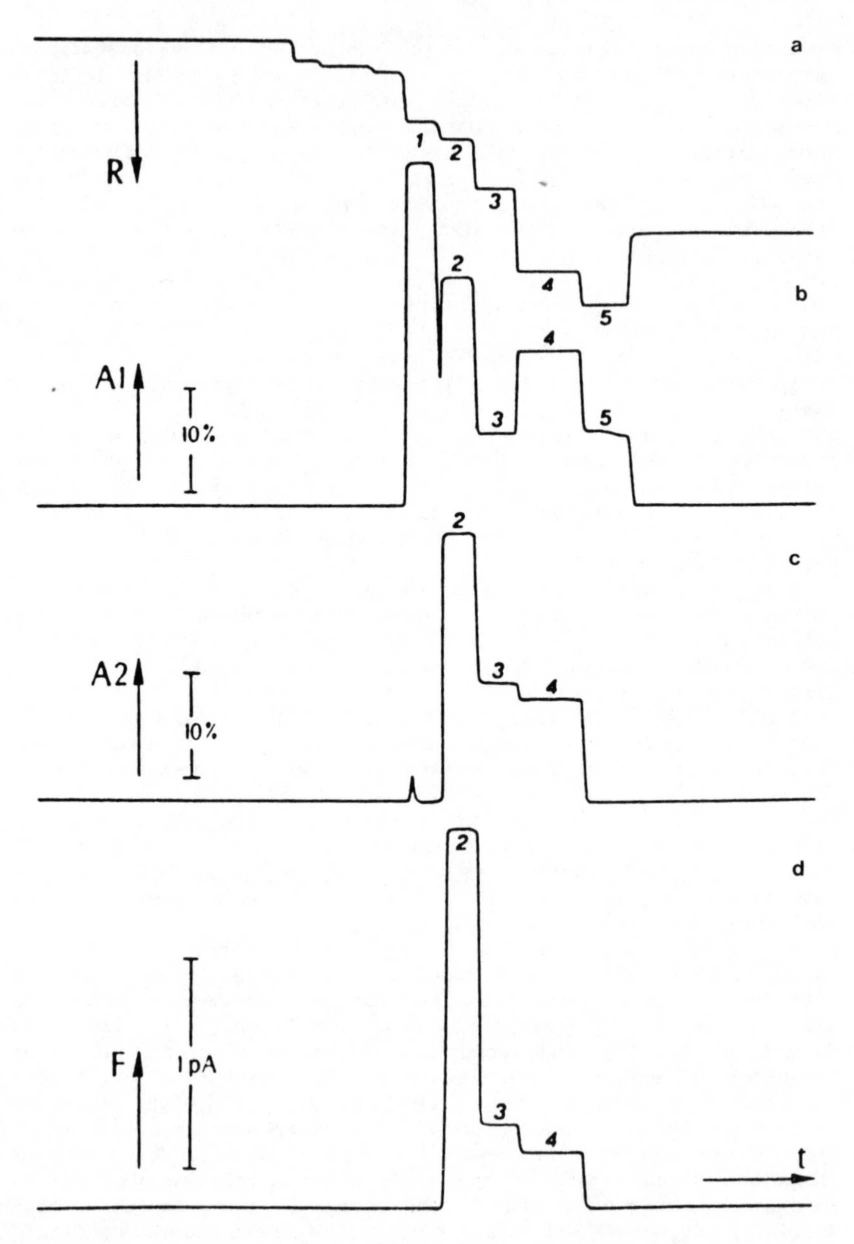


Figure 3. Selective fluorescence of vitamin B constituents, analyzed as cations at pH 5.0 with 0.01 M potassium/acetate as leading electrolyte and $H^-$ as terminator. (a) Conductivity trace, (b) UV-trace at 254 nm, (c) UV-trace at 340 nm, the wavelength of excitation, and (d) fluorescence emission above 350 nm. Approximately 1 nmole each of the following separands were injected: 1 = thiamine B1, 2 = pyridoxamine B6, 3 = pyridoxine B6, 4 = pyridoxal B6 and 5 = nicotinic acid amide. Operational conditions appear in Table II. Reproduced with permission from Ref. 25. Copyright 1983 Elsevier.

with the experiments on disc-electrophoresis, performed in 1964 by Ornstein [13] and Davis [14].
Further lowering of the minimum detectable amount has been achieved by decreasing the inner diameter of the separation compartment to 0.1 mm (Verheggen [15]), by using volume-coupling (Verheggen [16], Shimadzu [8]), and by two-dimensional column-coupling (Verheggen [17], Kaniansky [18], Eriksson [19]). These developments have enabled the minimum detectable amount to be decreased by a factor of ca. 100 compared to commercially available standard isotachophoretic equipment.
With the conductivity or potential gradient detector [19b], the adjoining zones can be resolved into zone volumes as small as 3 nl. Under standard operational conditions this is equivalent to 30 pmols of an electrolyte. The nature of the universal detector signal in ITP, however, makes signal processing by commercially available equipment (chromatographic peak integrators) impossible. The amplitude of the signal provides only qualitative information, whereas the time axis contains both qualitative (sequence of zones) and quantitative (length of zones) information. The differential of the signal is widely used for measuring zone-lengths manually and attempts at automation have thus far not been successful, with exception of Stover et al. [19a]. The only signal processor commercially available for ITP (type I-EIB Shimadzu) is, in fact, a modified integrator for chromatography and makes use of the differential of the isotachopherogram for the detection of the zone transitions. Failure to detect a zone transition obscures the quantitative results of other zones, whereas the qualitative accuracy is determined by the stability of the universal detector.
Reijenga [20] has introduced a signal processing method for ITP which converts the linear trace of the isotachopherogram to a signal with chromatographic properties (Fig. 4b) which is then treated as such. The amplitude of the converted signal provides quantitative information. Thus, a great deal of software and hardware developed for chromatography can be used for ITP.
A computer programm for the conversion of ITP signals, written in BASIC, can be used on any microprocessor with an 8 bit ADC and c. 10 kbyte of RAM. With this programm it is possible to resolve zones, e.g. in trace analysis, which approach the theoretical minimum detectable volume in the detector probe used. Quantitative accuracy in ITP, with a well-defined leading electrolyte transport number, is determined only by the stability of the driving current and the accuracy with which the zone-lengths are measured. The method described takes both these effects into account, as the microprocessor also measures the driving current with an absolute accuracy of 0.1%. It has been found sufficient to measure the qualitative information with a resolution of 0.5%. 200 stepheight intervals are available with the microprocessor, which means that, in principle, one can qualitatively identify 199 separands between the leading electrolyte and the terminating electrolyte.
The use of specific detectors, such as UV-absorption or fluorescence detectors, has provided useful additional information in isotachophoresis, especially since at the steady-state the separand zone in mixed only with the counter ion. The concentration is

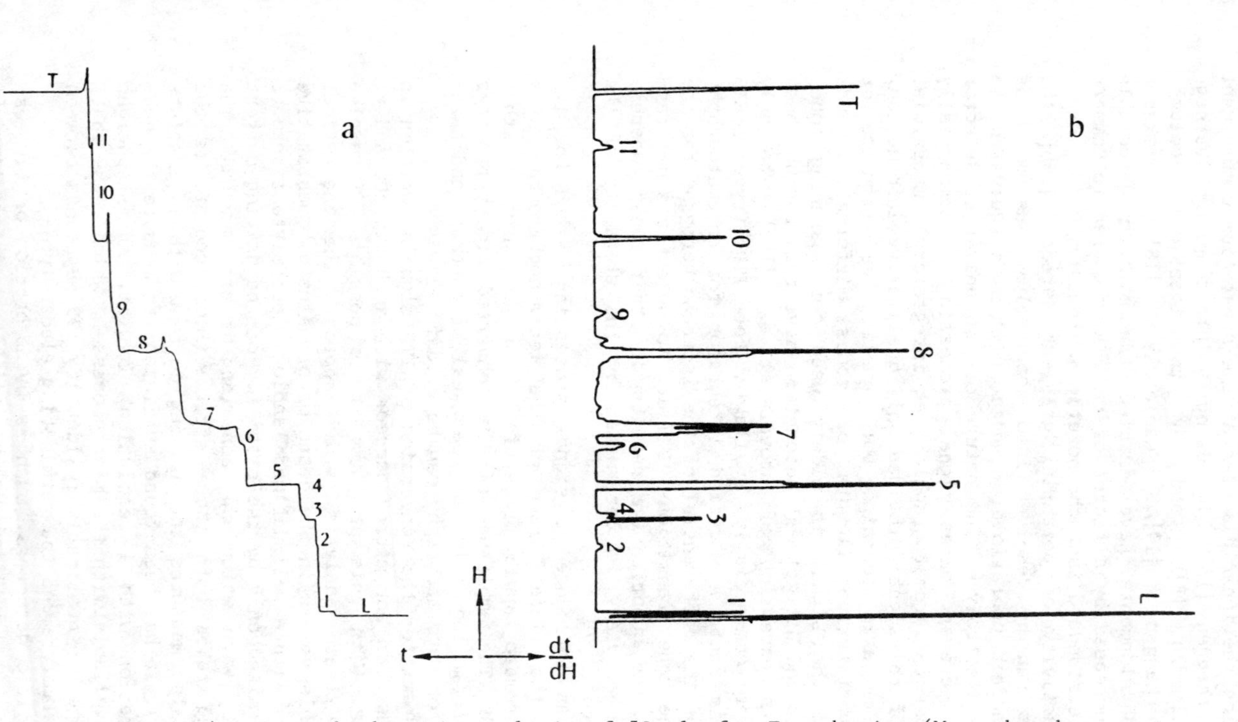


Figure 4. Isotachophoretic analysis of 50 nl of a French wine (Muscadet de Sevre et Maine) at pH 3.0 with 0.01 M chloride/β-alanine as leading electrolyte (L) and acetate as terminator (T). (a) Trace of conductivity detector, (b) the computer-converted isotachopherogram, which has the properties of a chromatogram, and is treated as such. 1 = sulphate, 2 = sulphite, 3 = phosphate, 4 = malonate, 5 = tartrate, 6 = citrate, 7 = malate, 8 = lactate, 9 = gluconate, 10 = succinate and 11 = dehydroascorbate. Operational conditions appear in Table III. Reproduced with permission from Ref. 26. Copyright 1984 Elsevier.

adjusted to the concentration of the leading electrolyte, which makes it necessary to use detector cell volumes less than 10 nl. The introduction of dual-wavelength detection, making use of such a measuring cell [21] with computerized signal processing, has been realized by the Eindhoven laboratory [22]. Scanning detectors are not yet available for ITP because the scan must be completed within 0.1s to allow the resolution of short zones. At the present time it is possible to choose two wavelengths from 206, 254, 280 and 340 nm with the plasma lamp/filter combinations commercially available. More wavelength-combinations are available. Making use of the UV-absorption (or absorbance) ratios, the method has been extremely useful for identification and quantification of steady-state zones, even where these were short (0.1s).

The detection unit developed for dual-wavelength UV-absorption detection has made it possible to apply fluorescence detection (see Fig. 3).

An even more specific detection method uses radioactivity, as introduced by Kaniansky [18]. As an example of dual-wavelength UV-absorption detection, an analysis will be given performed with the computerized dual-wavelength photometric detector.

The mutual interference of the two UV-light beams perpendicular to one another was determined with detection at 254 nm for both channels, with the filter placed at the detector side, amounted to less than 1%. This interference was completely eliminated when two different filters were placed before each detector. The noise level of each channel was measured with the DACs at 12 bits (0.025%). As the 1-Hz region of the noise spectrum of the detector signal is most important in isotachophoresis (with respect to the detector response time required), the amplitude of both detector signals was measured ten times at 1-sec intervals. From these values the average baseline (offset) and noise were calculated. The average noise level was ca. 0.1% (0.0004 a.u.) for 206, 254 and 280 nm.

The detector signals were then continuously monitored at 12 bits, 2 Hz, where the baseline values were updated for drift. If the signal-to-noise ratio of one channel exceeds 4, signal storage will commence. Now the full-scale resolution was 8 bits (0.4%) so that the detector noise was filtered out. Baseline offset correction was applied simultaneously. The sampling frequency could be chosen up to 59 Hz in the BASIC program, depending on the time required. However, there was a limit to the number of data points that could be stored in the available random access memory (RAM).

At the end of the run a choice can be made from a number of output facilities: (a) visual display on the terminal of the ratio of absorption plus channel 1 of the entire run; (b) plotting with a two-pen recorder of channel 1 plus channel 2; (c) plotting of channel 1 plus the ratio of channels 1 and 2; (d) plotting of channel 1 plus those data points of channel 1 that comply with a certain ratio. Further, all signals can be plotted against each other with an X-Y recorder. In all instances the output frequency can be optimally chosen to match the response of the recording instrument.

No logarithmic conversion of the transmission signals was applied.

When calculating the UV-ratio of a spike, the concentration distribution will cause some non-linearity above 50% absorption. However, this poses no problems when using the ratio in a qualitative sense for identification or determination of the purity of a spike or zone. A threshold value for ratio calculation is chosen for both channels. Insignificant variations in the ratio at low signal amplitudes are thus deleted. In that case the ratio is taken as zero. A ratio of greater than 10 is considered to be off-scale, so that the resolution is 0.04 ratio units when using the DAC at 8 bits.

## SELECTION OF ZONES

A standard mixture of nucleotides detected simultaneously at (a) 254 and (b) 280 nm is shown in Fig. 5. It is known from the literature [23,24] that the different classes of nucleotides can be characterized by a certain ratio of absorption at these two wavelengths. This ratio is more specific for a particular class of nucleotide than just the absorption at any of the two wavelengths. A ratio plot (Fig. 5c) illustrates this. The importance of dual-wavelength detection in verifying the purity of zones is also shown.

Whereas detection at one wavelength may suggest a pure zone, an interference will be detected only at the other wavelength. If no choice can be made as to which wavelength is best for a certain class of compounds, such as nucleotides, a ratio plot includes the information of both wavelengths. Quantitation by measuring zone lengths can also be applied to the ratio plot. Further data reduction by the micro computer is possible.

The entire dual-wavelength isotachopherogram can be plotted from the memory, but is can also be limited to those zones which comply with a certain ratio. This is illustrated in Fig. 5, in which the original 254 nm isotachopherogram is shown together with a reduced isotachopherogram (Fig. 5d), where only the zones with a ratio of 3.0 are seen. For this facility, a ratio window for recognition has to be used. The width of this window (20% in this instance) must be greater than the variation of the ratio of the zones to be selected.

Because of the sequence of the zones, the qualitative information from the time axis is more pronounced. Alternatively, the output can be limited to zones with a ratio above or below (Fig. 5e) a pre-set value. A threshold value for calculation of 5% absorption was used in this instance.

The method of selection of zones on the basis of the UV-ratio was applied to the analysis of an extract of nucleotides from a sea snail (Nassarius reticularis) eggs [24].

## RECENT DEVELOPMENTS IN INSTRUMENTATION

Column coupling [17,18,19] equipped with a microprocessor for handling the system, for controlling various operations and for stabilizing the electric driving current, enhances the versatility

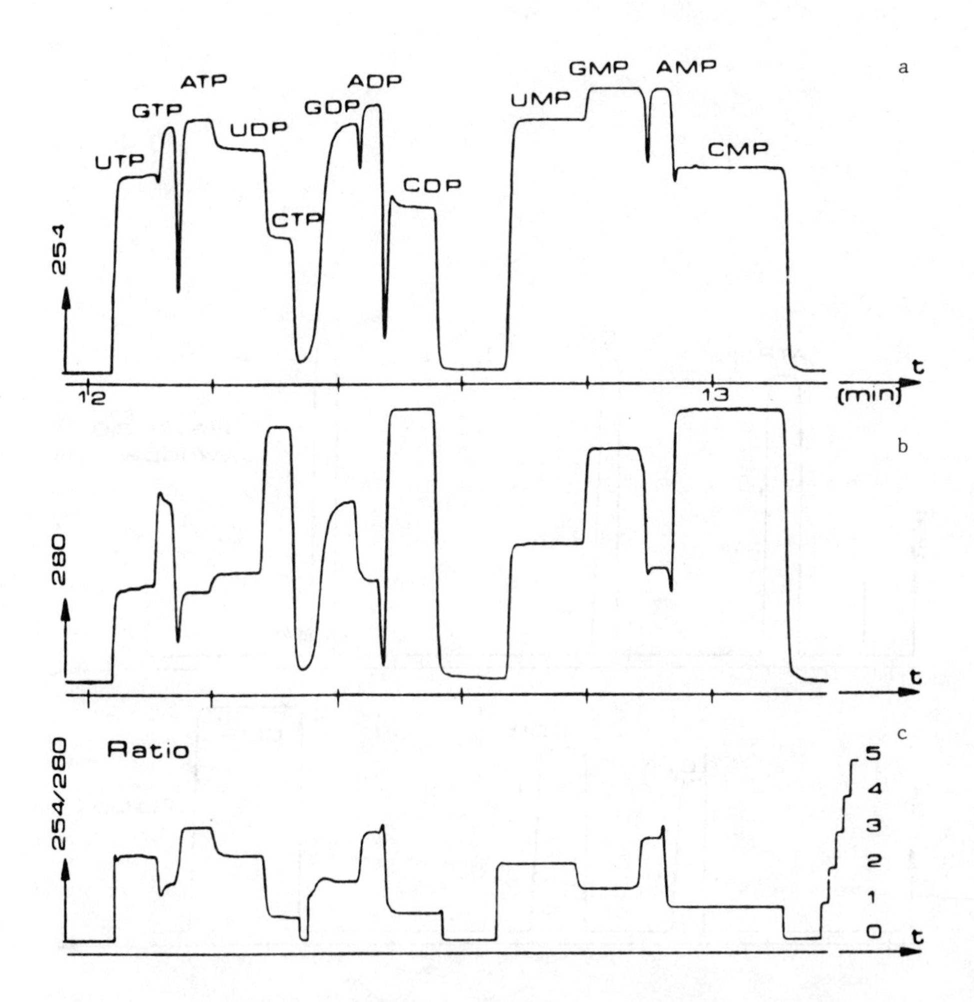


Figure 5. Analysis at pH 3.9 (Table IV) of a standard mixture of nucleotides, detected simultaneously at (a) 254 nm and (b) 280 nm. The 254/280 nm absorption ratio (c) can be plotted from the computer memory. Each class of nucleotides is characterized by a distinct ratio: cytidine ca. 0.8, guanosine ca. 1.5, uridine ca. 2.4 and adenosine ca. 3.0. Selected output of the isotachopherogram at 254 nm is possible on the basis of this ratio. A ratio of 3.0 with a 20% window will select the zones of adenosine nucleotides (d). In this sample only the cytidine nucleotides have a ratio smaller than 1 (e). The threshold value for ratio calculation was 5%. Reproduced with permission from Ref. 22. Copyright 1983 Elsevier.

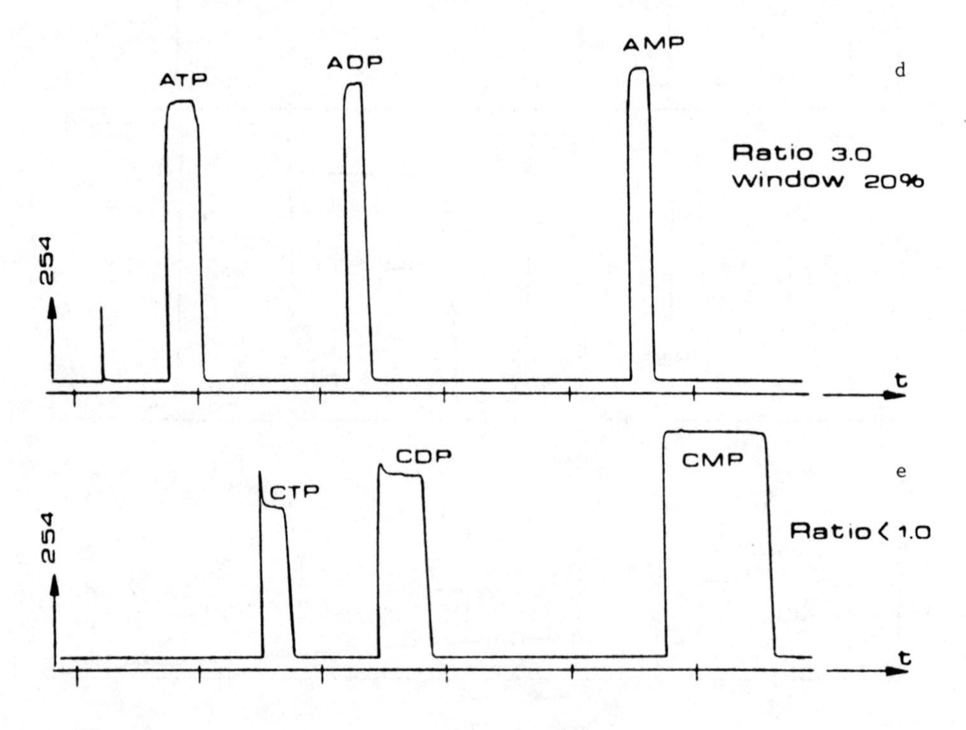


Figure 5. Continued.

of isotachophoresis without requiring more complex equipment. Column coupling makes use of two PFTE-tubes with different internal diameters. In the pre-separation tube, which has the larger internal diameter, a high pre-separation current is permitted. At a well defined distance from a conductivity detector - a 'telltale detector' - the final separation compartment is coupled to the pre-separation capillary in the bifurcation block [17,18,19]. The zones of interest can easily be selected from the sample train, migrating isotachophoretically in the pre-separation compartment via the tell-tale detector. The smaller internal diameter of the final separation compartment permits a higher current density during detection by means of the high resolution detectors described earlier. This system possesses several advantages over conventional isotachophoretic equipment:

- A higher sample load can be handled in the same analysis time;
- Higher concentration ratios of separand are permitted;
- Different operational systems can be applied in the two separation compartment (multidimensional isotachophoresis);
- Various electrophoretic separation principles can be combined, e.g. isotachophoresis followed by zone-electrophoresis.

Such an equipment is produced by the Institute of Radio-ecology and Applied Nuclear Techniques (Column-coupling isotachophoretic analyser ZKI-001; Pzo Kovo, Jankovcova 2, 17088 Praha, CSSR).

## CONCLUSIONS

Capillary isotachophoresis makes it possible to analyse both low and high molecular-weight charged substances with a minimum of sample pretreatment. A survey of recent ITP literature [7,8,9] indicates that there is a considerable overlap in applications with HPLC. Modern developments in isotachphoretic equipment and detection systems, combined with the use of microprocessors for equipment handling and signal processing make this analytical separation technique attractive because of its flexibility, reproducibility, accuracy and its extremely low running costs.

## OPERATIONAL CONDITIONS

Operational conditions used for the experiments given in this chapter.
For the analysis as shown in Fig. 1 the Table I is given; for the analysis of Fig. 3 the Table II is given; for the analysis of Fig. 4 the Table III is given; for the analysis of Fig. 5 the Table IV is given.

TABLE I

Operational system for isotachophoretic analysis of anions. The current was 80 µA in a PTFE capillary (I.D. 0.45 mm), PVA = polyvinyl alcohol (Hoechst Frankfurt, GFR), HEC = hydroxyethylcellulose (Polysciences, Warrington, P.A., U.S.A.).

| Parameter | pH = 4.03 |
|---|---|
| Leading ion | chloride |
| Concentration | 0.01 M |
| Counter ion | γ-aminobutyric acid |
| Additive | 0.05% PVA |
| | 0.2% HEC |
| Terminating ion | propionate |
| Concentration | ca. 0.005 M |

TABLE II

Operational system for isotachophoretic analysis of cations. The current was 30 µA in a PTFE capillary (I.D. 0.2 mm)

| Parameter | pH = 5.0 |
|---|---|
| Leading ion | potassium |
| Concentration | 0.01 M |
| Counter ion | acetate |
| Additive | none |
| Terminating ion | $H^+$ |
| as acetic acid pH | ca. 3.5 |

TABLE III

Operational system for isotachophoretic analysis of anions.
The current was 25 µA in a PTFE capillary (I.D. 0.2 mm). PVA = polyvinyl alcohol (Hoechst Frankfurt, GFR), HEC = hydroxyethylcellulose (Polysciences, Warrington, P.A., U.S.A.).

| Parameter | pH = 3 |
|---|---|
| Leading ion | chloride |
| Concentration | 0.01 M |
| Counter ion | β-alanine |
| Additive | 0.05% PVA |
| | 0.2% HEC |
| Terminating ion | acetate |
| Concentration | ca. 0.005 M |

TABLE IV

Operational system for isotachophoretic analysis of anions.
The current was 25 µA in a PTFE capillary (I.D. 0.2 mm) CTAB = cetyltrimethylammonium bromide (Merck, Darmstadt, GFR)

| Parameter | pH = 3,9 |
|---|---|
| Leading ion | chloride |
| Concentration | 0.01 M |
| Counter ion | β-alanine |
| Additive | 0.05% PVA |
| | 0.2 mM CTAB |
| Terminating ion | caproate |
| Concentration | ca. 0.005 M |

REFERENCES


1. Kohlrausch F. (1897), Ann.Phys. (Leipzig) 62, 209.
2. Martin A.J.P. (1942), Unpublished results..
3. Everaerts F.M., Beckers J.L. and Verheggen Th.P.E.M., (1976) Isotachophoresis, Theory, Instrumentation and Applications, Journal of Chromatography Library, vol 6, Elsevier Publishing Company, Amsterdam, Oxford and New York.
4. Deyl Z., Everaerts F.M., Prusik, Z. and Svendsen P.J. (Eds) (1979), Electrophoresis (part A: Theory), Journal of Chromatography library, vol 18A, Elsevier Scientific Publishing Company Amsterdam, Oxford and New York.
5. Deyl A., Chrambach A, Everaerts F.M. and Prusik, Z. (Eds) (1983), Electrophoresis (part B: A Survey of Techniques and Applications), Journal of Chromatography library, vol 18B. Elsevier Scientific Publishing Company Amsterdam, Oxford and New York.

5a. Mikkers F.E.P., Everaerts F.M. and Peek P.A.F., J. Chromatogr., 168 (1979) 293-315.

5b. Mikkers F.E.P., Everaerts F.M. and Peek P.A.F., J. Chromatogr., 168 (1979 317-332.

6. Lederer M. (Ed), Macek K. (Ass Ed) and Heftmann E. (Ed Symposium volumes), Journal of Chromatography, vol 320, no. 1, February 1985 (Special Issue): 4th International Symposium on Isotachophoresis Hradec Kralové, Sept. 2-6, 1984.
7. Application notes on Isotachophoresis, LKB Produkter AB, Box 305, S-161 26 Bromma, Sweden.
8. Application notes on Isotachophoresis, Shimadzu Corp. R8D Eng. Dept, Analyt.Inst.Div., Kyoto, Japan.
9. Acta Isotachophoretica, LKB Produkter AB, Box 305, S-161 26 Bromma, Sweden.

9a. Bocek P., Gebauer P., Dolink V. and Foret F., J. Chromatogr., 334 (1985) 157.

10. Everaerts F.M., (1964) Graduation Report, University of Technology, Eindhoven, The Netherlands.
11. Kenndler E. and Kaniansky D. (1981) J. Chromatog. 209, 306.
12. Schoots A.C. and Everaerts F.M. (1983) J. Chromatog. 277, 328.
13. Ornstein L. (1964), Ann. NY Acad.Sci. 121, 321.
14. Davis B.J. (1964), Ann. NY Acad.Sci. 121, 404.
15. Verheggen, Th.P.E.M., Mikkers F.E.P. and Everaerts F.M. (1977), J. Chromatog. 132, 205.
16. Verheggen, Th.P.E.M. and Everaerts F.M. (1982), J. Chromatog. 249, 221.
17. Verheggen, Th.P.E.M., Mikkers F.E.P., Kroonenberg D.M.J. and Everaerts F.M. (1980) in Biochemical and Biological Applications of Isotachophoresis (Adam A and Schots C, eds) Elsevier Publishing Company pgs 41-46.
18. Kaniansky D., (1982), Thesis, Comenius University of Bratislava, CSSR.
19. Eriksson G., (1983), Thesis, University of Lund, Sweden.

19a. Stover F., Deppermann K.L., and Grote W.A., J. Chromatogr., 269 (1983) 198.

19b. Verheggen Th.P.E.M., van Ballegooijen E.L., Massen C.H. and Everaerts F.M., J. Chromatogr., 64 (1972), pp. 187.

20. Reijenga J.C., van Iersel W., Aben G.V.A., Verheggen Th.P.E.M. and Everaerts F.M., (1984) J. Chromatog., 292, 217.
21. Verheggen Th.P.E.M., Oral presentation at the 3rd International Symposium on Isotachophosesis, (1982), Goslar, Germany.
22. Reijenga, J.C., Verheggen Th.P.E.M. and Everaerts F.M., (1983), J. Chromatog. 267, 75.
23. Sahota A., Simmonds H.A. and Payne R.H. (1979), J. Pharm. Methods 2, 303.
24. Van Dongen C.A.M., Mikkers F.E.P., De Bruijn Ch. and Verheggen Th.P.E.M., (1981), in Analytical Isotachophoresis (Everaerts F.M. ed.), Analytical Chemistry Symposia Series, volume 6. Elsevier Scientific Publishing Company, Amsterdam, Oxford and New York.
25. Everaerts, F. M., Verheggen, Th. P. E. M., and Reijenga, J. C. Trends Anal. Chem. 1983, 2(9), 189.
26. Reijenga, J. C., Verheggen, Th. P. E. M., and Everaerts, F. M. J. Chromatogr. 1984, 283, 105.

RECEIVED July 24, 1986

# Chapter 15

# Isotachophoresis of Synthetic Ion-Containing Polymers

**L. Ronald Whitlock**

**Commercial and Information Systems Group, Research Laboratories, Eastman Kodak Company, Rochester, NY 14650**

Analytical capillary isotachophoresis offers an alternative approach to chromatographic methods for the separation and characterization of high-molecular-weight synthetic copolymers containing ionizable functional groups. Polymer mixtures are separated into discrete zones inside a small capillary without the aid of stabilizing media according to the electrophoretic mobility of the individual polymer chains. Polymer mobility is independent of molecular weight, eliminating the influence of chain-length polydispersity on the separation process. Electrophoretic mobility of charged polymers is governed by the ratio of ionic and nonionic repeat units in the chain, degree of ionization of the ionic groups, and extent of counterion binding. Resolution of copolymers has been achieved when there is at least a 0.05 mole fraction difference in ionic group and when at least one of the copolymers has <0.5 mole fraction ionic repeat unit content.

Chemical heterogeneity in synthetic polymers offers a challenge to the analytical chemist to devise sensitive techniques for the characterization of these chemical distributions. It is well known that many synthetic copolymers consist of a collection of polymer chains that differ in their individual compositions. This distribution of repeat-unit composition from chain to chain can influence the physical properties of synthetic polymers significantly. Consequently, a thorough characterization of a copolymer sample would include a description of the average composition and its compositional distribution.

Size-exclusion chromatography and, to a lesser extent, reversed-phase and adsorption high-performance liquid chromatography are recognized separation methods for synthetic polymers. Their application to ion-containing, water-soluble polymers has been troublesome because of unpredictable influences of the charge-bearing groups on the separation process. In addition, measurements of chemical heterogenity accomplished with a chromatographic separation or solvent fractionation are sensitive to molecular-weight heterogeneity.

0097-6156/87/0335-0222$06.75/0

Separations of copolymers by isotachophoresis will lack this molecular weight dependence, which considerably simplifies their interpretation.

Isotachophoresis (ITP) offers an alternative to these chromatographic methods for the separation and chemical characterization of high-molecular-weight synthetic polymers containing an ionizable functional group. The separations occur because of differences in the effective electrophoretic mobilities of the macromolecules.

Isotachophoresis is now a fairly advanced microseparation method predominantly used for small, ionic molecules. The use of ITP for the separation of macromolecules has been limited until recently to biopolymer applications such as the separation of peptides, the profiling of protein mixtures (1-3), and the analysis of enzymes. These applications have been reviewed by Bocek (4) and Hjalmarsson and Baldesten (5). These successful applications of ITP to biomacromolecules can be attributed in large part to the predominance of electrokinetic separation techniques in the characterization of biochemical systems.

Yet many of the same features of biomacromolecules that lend themselves to electrophoretic separations, such as water solubility and ionic charge, are shared with certain classes of synthetic polymers including polyelectrolytes and polyionomers. The objective of recent work by this author has been to investigate the capabilities of ITP as a new approach to the characterization of synthetic polymers containing an ionizable functional group as a fraction of its chain repeat units.

ITP offers a number of potential advantages over other techniques for examining heterogeneity in ion-containing synthetic polymers. The separation occurs without the aid of a support medium or column packing. This eliminates the source of secondary polymer-substrate interactions that can complicate the interpretation of chromatographic separations. The ion-containing polymer does not have to be chemically treated before the separation. The self-sharpening nature of the boundary between migrating zones ensures resolution of components that is not compromised by molecular diffusion effects. Each zone is detected, regardless of its composition, with a potential-gradient detector. Additional selectivity is available using dual detectors such as UV and electrochemically based detection. The separated components can be quantitated from zone-length measurements and standard calibration techniques.

Isotachophoresis theory and practice have been described in several review articles (6-9) and in a book by Everaerts (10). Briefly, a solution of the polymer sample is introduced at the interface between a leading electrolyte of high effective mobility and a trailing electrolyte of low effective mobility. Sample ions of different electrophoretic mobilities separate into individual zones with the zone order in direct relation to their effective mobilities. The zones migrate past an on-column potential-gradient or conductivity detector with a response that is proportional to the effective mobility and amount of ion present.

In this paper we have described the separation and characterization of four polymers by ITP: carboxymethylcellulose (CMC) (11) with degree of substitutions of 0.4-1.2 carboxymethyl group; poly(2-hydroxyethylmethacrylate-co-2-acrylamido-2-methylpropanesulfonate) (12) containing >20% of the sulfonated repeat unit in the polymer chain; a series of sulfonated polystyrenes of different molecular weight (12); and a series of compositions of well-characterized samples of poly(acrylamide-co-acrylic acid) obtained by alkaline hydrolysis of polyacrylamide (13).

Characterization of the Carboxymethyl Distribution in Carboxymethylcellulose

Isotachophoresis was used to evaluate the degree of substitution (DS) and the distribution of substitution of carboxymethyl groups in carboxymethylcellulose (CMC). Commercial samples can range from very homogeneous to heterogeneous with different degrees of substitution. CMC is prepared from a homogeneous natural polymer, cellulose. However, the derived polymer can be heterogeneous in substitution of carboxymethyl groups, because cellulose contains both amorphous and crystalline domains. Crystalline regions in the original cellulose sample might be less highly substituted, whereas the amorphous material is likely to be more rapidly swollen and receive a greater degree of substitution, leading to nonuniformity of carboxymethylation. A wide variety of processes are used to prepare CMC; some are crude and give nonuniform substitution, whereas other processes give

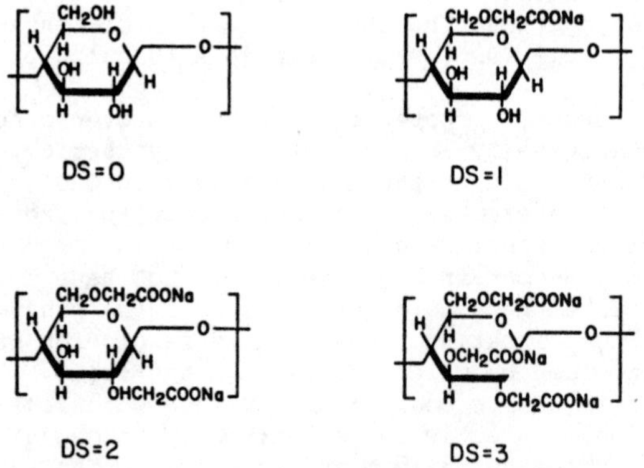

DS=0

DS=1

DS=2

DS=3

more uniform substitutions. The degree of substitution DS can vary between zero and 3.0 (all three hydroxyl groups substituted).

Samples of CMC were obtained from Hercules, Inc. (Wilmington, DE) and from Polysciences, Inc. (Warrington, PA). Table I summarizes the sample designations, nominal degree of substitution, and actual degree of substitution (determined by potentiometric titration).

Experimental. The separations were performed on a Shimadzu IP-2A isotachophoretic analyzer (Shimadzu Scientific Instruments, Inc., Columbia, MD) equipped with a 60-mm OD x 1.0-mm ID Teflon first-stage capillary tube and a 100-mm OD x 0.5-mm ID fluorinated poly(ethylene-propylene) second-stage capillary tube. The compartment holding the capillaries and the potential-gradient detector cell was kept at 25°C by Peltier elements using a fluorinated-hydrocarbon cooling fluid. Zones were detected by a potential-gradient detector. Instrumental operating pararmeters are given in the figures with the isotachopherograms.

Table I. Description of CMC Materials

| Designation | Degree of Substitution Nominal | Degree of Substitution Analysis | Molecular weight |
|---|---|---|---|
| Hercules Chemical (samples dialyzed) | | | |
| 4M6SF | 0.4 | 0.44 | |
| 7M1 | 0.7 | 0.79 | |
| 7H | 0.7 | 0.74 | |
| 7H3S | 0.7 | 0.74 | |
| 7M8SF | 0.7 | 0.67 | |
| 9M8 | 0.9 | 0.66 | |
| 9M31 | 0.9 | 0.72 | |
| 12M8 | 1.2 | 1.20 | |
| 12M31 | 1.2 | 1.15 | |
| Polysciences, Inc. (samples not dialyzed) | | | |
| 6140 | 0.7 | 0.60 | 80,000 |
| 6138 | 0.7 | 0.70 | 250,000 |
| 6139 | 0.7 | 0.72 | 700,000 |

SOURCE: Reproduced with permission from Ref. 11. 

The leading electrolyte was 0.01 M chloride, buffered at pH 7.0-7.5 with imidazole (Kodak Laboratory and Research Products, Rochester, NY) (recrystallized three times from ethanol) and contained 0.5% Triton X-100 surfactant (Rohm and Haas, Philadelphia, PA). The surfactant minimized electroosmotic flow. The terminating electrolyte was 0.01 M n-hexanoic acid (Kodak) and contained 0.5% Triton X-100 surfactant (purified by fractional distillation). The Triton X-100 surfactant was purified by dialysis to remove low-molecular-weight anionic impurities.

The CMC samples were analyzed for actual DS by potentiometric titration of the sodium carboxylate group. Samples were prepared for injection by dissolving 10 or 50 mg of CMC in 10 ml of distilled water. No significant advantage in the appearance of zone shapes was observed when samples were dissolved in leading electrolyte as is sometimes recommended.

CMC with Various Degrees of Substitution. Isotachopherograms obtained on CMC with DS of 0.4, 0.7, and 1.2 are shown in Figure 1. The effective mobility of these materials is governed by three factors: the actual number of carboxylic acid groups per polymer chain, the degree of acid-group ionization $\alpha$, and the extent of counterion binding that occurs with most high-charge-containing polyelectrolytes in water. This third factor, the extent of counterion binding for CMC, is discussed here and in a later section since this influences the interpretation of the isotachopherograms and ITP provides a convenient and accurate method for its measurement. The value of $\alpha$ is controlled by the pH of the leading electrolyte through which the polymer migrates. By potentiometric titration, the apparent $pK_a$'s of CMC in water were found to be 4.0, 4.3, and 4.6 for DS of 0.4, 0.7, and 1.2, respectively. At pH 7-7.5, ionization of these groups is essentially complete.

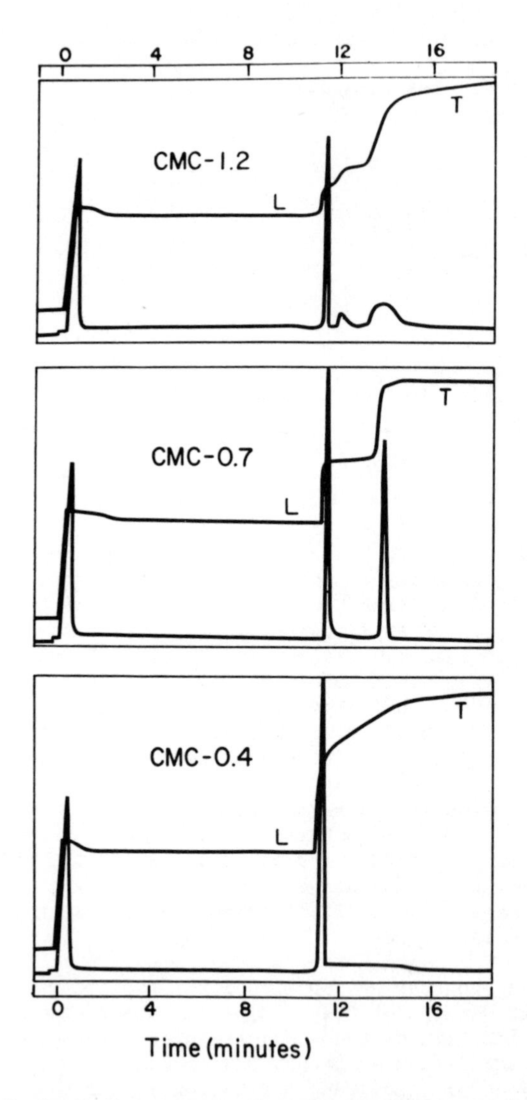


Figure 1. Isotachopherograms of carboxymethyl cellulose with DS of 0.4, 0.7, and 1.2. Leading electrolyte (L): 0.005 M $Cl^-$, pH 7.0 imidazole buffer, 0.5% Triton X-100 surfactant; terminal electrolyte (T): 0.01 M n-hexanoic acid, 0.5% Triton X-100 surfactant; migration current: 150 μA; sample: 30 μl of 1000 ppm solutions in water; temperature: 25°C. Reproduced with permission from Ref. 11. Copyright 1986, Elsevier.

The appearance of each isotachopherogram is diagnostic of both DS and the distribution of substitution. The 1.2-DS sample has the highest effective mobility and, in addition, two zones are evident, suggesting that the sample is a mixture of two materials with different DS. A second lot of DS-1.2 material did not show two zones (Figure 5). The 0.7-DS sample has a single, intermediate mobility with sharp zone boundaries, suggesting it is the most homogeneous of the three samples. The 0.4-DS sample has the lowest mobility and a much broader range of substitution, as evidenced by the gradually sloping potential-gradient detector trace towards lower-mobility polymer.

Comparison of Grades of CMC. The isotachopherograms of four grades of CMC with nominal DS of 0.7 examined by ITP for differences in substitution are shown in Figure 2. The detector traces show that differences in distribution exist among the grades although the major portion of each sample has the same mobility and, consequently, a similar DS.

Two grades of CMC with reported DS of 0.9 gave isotachopherograms that were identical with that of the DS = 0.7 CMC (Figure 5), suggesting no differences in substitution existed. This was confirmed by the potentiometric titration analysis for average DS, which was 0.7 ± 0.1 for the nominal DS = 0.9 materials, consistent with the ITP data.

## Effect of CMC Molecular Weight on the Isotachophoretic Separation

The molecular-weight dependence on the separation was examined using samples of CMC with nominal molecular weights of 80,000, 250,000, and 700,000 daltons. By potentiometric titration, the DS values were 0.7 ± 0.1. The isotachopherograms are shown in Figure 3. No significant difference in apparent mobilities was observed for this range of molecular weights. This behavior is consistent with a lack of molecular-weight dependence on polyelectrolyte mobility determined by other electrophoretic techniques (14-16).

Quantitation of Isotachopherograms. For quantitative evaluation of an isotachopherogram, zone lengths displayed on the recorder tracing are measured. The zone lengths are taken as the distance in millimeters between peak maxima of the differential potential-gradient detector signal. The relationship between amount of CMC (DS-0.7) and zone length was evaluated from a calibration curve based on injections of 5-40 μg of CMC with molecular weights of 250,000 and 700,000. The zone lengths are plotted *vs.* equivalents of carboxyl group in Figure 4, showing excellent linearity of the calibration curve for CMC with no dependence on CMC molecular weight.

## Effect of DS on CMC Zone Length and Counterion Binding

We have evaluated the effect of DS on zone length for CMC with DS = 0.4, 0.7, and 1.2 and sodium acetate at the same molar concentration of carboxyl group. The isotachopherograms are shown in Figure 5. With the higher-charge-density polymer, more of the carboxyl groups appear as if they are not ionized. We determined that this behavior was not caused by incomplete neutralization of the carboxyl groups by obtaining the isotachopherograms in a leading electrolyte buffered at

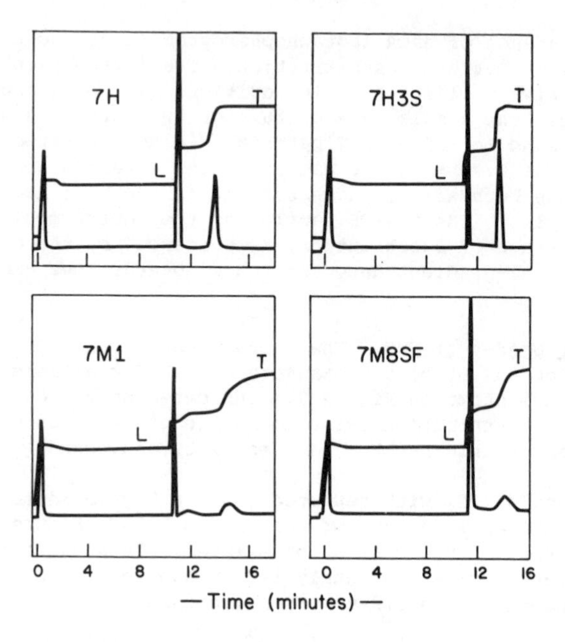


Figure 2. Isotachopherograms of four grades of CMC with DS of 0.7. Conditions same as in Figure 1. Reproduced with permission from Ref. 11. Copyright 1986, Elsevier.

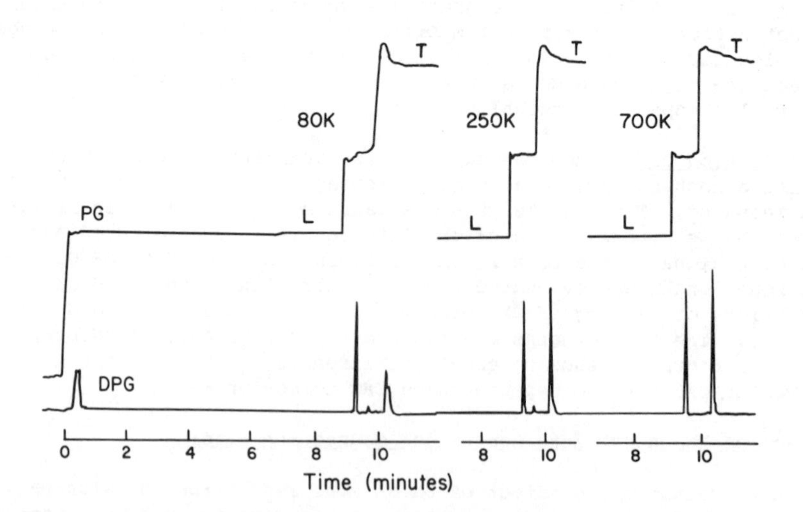


Figure 3. Isotachopherograms of CMC of different molecular weights. Conditions same as in Figure 1 except: leading electrolyte: 0.01 M $Cl^-$, pH 7.3 imidazole buffer, 0.5% Triton X-100 surfactant; migration current: 125 $\mu$A; temperature: 15°C; sample: 15 $\mu$l of 1000 ppm solutions in water. Reproduced with permission from Ref. 11. Copyright 1986, Elsevier.

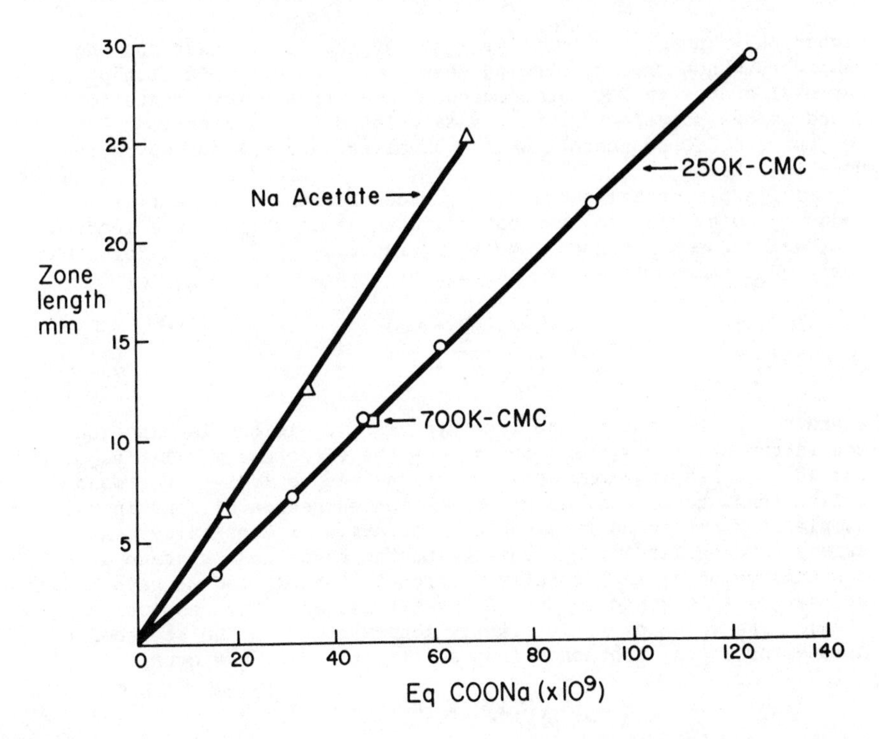


Figure 4. Zone-length calibration of CMC and sodium acetate. Reproduced with permission from Ref. 11. Copyright 1986 Elsevier.

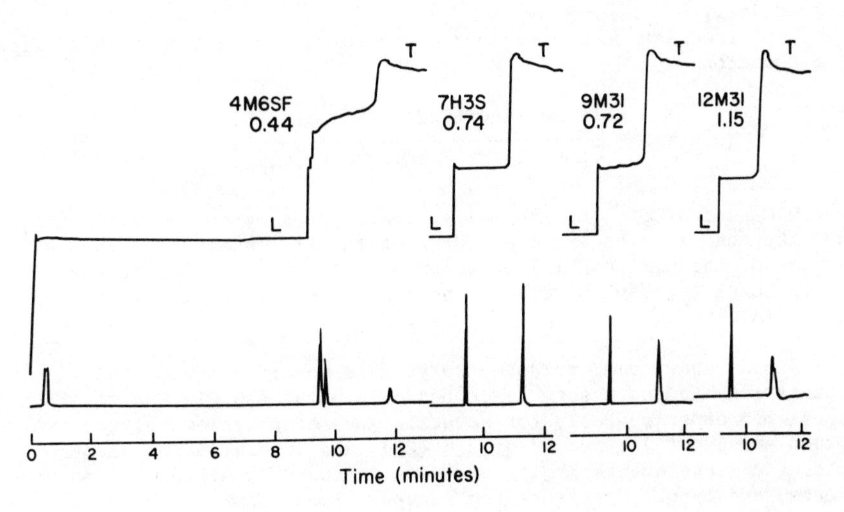


Figure 5. Isotachopherograms of CMC with different DS for evaluation of zone-length dependence on DS. Conditions same as in Figure 3 except: leading electrolyte: pH 7.5 imidazole buffer, 0.5% Triton X-100 surfactant; sample: 20 μl of CMC at 0.005 eq/ℓ carboxyl group. Reproduced with permission from Ref. 11. Copyright 1986, Elsevier.

higher pH values. The relative zone lengths for acetate and the CMC samples did not change, showing that no increased ionization of the carboxyl groups in CMC had occurred. The fraction of counterion bound to the polyelectrolyte was calculated from the ratio of zone lengths for CMC to acetate when equivalent carboxyl concentrations were injected.

The experimental sample zone concentrations must be first adjusted to have an ion concentration equal to that of the leading ion, using the Kohlrausch regulating function (10), which forms the basis of isotachophoresis. The equation is

$$C_i = C_L \frac{m_i(m_L + m_c)}{m_L(m_i + m_c)} \tag{1}$$

where $C_i$ is the sample zone concentration, $C_L$ is the leading ion concentration, and $m_i$, $m_L$, and $m_c$ are the effective mobilities of sample ion, leading ion, and counterion, respectively. For monovalent ions, this equation gives the concentration of ions in the sample zone resulting from the concentration discontinuity that exists between leading and sample ions as these ions migrate at constant velocity in a constant current. Both $m_L$ and $m_c$ have known values ($79 \times 10^{-5}$ and $29.5 \times 10^{-5}$ $cm^2/V$ sec at 25°C, respectively (17)). The value of $m_i$ is readily determined from the step height dimensions in the isotachopherogram (17,18) using the equation,

$$m_i = m_L(E_L/E_i) = m_L(h_L/h_i) \tag{2}$$

where $E_L$ and $E_i$ are the potential gradients and $h_L$ and $h_i$ are the step heights in millimeters of the leading and sample zones, respectively.

The fraction of counterions bound to CMC, $\phi$, was calculated from the equation

$$\phi = 1 - \frac{(\text{zone length}/\mu\text{eq CMC})C_{cmc}}{(\text{zone length}/\mu\text{eq acetate})C_{acetate}} \tag{3}$$

where $C_{cmc}$ and $C_{acetate}$ are moles/ℓ calculated by using Equation 1, and the zone lengths are the distances in millimeters between peak maxima of the differential detector signal trace divided by the μeq of carboxylic group injected. The results of these calculations are given in Table II.

This behavior of polyelectrolytes is consistent with the counterion-binding theory described by Manning (19-21) and others and confirmed experimentally for several types of polyelectrolytes by Okubo and Ise (22), Record *et al.* (23), and others, using electrophoretic light scattering, tracer-ion diffusion-coefficient measurements, and osmotic-pressure techniques. Specifically, the high charge density on polyelectrolytes is "self-lowered" through binding by site-bound or territorially bound counterions that exist near the fixed charges on the polymer. These counterions move with the polyion in an electric field. The theory suggests that the extent of counterion

Table II. Fraction of Counterion Bound to CMC

| Sample | Fraction of Counterion Bound pH 7.5 (Imidazole Buffer) |
|---|---|
| Acetate, Na salt | 0 |
| Methoxyacetate, Na salt | 0 |
| CMC | |
| DS-0.4 | 0.22 |
| DS-0.7 | 0.27 |
| DS-1.2 | 0.43 |

SOURCE: Reproduced with permission from Ref. 12. Copyright 1986 Elsevier.

binding is dependent on several parameters, the most important being the charge spacing along the polymer chain and the solvent dielectric constant. Counterion binding has considerable influence on the interpretation of isotachopherograms of ion-containing polymers.

Separation of Copolymer Mixtures of Poly(2-hydroxyethylmethacrylate-co-2-acrylamido-2-methylpropanesulfonate)

Chemically heterogeneous copolymers prepared from 2-hydroxyethylmethacrylate (HEMA) and 2-acrylamido-2-methylpropanesulfonate (AMPS) have been characterized by ITP by separation of the copolymer mixture into discrete zones. The separations were achieved from differences in electrophoretic mobilities of individual copolymer chains. Mobility is governed by the ratio of ionic to nonionic repeat units in the chain for mole fractions of 0-0.6 ionic repeat units and by the extent of binding of counterions to the ionic groups for mole fractions of 0.6-1.0 ionic groups. Compositions of several copolymer samples were calculated from ITP zone dimensions and chemical analysis data.

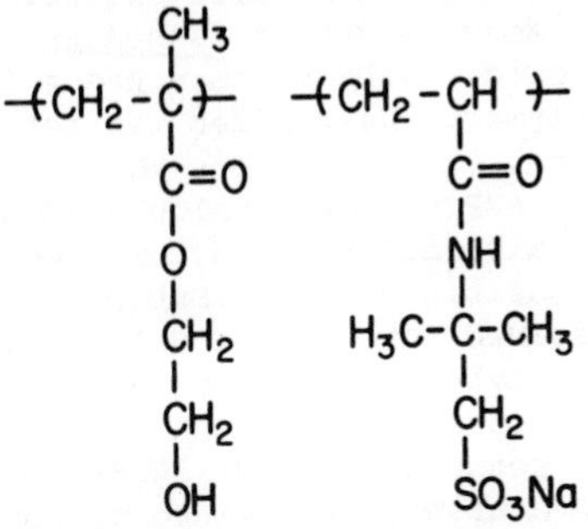

HEMA: $-(CH_2-C(CH_3)(C(=O)-O-CH_2-CH_2-OH))-$

AMPS: $-(CH_2-CH(C(=O)-NH-C(CH_3)_2-CH_2-SO_3Na))-$

HEMA AMPS

Copolymer Synthesis. Copolymers were prepared with HEMA (Kodak Laboratory and Research Products, Rochester, NY) and AMPS (Lubrizol Chemical Corp., Wickliffe, OH) at weight ratios of 78/22, 60/40, 40/60, and 20/80 by free-radical polymerization in water/ethanol (80/20) under nitrogen at 60°C. The reaction time was 20 h. Two of the five polymer solutions showed phase separation at polymer concentrations >10% in water. For the phase-separated samples, ITP analyses were performed on each phase.

The total AMPS content in each sample was determined by potentiometric titration of the sulfonate group with sodium hydroxide after ion exchange of the copolymers using Amberlite IR-10 ion-exchange resin.

Experimental. The separations were performed on a Shimadzu IP-2A isotachophoretic analyzer. Quantitative evaluation of the isotachopherograms was performed by measuring zone lengths between peak maxima of the differential potential-gradient detector signal. The relationship between the amount of AMPS homopolymer in the copolymer samples and the zone length was established from a calibration curve constructed from injections of the polymer (10-60 μg) dissolved in the leading electrolyte.

The leading electrolyte was 0.01 M HCl buffered to pH 3.8 with β-alanine (Kodak Laboratory and Research Products, Rochester, NY) and contained 0.5% Triton X-100 surfactant (Rohm and Haas, Philadelphia, PA). The terminal electrolyte was 0.01 M n-hexanoic acid (Kodak Laboratory and Research Products) and contained 0.5% Triton X-100 surfactant. Copolymer samples were prepared at 2000 ppm by dissolving 20 mg of the freeze-dried polymer in 10 ml of the leading electrolyte or water.

Separations of HEMA-AMPS Copolymers. Isotachopherograms were obtained on copolymers synthesized with HEMA-AMPS monomer feed ratios (wt.) of 78/22, 60/40, 40/60, and 80/20. Copolymer samples prepared at 60/40 and 40/60 monomer ratios were phase separated and are referred to as upper and lower phases. The phases are treated separately for quantitative analysis, but were recombined in dilute solution for the analysis of a copolymer mixture by ITP that is described in a later section.

Isotachopherograms of the seven polymer solutions are shown in Figures 6-8. Copolymer 78/22 and both the upper and lower phases of copolymers 60/40 and 40/60 contain two distinct zones. One component in each copolymer had an effective mobility identical with that of AMPS homopolymer and its zone length increased proportionately when authentic AMPS-100 was added. AMPS monomer has a lower effective mobility than the homopolymer. No monomer was detected in these dialyzed copolymer samples.

The concentration of AMPS-100 was calculated using its zone length and the AMPS-100 calibration curve. The copolymer component was of rather narrow compositional distribution, giving a zone that could not be further resolved into additional components using longer capillaries, other electrolyte systems, or injection of larger amounts of the copolymer. The compositions of each copolymer zone were calculated by subtracting the concentration of AMPS-100, determined by ITP, from the total AMPS content, determined by potentiometric titration. The HEMA content was calculated by the difference.

The copolymer compositions are summarized in Table III. The phase-separated preparations consist of three distinct components. The lower phases were homopolymer and copolymer with an AMPS composition less than the monomer feed ratio. The upper phases were homopolymer and copolymer with AMPS composition greater than the monomer feed ratio. The two copolymers formed in each preparation were of such widely differing composition that they are incompatible

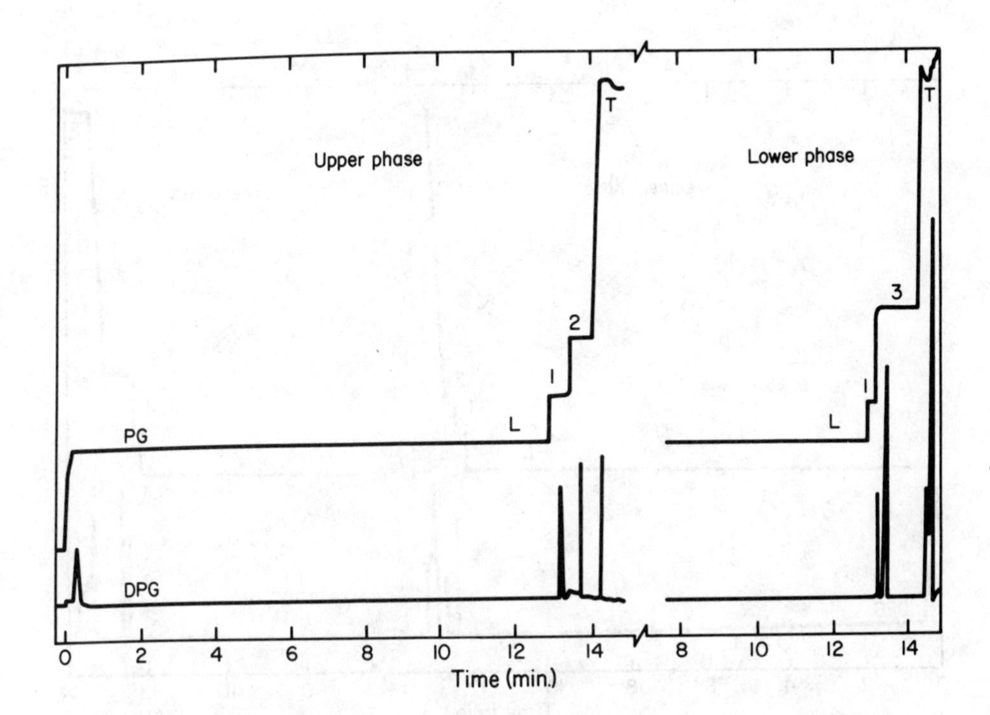


Figure 6. Isotachopherograms of HEMA-AMPS copolymer 60/40, upper phase and lower phase. 1 = AMPS (100); 2 = HEMA-AMPS (65/35); 3 = HEMA-AMPS (80/20). Leading electrolyte (L): 0.01 M $Cl^-$, pH 3.8, β-alanine buffer, 0.5% Triton X-100 surfactant; terminating electrolyte (T): 0.01 M hexanoic acid, 0.5% Triton X-100 surfactant; PG-potential gradient; DPG-differential potential gradient.

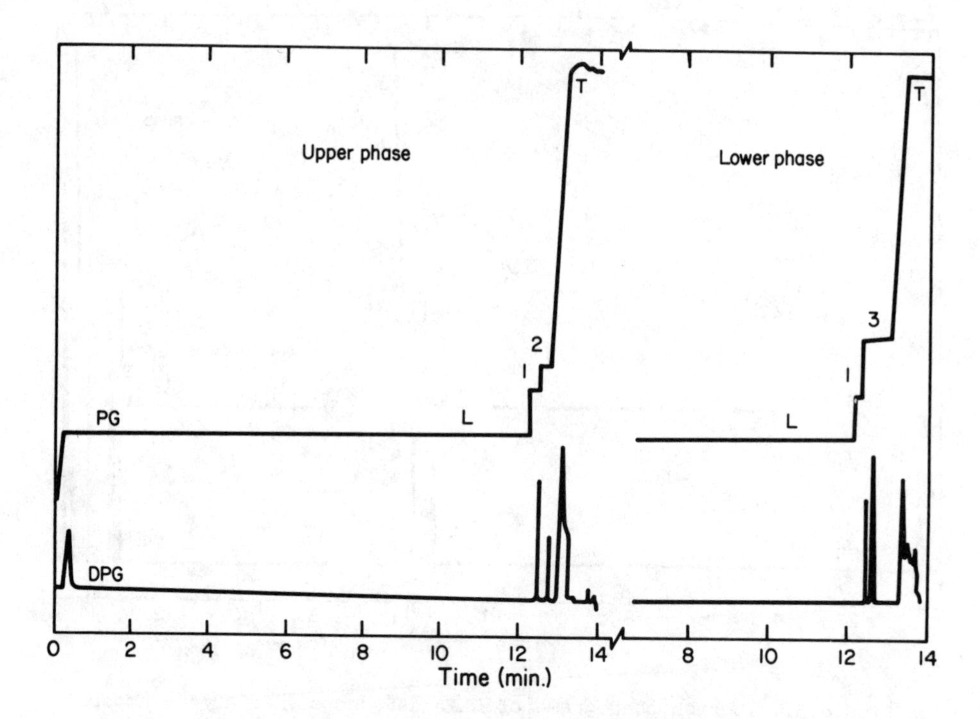


Figure 7. Isotachopherograms of HEMA-AMPS copolymer 40/60, upper phase and lower phase. Electrolytes, see Figure 6. 1 = AMPS (100); 2 = HEMA-AMPS (46/54); 3 = HEMA-AMPS (66/34).

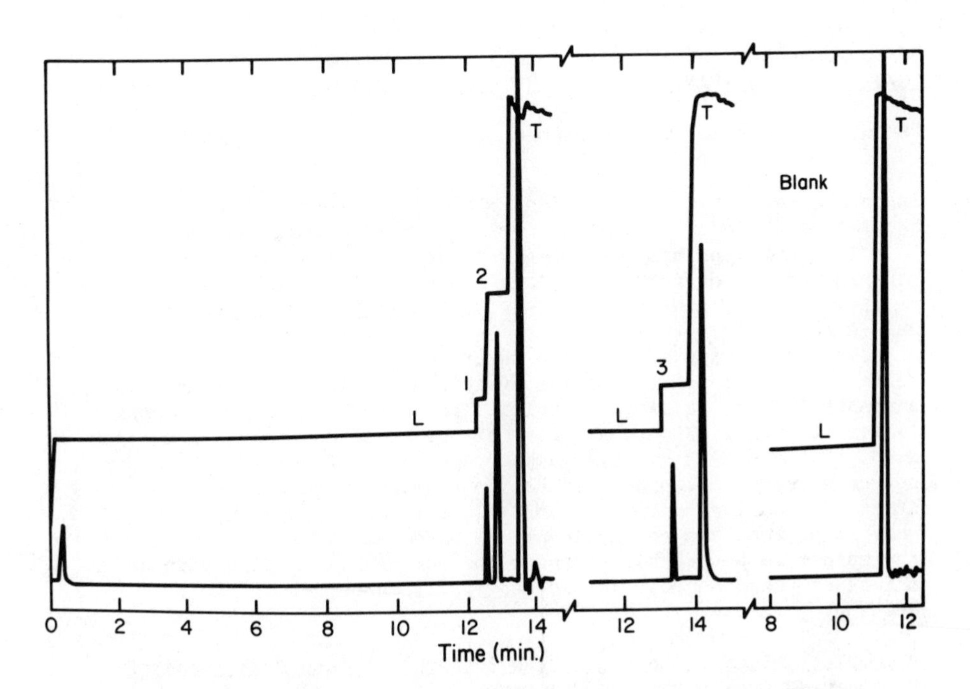


Figure 8. Isotachopherograms of HEMA-AMPS copolymers 78/22, 20/80, and electrolyte blank. Electrolytes, see Figure 6. 1 = AMPS (100); 2 = HEMA-AMPS (84/16); 3 = HEMA-AMPS (25/75).

Table III. Composition and Mobility of HEMA-AMPS Copolymers

| Polymer Designation (Wt. Ratio) | Total Wt. % AMPS (Titration) | % (Rel.) AMPS as Homopolymer (ITP) | Wt. Ratio HEMA/AMPS | Apparent Mobility ($X\ 10^5\ cm^2/V\ sec$) Zone A | Zone B |
|---|---|---|---|---|---|
| AMPS-100 | 100 | 100 | 0/100 | 53.1 | a |
| HEMA/AMPS | | | | | |
| 20/80 | 75.4 | a | 25/75 | 52.5 | a |
| 40/60 upper | 60.4 | 24.5 | 46/54 | 53.8 | 46.9 |
| 40/60 lower | 36.4 | 9.6 | 66/34 | 53.6 | 38.3 |
| 60/40 upper | 42.0 | 27.1 | 65/35 | 53.8 | 39.1 |
| 60/40 lower | 22.4 | 14.3 | 80/20 | 55.4 | 32.5 |
| 78/22 | 20.8 | 27.9 | 84/16 | 54.0 | 30.5 |

[a]Shows single zone in isotachopherogram

in concentrated aqueous solution. This results in the formation of two liquid phases with unequal polymer concentration and composition.

This behavior has been observed for other ion-containing copolymers including poly(ethyl acrylate-co-methacrylic acid) (24), poly(methyl methacrylate-co-methacrylic acid) (25), and poly(styrene-co-maleic acid) (26).

The capability of ITP to separate complex mixtures of high-molecular-weight copolymers was evaluated by blending six copolymers and AMPS-100 into a single sample. The isotachophoretic separation is illustrated in Figure 9. Five of the seven components are separated. Two copolymers, 60/40 upper phase and 40/60 lower phase were not resolved because their actual compositions and effective mobilities are nearly identical. AMPS-100 and copolymer 20/80 have mobilities that are nearly identical as a result of counterion binding and cannot be separated. Reasons for the inability to obtain separations of these components are more fully described in the last section.

## Molecular Weight Influence on the Isotachophoretic Separation of Sulfonate-Group-Containing Polymers

The molecular weight dependence on the ITP separation of sulfonated polymers was examined using six sulfonated polystyrenes with narrow-molecular-weight distributions (Pressure Chemical Co., Pittsburgh, PA) with weight-average molecular weights of 16,000 to 780,000 daltons. These polymers were prepared commercially by sulfonation of anionically polymerized styrene. The degree of sulfonation was determined by sulfur analysis on dialyzed samples and by potentiometric titration of the sulfonate group after ion exchange to the sulfonic acid. Data for these polymers are summarized in Table IV. The isotachopherograms in Figure 10 show no significant variation in the observed effective electrophoretic mobilities. This lack of molecular weight influence for polyelectrolyte mobility is consistent with our reported values for CMC (11) and the work of Rice and Nagasawa (14).

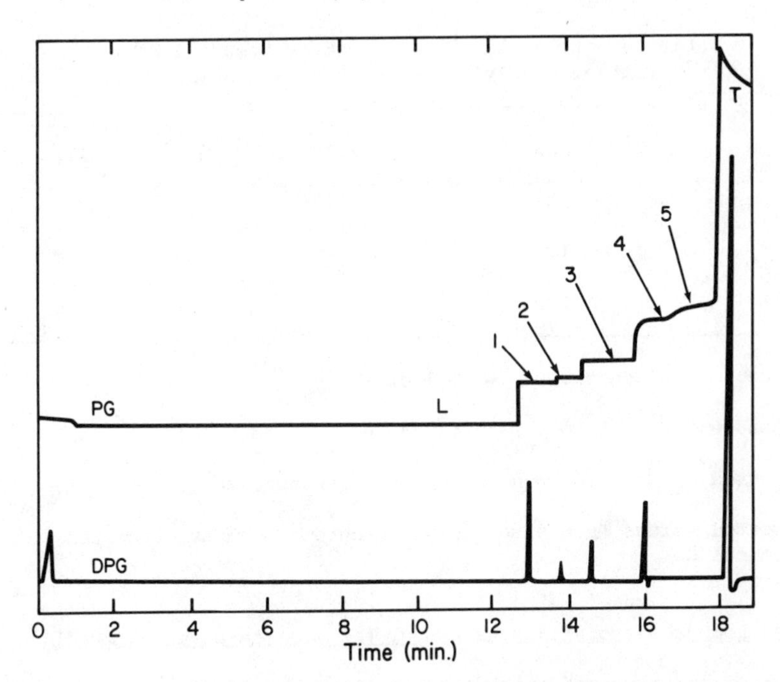


Figure 9. Isotachopherograms of a mixture of HEMA-AMPS copolymers. 1 = AMPS (100) and 20/80; 2 = 40/60 upper phase; 3 = 60/40 upper and 40/60 lower phases; 4 = 60/40 lower phase; 5 = 78/22. Reproduced with permission from Ref. 12. Copyright 1986, Elsevier.

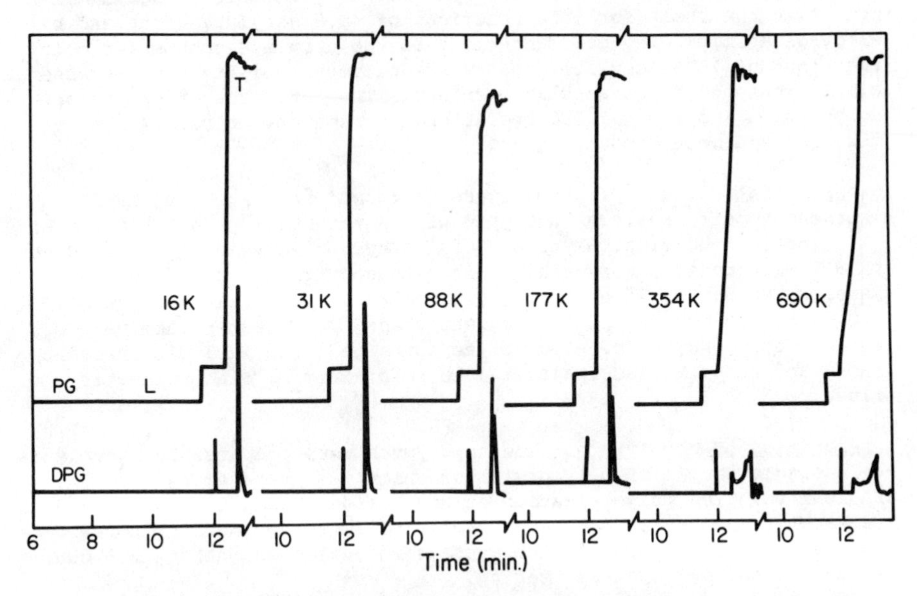


Figure 10. Isotachopherograms of poly(styrenesulfonate) Na salt of different molecular weights. Polymer descriptions in Table IV.

Table IV. Molecular Weight, Composition and Mobility Data for Poly(styrenesulfonate) Na Salts[a]

| Lot No.[b] | $\bar{M}w$[c] ($\times 10^{-3}$) | $\bar{M}w/\bar{M}n$[d] | Effective Mobility ($\times 10^5$ $cm^2/V$ sec) |
|---|---|---|---|
| 20 | 18 | 1.06 | 52.6 |
| 11 | 35 | 1.06 | 51.5 |
| 25 | 100 | 1.05 | 53.0 |
| 26 | 200 | 1.05 | 51.1 |
| 12 | 400 | 1.05 | 52.0 |
| 16 | 780 | 1.05 | 52.3 |

[a]Degree of sulfonation - 94 to 100%.

[b]Pressure Chemical Co., Pittsburgh, PA.

[c]Reported $\bar{M}w$ from membrane osmometry, Pressure Chemical Co.

[d]Reported values from size-exclusion chromatography, Pressure Chemical Co.

## Isotachophoresis of Mixtures of Poly(acrylamide-co-acrylic Acid)

High-molecular-weight synthetic copolymers of acrylamide (A) and acrylic acid (C) can be separated into pure zones by ITP based on differences in electrophoretic mobility of the polymer chains. Copolymers with compositions of 1.8 to 61 mol% acrylic acid were prepared by alkaline hydrolysis of polyacrylamide. Copolymer mobility is governed by the mole ratio of acrylic acid to acrylamide groups in the chain for mole fractions of <0.6 acrylate group and by electrostatic binding of counterions to the acrylate groups for mole fractions of >0.6 acrylate group where polymer mobility becomes essentially independent of acrylate content. The influence of copolymer composition and several ITP operational parameters on resolution of mixtures has been reported (13).

Experimental. The separations were performed on a Shimadzu IP-2A isotachophoretic analyzer equipped with a potential gradient detector. The leading electrolyte was 0.01 M hydrochloric acid buffered to pH 8.2 with tris(hydroxymethyl)aminomethane (Kodak Laboratory and Research Products, Rochester, NY). The terminating electrolyte was 0.01 M sodium borate (pH 9) (reagent grade, J. T. Baker Chemical Co., Phillipsburg, NJ). The electrolyte contained 0.5% TX-100 surfactant (Rohm and Haas, Philadelphia, PA) to help minimize electroosmotic flow.

Preparation of Copolymers. The copolymers were prepared by hydrolysis of a single sample of polyacrylamide which was prepared by free-radical solution polymerization of acrylamide (Kodak Laboratory and Research Products) in water/methanol at 60°C using ammonium persulfate initiator (0.35%). The weight average molecular weight was 285,000, as determined by light-scattering.

The copolymers were obtained by alkaline hydrolysis of 2% solutions of polyacrylamide at 60°C in either 0.1 M or 1 M sodium hydroxide using hydrolysis times of 10 min to 24 h. The hydrolysis reaction products were neutralized to pH 7.5 with hydrochloric acid and dialyzed.

Copolymer samples were prepared for analysis by dissolving sufficient copolymer (20-500 mg) in 10 ml distilled water or leading electrolyte to obtain an acrylate concentration of 0.02 meq/ml. Isotachopherograms were obtained by injecting 10-30 µl of sample solution.

Copolymer Characterization. The base-catalyzed hydrolysis reaction of polyacrylamide has been studied by several workers (27-29). Halverson, et al. (29) concluded from interpretations of $^{13}C$ NMR spectra that a well-spaced distribution of acrylate groups is produced along the polymer chain. This hydrolysis procedure offered an opportunity to prepare nearly model copolymers free of chemical sequence heterogeneity.

The extent of hydrolysis of eight copolymers was determined by potentiometric titration. The $pK_a$ of the copolymer is an important parameter used in making the choice of leading ITP electrolyte. The $pK_a$ of the acrylic acid group increases slightly with extent of copolymer hydrolysis due to electrostatic effects associated with ionized carboxyl groups. With a leading electrolyte of pH 8.2 (used in this work) the carboxyl groups are >99% ionized for values of $pK_a$'s of 6.0 or less as occurred for these copolymers.

Listed in Table V are the sample designation code, mole percent acrylate, apparent $pK_a$, average number of repeat units per acrylic acid group (n), and average distance between ionic groups assuming a fully extended polymer chain and a length of 2.52 angstroms between repeat units. The molecular weight determined by light scattering of several copolymers was 280,000 ± 10% and was not affected by the alkaline hydrolysis.

Table V. Composition, Structural Dimensions and Acidity of Poly(acrylamide-co-acrylate)

| Copolymer Designation | Mole % Acid | n[a] | b,[b] Å | $pK_a$ |
|---|---|---|---|---|
| AC-1.8 | 1.8 | 55 | 140 | 4.6 |
| AC-5.0 | 5.0 | 20 | 50 | 4.7 |
| AC-6.3 | 6.3 | 16 | 40 | 4.8 |
| AC-9.4 | 9.4 | 11 | 27 | 4.8 |
| AC-17 | 17.4 | 6 | 15 | 5.1 |
| AC-31 | 30.9 | 3.2 | 8.2 | 5.2 |
| AC-53 | 53.3 | 1.9 | 4.7 | 5.6 |
| AC-61 | 61.4 | 1.6 | 4.1 | 5.7 |
| C-100 | 100 | 1 | 2.5 | 6.1 |

[a]n is average number of repeat units per acrylate group.

[b]b is the average linear distance (angstroms), assuming a fully extended polymer chain, per acrylate group.

Isotachophoretic Behavior of PAM-PAA Copolymers. Isotachopherograms were obtained for eight AC copolymers and the C-100 homopolymer in pH 8.2 electrolyte, which ensured complete ionization of the carboxylic groups. The isotachopherograms, shown in Figure 11, have well-formed zones with sharp boundaries between leading electrolyte, sample zone, and terminating electrolyte. The apparent mobility of each copolymer, $m_i$, was calculated from step height dimensions using Equation 2.

The effective mobilities are plotted vs. mole fraction of acrylate in these copolymers in Figure 12. Copolymer mobility increases nearly linearly for mole fractions <0.5-0.7 acrylate, beyond which mobility remains nearly constant. The lack of dependence of mobility on composition for the higher charge density copolymers is principally in response to electrostatic binding of counterions to the otherwise fully neutralized carboxyl groups and occurs only when the ionized acrylate groups are spaced very closely (separations of <10 A) along the extended polymer chain. This close charge-group spacing results in a high local electrostatic charge. The high charge density on the polymer becomes "self-lowered" through counterion condensation near the groups on the polymer chain. Since these condensed counterions move with the polyion in an electrophoresis experiment, polymer mobility becomes independent of composition beyond 0.5-0.7 mole fraction of acrylate repeat unit.

An observation of the transition from nonassociated counterions to counterion binding can be seen from this work by plotting the ITP mobility data for these copolymers vs. the dimensionless parameter $\xi$. First introduced by Manning (19), this parameter is a measure of the charge density along the polymer chain,

$$\xi = e^2/\epsilon kTb \tag{4}$$

where e is the electronic charge, $\epsilon$ is the bulk dielectric constant of the solvent, k is the Boltzmann constant, T is the Kelvin temperature, and b is the average axial spacing between charged groups. For acrylate polymers, the value of b is approximately 2.52 A and for copolymers of acrylamide/acrylate the value of b is n(2.52) A where n is the average number of repeat units between ionized acrylate groups. Values for b are given in Table V; for aqueous solutions at 25°C, the value of $\epsilon$ is 80.4 and $\xi = 7.12/b$. The value of $\xi$ is nearly independent of temperature since for water the product of $\epsilon T$ has only a slight temperature dependence. The data are plotted in Figure 13 and provide new experimental evidence of counterion binding near $\xi = 1$, where the mobility undergoes a transition from a relation nearly linear with charge density to independence of charge density. For copolymer compositions with $\xi > 1$, counterions are electrostatically attracted and move with the polyion causing mobility to become independent of polymer ionic group content.

Separation of Acrylamide-Acrylic Acid Copolymer Mixtures. Resolution and separation capacity of simple ionic compounds by ITP has been described by Mikkers (30,31) and Bocek (32) to be primarily governed by the mobility ratio of the mixed components, their relative amounts, and the total electric current applied (i.e., the current-time integral). Resolution in ITP has been defined by Everearts (10) most simply as the fractional amount of separated constituent in a mixture and its numerical value may vary between zero (no separation) and

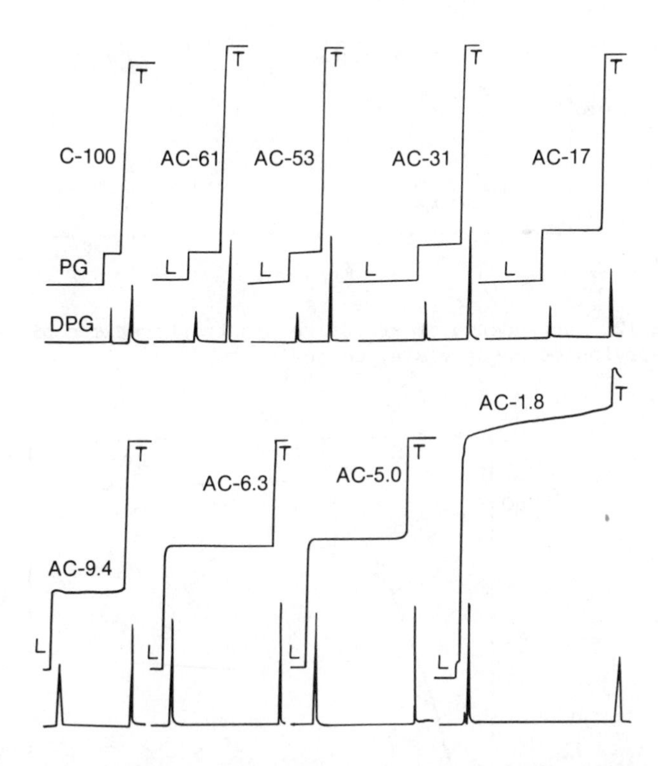


Figure 11. Isotachopherograms of poly(acrylamide-co-acrylate) of different compositions and injected at identical acrylate concentration of 0.02 μg/μl acrylate. Copolymer descriptions given in Table V. Leading electrolyte (L): 0.01 <u>M</u> $Cl^-$, pH 8.2 Tris buffer, 0.5% Triton X-100 surfactant; terminating electrolyte (T): 0.01 <u>M</u> sodium borate, pH 9, 0.5% Triton X-100 surfactant.

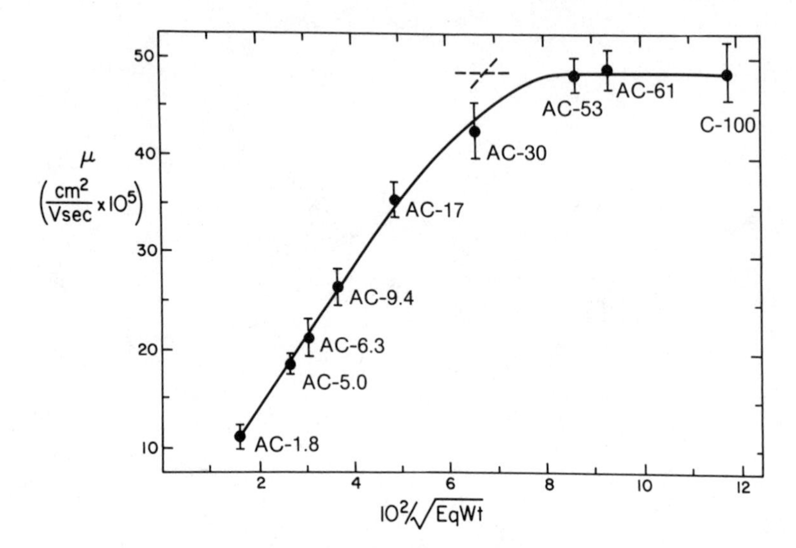


Figure 12. Dependence of effective electrophoretic mobility on poly(acrylamide-co-acrylate) composition.

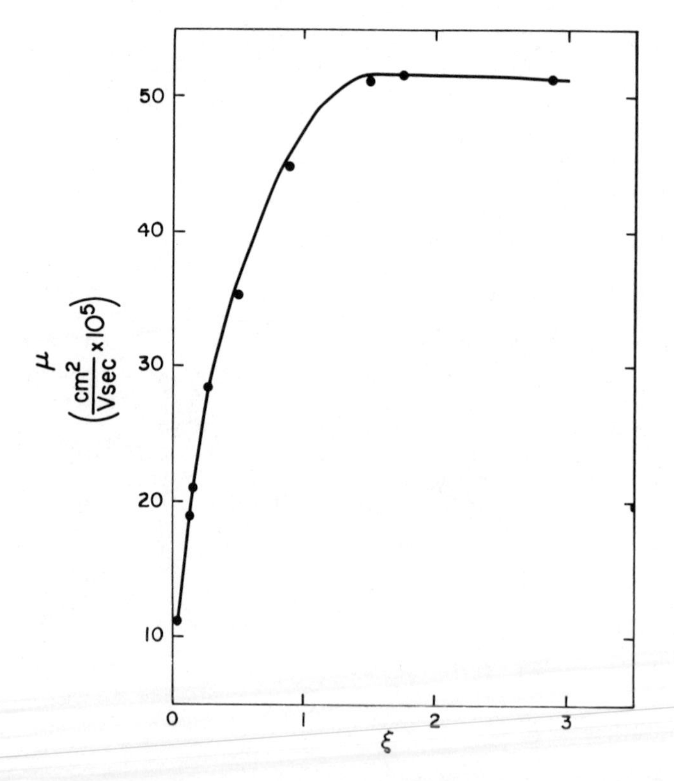


Figure 13. Dependence of effective electrophoretic mobility on charge density parameter $\xi$.

unity (complete separation). In principle if partial resolution is achieved for a mixture at a given set of operational conditions, unit resolution can be obtained with a sufficiently larger current-time integral. This is achieved most easily by increasing capillary length, capillary diameter (although convective disturbances can become severe), or by increasing leading ion concentration.

The resolution of mixtures will occur when their mobility ratio is sufficiently large. For simple organic ions this ratio has been reported to be as small as 1.05 (32). Counterion binding is not observed in simple anions but is shown in this work to influence the effective charge and mobility of polyions and, therefore, can have an influence on the resolution of copolymer mixtures.

Several two-component mixtures of copolymers were used to evaluate the separation capability of ITP where differences in acrylate content were as small as 3 mol% and as large as 50 mol%. Isotachopherograms of several mixtures are shown in Figure 14 where both complete and partial copolymer separations are shown. The AC-5 and AC-6.3 copolymers are separated completely from AC-53.3 copolymer. These copolymer mixtures have effective mobility ratio values of >2.0. Copolymers AC-5 and AC-6.3 are partially resolved from AC-17.4. The incomplete separation is evidenced by the appearance of a mixed zone of intermediate mobility and located between the two pure copolymer zones. The mixed zones are even longer for AC-5 and AC-6.3 when mixed with AC-9.4 copolymer, as is expected because the mobility ratios are <1.5.

The amount of separation of these copolymers into pure zones was increased by increasing capillary length, by decreasing the amount of copolymer injected, and by increasing leading ion concentration at a constant capillary length. The separations achieved at these operating conditions are illustrated in Figure 15. The mixture of AC-9.4 and AC-53.3 is partially resolved with a 8-cm long first-stage capillary and completely resolved with a 20-cm long first-stage capillary. Alternately, complete resolution was achieved with the AC-9.4/AC-53.3 mixture with the 8-cm capillary after increasing the leading ion concentration to 0.02 M or by decreasing the amount of copolymer mixture injected into the capillary by one half.

## Conclusions

Isotachophoresis provides a rapid and sensitive analytical separation of synthetic, charge-bearing polymers that is useful for the characterization of compositional mixtures. ITP, therefore, offers an alternative to chromatographic methods for the separation and chemical characterization of high-molecular-weight synthetic polymers containing ionizable functional groups. Chromatographic separations and solvent fractionations of synthetic polymers often are difficult to interpret because the separation mechanisms are sensitive to both molecular weight distributions and composition distributions. Separations by ITP are free of the influence of polymer molecular weight distribution because their chain mobilities are governed by the ratio of ionic to nonionic repeat unit, which is a property independent of chain length. This simplifies considerably the interpretation of an isotachophoretic separation when compared to chromatographic processes. The separation is obtained without the aid of capillary packing materials or stabilizing media. This eliminates the source

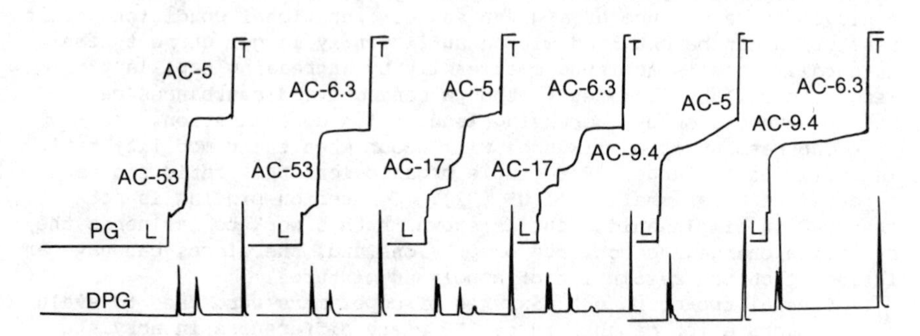


Figure 14. Isotachopherograms of two-component mixtures (1:1 in acrylate) of poly(acrylamide-co-acrylate) showing the influence of relative mobilities on resolution. Electrolyte same as in Figure 11. 1 = AC-53; 2 = AC-5; 3 = AC-6.3; 4 = AC-17; 5 = AC-9.4; 6 = mixed zone; L = leading electrolyte; T = terminating electrolyte.

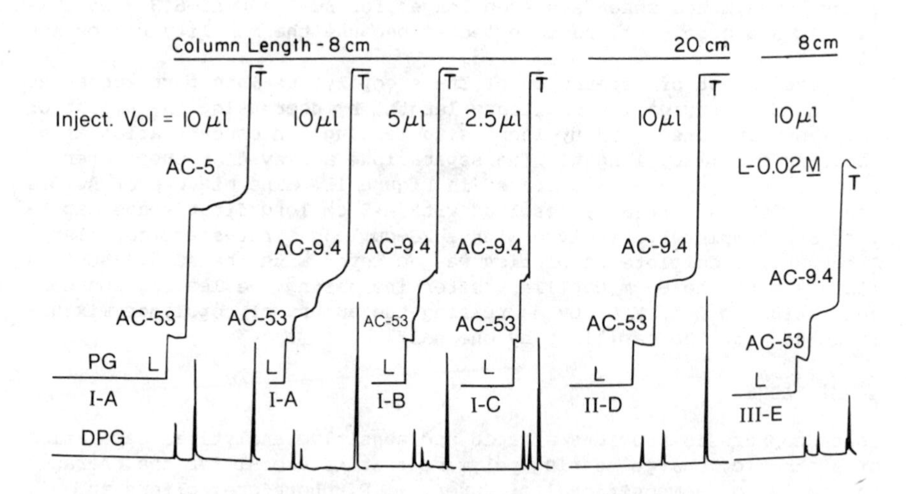


Figure 15. Isotachopherograms of AC-9.4 and AC-53 showing the influence of various ITP operational variables on resolution. (I) 8-cm first stage capillary, 0.01 M $Cl^-$ leading electrolyte; (A) = 10 μl of mixture (0.2 μeq total acrylate); (B) = 5 μl of mixture (0.1 μeq total acrylate); (C) = 2.5 μl of mixture (0.05 μeq total acrylate). (II) 20-cm first stage capillary, 0.01 M $Cl^-$ leading electrolyte; (D) = 10 μl of mixture (0.2 μeq total acrylate). (III) 8-cm first stage capillary, 0.02 M $Cl^-$ leading electrolyte; (E) = 10 μl of mixture (0.2 μeq total acrylate). Zones: 1 = AC-53; 2 = mixed zone; 3 = AC-9.4; L = leading electrolyte; T = terminating electrolyte.

of polymer-substrate interactions that can also complicate the interpretation of polymer separations obtained from packing columns.

Zone detection of the charge-bearing polymeric components is achieved regardless of composition using the potential gradient detection method. Zone lengths are proportional to the amount of polymer introduced into the capillary and are used in the calculation of copolymer compositions.

The capability of ITP to resolve copolymer mixtures is strongly influenced by the relative difference in concentration of ionic repeat unit in the chains. The separation of copolymers into discrete zones has been achieved when the copolymer mixtures contained at least a 5 mol% difference in ionic group. Resolution is highest and the separation is fastest for copolymer mixtures with large mobility ratios (>1.5). Resolution is possible for mixtures where at least one of the copolymers has <0.5 mole fraction ionic group content. Resolution of copolymer pairs is not possible when both copolymers contain >0.6 mole fraction of the charge-bearing group owing to significant counterion binding to the charged groups in these high-charge-density copolymers. This changes their effective electrophoretic mobility to similar or identical values.

Acknowledgments

The technical assistance of Valeri Childs and Louise Wheeler during this work is gratefully acknowledged.

Literature Cited

1. Schmitz, G.; Borgmann, A.; Assmann, G. J. Chromatogr. 1985, 320, 253.
2. Delmatte, P. Sep. Purif. Methods 1981, 10, 29.
3. Lange, P. In "Biochemical and Biological Applications of Isotachophoresis"; Adam, A.; Shots, C., Eds.; Elsevier: Amsterdam, 1980; p. 187.
4. Bocek, P.; Gebauer, P.; Dolnik, V.; Foret, F. J. Chromatogr. 1985, 334, 157.
5. Hjalmarsson, S.; Baldesten, A. Crit. Rev. Anal. Chem. 1981, 11, 261.
6. Everaerts, F. M.; Mikkers, F. E. P. Sep. Purif. Methods 1977, 6(2), 287.
7. Bocek, P. Top. Curr. Chem. 1981, 95, 131.
8. Everaerts, F. M.; Verheggen, T. P. E. M.; Reijenga, J. C. Trends. Anal. Chem. 1983, 2, 188.
9. Thormann, W. Sep. Sci. Technol. 1984, 19, 455.
10. Everaerts, F. M. "Isotachophoresis"; J. CHROMATOGR. LIBRARY, Elsevier Publishing: New York, 1976, p. 6.
11. Whitlock, L. R. J. Chromatogr. 1986, 363, 267.
12. Whitlock, L. R.; Wheeler, L. M. J. Chromatogr. 1986, 368, 125.
13. Whitlock, L. R. J. Chromatogr. 1986, in press.
14. Rice, S.; Nagasawa, M. "Polyelectrolyte Solutions"; Academic Press: New York, 1961, p. 533.
15. Olivera, B. M.; Baine, P.; Davidson, N. Biopolymers 1964, 2, 245.
16. Meullenet, J. P. J. Phys. Chem. 1979, 83, 1924.
17. Hirokawa, T.; Kiso, Y. J. Chromatogr. 1982, 252, 33.

18. Hirokawa, T.; Kiso, Y. J. Chromatogr. 1982, 242, 227.
19. Manning, G. S. Acc. Chem. Rev. 1979, 12, 443.
20. Manning, G. S. J. Phys. Chem. 1981, 85, 1506.
21. Manning, G. S. Q. Rev. Biophys. 1978, 11, 179.
22. Okubo, T.; Ise, N. Macromolecules 1978, 11, 439.
23. Record, M. T.; Anderson, C. F.; Lohman, T. M. Q. Rev. Biophys. 1978, 11, 103.
24. Hughes, L. J.; Britt, G. E. J. Appl. Polym. Sci. 1961, 5, 337.
25. Koleske, J. V.; Lundberg, R. D. J. Polym. Sci., Part A-2 1969, 7, 795.
26. Kosai, K.; Higashino, T. Nippon Setchaku Kyokai Shi. 1975, 11, 2.
27. Kulicke, W. M.; Horl, H. H. Coll. Polym. Sci. 1985, 263, 530.
28. Truong, N. D.; Galin, J. C.; Francois, J.; Pham, Q. T. Polymer 1986, 27, 459.
29. Halverson, F.; Lancaster, J. E.; O'Connor, M. N. Macromolecules 1985, 18, 1139.
30. Mikkers, F. E. P.; Everaerts, F. M.; Peek, J. A. F. J. Chromatogr. 1979, 168, 293.
31. Mikkers, F. E. P.; Everaerts, F. M.; Peek, J. A. F. J. Chromatogr. 1979, 168, 317.
32. Bocek, P.; Deml, M.; Kaplanova, B.; Janak, J. J. Chromatogr. 1978, 160, 1.

RECEIVED January 23, 1987

Chapter 16

# Recent Advances in Preparative Electrophoresis

**Richard A. Mosher, Wolfgang Thormann, Ned B. Egen, Pascal Couasnon, and David W. Sammons**

**Center for Separation Science, University of Arizona, Tucson, AZ 85721**

The renewed interest in the preparative applications of electrophoretic techniques is due to the demands of the rapidly emerging biotechnology industry for large scale protein purification methods. This interest is exemplified by NASA's recent establishment of two university based centers of excellence in separation science as well as by the Electrophoresis Operations in Space program of the McDonnell Douglas Corporation. A variety of preparative electrophoretic methodologies are briefly examined from the perspective of the methods used to establish fluid stability. Three instruments available at the Center for Separation Science, which have exceptionally large throughputs, are described and their performance evaluated.

The recent technological advances in molecular biology have permitted the production of a variety of important molecules on a very large scale. In most cases the compounds of interest are produced in a broth with a large number of similar species and require purification. Compounds to be employed for therapeutic purposes require a high degree of purity. This is especially true if the molecule is the product of a transformed cell, as are monoclonal antibodies. The most common large scale purification techniques for proteins are chromatographic. Some of the limitations of these methods include resolution that suffers when scale-up is attempted, the need for more than one separation principle to accomplish a purification, and a batch rather than a continuous mode of operation.

The most powerful methods for the analysis of protein mixtures are electrophoretic. Isoelectric focusing in polyacrylamide gels is capable of resolving proteins which differ in pI by 0.001 pH units (1). As many as ten thousand components can be resolved by two dimensional techniques (2). It is because of this power that analytical electrophoretic methods are used in numerous labs throughout the world. As a preparative methodology however, these techniques are

0097-6156/87/0335-0247$06.00/0

not as popular. There are problems associated with scale-up: the fluid in which the separation occurs must be stabilized against flows due to convection and electroosmosis, and the heat generated by the current must be dissipated. The difficulties in surmounting these impediments are illustrated by the paucity of commercially available preparative electrophoretic devices. It is because of the deleterious effects of gravity driven convective flows that the National Aeronautics and Space Administration has identified electrophoresis as a method which will likely benefit from operation in microgravity. The leader in the field of microgravity electrophoresis is the McDonnell Douglas Corporation with its ambitious Electrophoresis Operations in Space (EOS) program. In addition to the experiments aboard the shuttle, NASA has established two centers for both ground based and microgravity research in separation processes. These are the Center for Separation Science (CSS) at the University of Arizona and the Bioprocessing and Pharmaceutical Research Center in Philadelphia.

The CSS is dedicated to the development of new methods and instrumentation for electrophoresis. It has the largest collection of preparative electrophoretic equipment in the world. This includes two commercially available devices, the Elphor VaP 21, a free flow instrument, and the shear-stabilized BIOSTREAM Separator. In addition, the CSS has developed two instruments for preparative isoelectric focusing based on a recycling principle. The Center is thus in a unique position to bring a variety of preparative electrophoretic methods to bear on separation problems and represents an important resource for the biotechnology industry. This existing instrumentation also provides a valuable comparison for the performance of newly developed devices which must offer higher throughput and increased resolution to meet the demands of the growing biotechnology industry.

Preparative electrophoresis is seen as an attractive alternative to the purification methods based on chromatography, aqueous two-phase extraction and filtration. The various approaches for preparative electrophoresis are briefly reviewed and three recently developed instruments for free fluid electrophoresis on an industrial scale are described.

## Fluid Stability in Electrophoretic Processes

Convective disturbances present serious problems in electrophoretic experiments. One illuminating way to examine preparative methodology is categorization by the way in which the fluid is stabilized. The most widely used methods by far employ coherent or granular gels. The former are most commonly polyacrylamide or agarose and the latter is usually Sephadex. Gels have been used in the preparative applications of each of the three common electrophoretic modes, isoelectric focusing (IEF), isotachophoresis (ITP) and zone electrophoresis (ZE). Regardless of the mode utilized, the procedure is most often a batch operation and employs either a cylindrical flat bed or annular separation chamber. Isotachophoresis has been applied to preparative scale fractionations in both polyacrylamide (3) and Sephadex (4). The latter method was capable of processing up to 2g of plasma protein in a 20 hour experiment. Zone electrophoretic methods in gels are suitable for the preparation of micro- to milligram quantities of proteins in batch mode. A large number of devices have been designed for this purpose (5). There have been relatively few reports of continuous zone electrophoresis in gels. In one of the more interesting, a slab gel is in constant motion with respect to the electric field (6) producing a two dimensional separation. The most popular method for preparative gel electrophoresis, in both polyacrylamide and Sephadex, is batch IEF. This subject has recently been reviewed by Radola (7). There have also been many attempts to perform continuous IEF in gel beds, granulated gels being most suitable (8). One apparatus was capable of processing 4g of hemoglobin a day, separating the A and A2 species which have pIs differing by 0.4 pH units (9). The most significant recent advance in preparative gel work utilizes immobilized pH gradients. Righetti has reported that these gels can support a protein load of several hundred milligrams in a single focused band (10) while maintaining a resolving power of 0.01 pH units.

There are several fundamental problems associated with preparative gel based methods. The manipulation and preparation of the support is time consuming and tedious. There is often a problem with adherence of sample components to the gel which not only results in a decreased recovery but can also cause electroosmosis which has deleterious effects on resolution. In addition, the matrix must be removed from the recovered sample, cannot generally be reused and therefore, becomes another expense. In general, gel based methods will be adequate for most research lab requirements. The practical upper limit of sample size is approximately a few grams per day which is inadequate for industrial purposes.

Density gradients, produced by concentration gradients of a neutral molecule such as sucrose or sorbitol, are also used to promote fluid stability. These are most often used for batch IEF separations using vertical columns such as those marketed by LKB Produkter and ISCO. The larger of the two LKB columns has a volume of 440 ml and can be used for sample sizes in the neighborhood of 1g protein (8). There is one report of 7.3g of a cytoplasmic extract of Cytophaga johnsonii focused in this column (11). There was a heavy precipitate formed in the bottom of the column, but minor components could be recovered in well focused bands at the top. Sample loads of this size are possible only in special cases. Many investigators have constructed columns for density gradient IEF (12-14) which employ slightly different designs. Cooling requirements are an impediment to scale-up. There have been some attempts to adapt the density gradient method to allow continuous operation in both IEF and ZE modes. Fawcett has reviewed these attempts and described a continous flow density gradient IEF method which will process 1-2g of protein per hour (15). The necessity of forming the gradient is a limitation of both the batch and continuous modes as is the need to remove the material used to form the gradient from recovered samples. This material is not easily reused and is an additional expense in large scale operations.

One method of dealing with the sedimentation of zones of concentrated protein is unique to IEF. This batch technique, termed zone convection, utilizes the sedimentation of zones of focused protein to assist sample collection. In one variant the separation chamber is a horizontal coil of glass tubing (6 mm i.d., 9 mm o.d., with a coil diameter of 4.5 cm) (16). The helix consists of 37 turns, each with a volume of 4.5 ml. When focusing is complete the proteins have sedimented to the bottoms of the coils. The tops of the coils have ports which allow collection. A small amount of buffer is extracted simultaneously from each port, interrupting the liquid continuity and leaving behind what is effectively a series of disconnected U-tubes from which the remaining fluid can be removed independently. The capacity per unit volume of these devices is much greater than that of density gradients, with one report of an initial sample load of 7 mg/ml (17, p.106). A good descriptive review of zone convection devices (17, pp. 104-112) and an extensive list of proteins which have been purified by this method are available (18). Fawcett (15) has suggested a means by which this method could be adapted to continuous flow using a modification of a device developed by Valmet (19). The authors are unaware of any implementation of this idea.

One of the most effective methods, in terms of sample capacity, for large scale isoelectric focusing utilizes membranes to define subcompartments in an electrolyzer. The membranes prevent bulk flow between adjacent compartments while allowing the free migration of proteins. Rilbe has described several devices based on this principle (20). The most recent is a 7.6 liter cell with 46 separation compartments (21). It has a cylindrical geometry with closed compartments. The contents of each compartment are effectively mixed and cooled by the slow rotation of the submerged apparatus in a tank of cold water. The device has fractionated 14g of whey protein into the major components, albumin (pI = 4.6), alpha-lactalbumin (pI = 5.01) and beta-lactoglobulin (pI = 5.13-5.23), the latter partially separated into A and B subcomponents. This represents quite high resolution, particularly when viewed in light of the sample load. Devices compartmented with membranes offer some of the highest sample loading capacities of all preparative electrophoresis instruments. Most, however, operate in a batch mode, a distinct disadvantage if industrial scale throughputs are necessary (100 to 1000 g/day). One serious drawback to the Rilbe instrument is the cost of the commercially prepared carrier ampholytes needed to establish the pH gradient. Binion and Rodkey have published a simple procedure by which carrier ampholytes can be synthesized for approximately 1% of the cost of commercial mixtures (22). These preparations have been characterized and found to be functionally equivalent to those which can be purchased (23). Martin and Hampson (24) have attempted to replace carrier ampholytes by using simple monovalent buffers to create the pH gradient in a membrane compartmented device. Their technique relies upon the synthesis of amphoteric membranes which are utilized in the pH gradient at their isoelectric points. The method has been recently reviewed (25). Only prepared mixtures have been separated so far.

Hjertén has devised a unique method for stabilizing zones of protein against sedimentation. ZE and ITP (26) or IEF (27,28) is carried out in a horizontal cylinder, with an internal diameter of 3 mm, which is rotated (40 rpm) about the electrophoretic axis. The rotation acts to resuspend zones of protein which would otherwise sediment due to their higher density. The technique is micropreparative at best with protein loads in the IEF mode of the order of 100 ug. If the diameter of the cylinder is less than 0.8 mm no rotation is necessary. In such capillary tubes the fluid is stabilized by its own viscosity. A preparative capillary ITP device has been described by Arlinger (29). The column eluent is collected on a moving cellulose acetate strip. This instrument is marketed by LKB Produkter under the name Tachofrac. Kobayashi et al. (30) have also described preparative capillary ITP experiments using a syringe to extract fractions after migration across a potential gradient detector. Hjertén (31) has described a micropreparative zone electrophoretic technique in which the separation takes place in a capillary filled with a polyacrylamide gel. A flow of buffer past the end of the capillary sweeps the samples through a UV detector after which collection is possible. This last method would seem to be the least useful because the smaller diameter tube limits the protein capacity to the ng to ug range.

## Large Scale Free Fluid Instrumentation

The Elphor VaP 21. The use of the inherent viscosity of a fluid to provide stability is the principle behind the free flow electrophoretic devices pioneered by Hannig (32). The separation chamber is a narrow gap between 2 flat plates. Carrier buffer and sample are admitted into the top of the chamber and exit through an array of outlet tubes at the bottom. The electric field is applied across the longer dimension transverse to the flow. Although the separation chamber volume is not large, the continuous nature of the process provides significant preparative potential. Transit time of the sample is a function of the flow rate and to a lesser degree the voltage gradient. Resolution is also affected by the rate of input of the sample, its concentration and composition and the number of fractions collected. An effective cooling method will allow a greater applied field strength, a longer residence time and therefore improve resolution. The electrophoretic mode utilized will also affect resolution.

One such device, developed by Wagner (33), is available from Bender & Hobein of Munich. A clear advantage of this instrument is the variety of operational modes. These include ZE, ITP, IEF, step isoelectric focusing and field step focusing. In ZE a small amount of sample is separated in a buffer of constant composition. This mode has a small throughput, about 0.1g/hr. The performance of conventional IEF, using carrier ampholytes to form the pH gradient, in a free flow, single pass device requires a sufficient residence time that the pH gradient can form and the sample can focus before exiting the cell. High field strengths are needed to accomplish this, which requires an efficient cooling mechanism. Although the cooling unit of the Elphor VaP is very efficient, the instrument can process only 0.05g protein per hour in this mode. It is the mandatory slow flow rate which limits the method. Step IEF attempts to correct this limitation. Several carrier buffers are used to preestablish a pH gradient. This gradient is a step function, in contrast to the continuous gradient of conventional focusing, with the number of plateaus equal to the number of input solutions. This allows much faster flow rates and it is possible to process 1.0g protein per hour in this mode. In ITP, the sample is introduced at the interface of a leading and a terminating buffer. The sample components separate according to differences in mobilities producing a steady state distribution migrating with constant velocity. In the ITP mode, the Elphor VaP 21 has processed 5g protein per hour. The highest throughput is achieved with the field step method. In its most common form this method uses 3 different buffer solutions. The center buffer is of low conductivity and is flanked by buffers of higher conductivities.

Sample is admitted to the chamber in the region containing the low conductivity buffer. The higher field strength in this area causes rapid sample migration to the buffer interfaces. At these points the migrating components encounter a region of low field strength, their velocity decreases and they accumulate. This method can process sample at rates of 30g per hour. The limitation is that only two fractions are obtained. An example of a fractionation using the Elphor VaP in ZE mode is presented in Fig. 1.

The BIOSTREAM Separator. A novel approach to very large scale electrophoretic separations has been developed at the Harwell Atomic Energy Institute utilizing a concept originated by J. St. L. Philpot and fully developed by Thompson (34). The separation takes place in an annulus between two vertically oriented concentric cylinders (see Fig. 2). The diameter of the outer cylinder is 9 cm and the annular space has a thickness of 3 mm. Carrier buffer is pumped into the chamber from the bottom. At the top is a stack of "maze plates" that divide the fluid into 29 fractions which in cross section are concentric rings. Fluid stabilization is achieved by rotation of the outer cylinder at 150 rpm, creating a velocity gradient across the annulus, which maintains a laminar flow profile. Sample is introduced through a thin circumferential slit at the base of the stator (inner cylinder). The walls of the rotor and stator which define the annular space are semipermeable, and isolate electrode chambers from the separation space. Samples are separated by zone electrophoretic principles. No internal cooling is provided. Instead, the sample, buffer and electrolyte streams are chilled(2°C) prior to being introduced into the instrument. Residence time is of the order of 15-60 sec. The operating conditions must be controlled so that the temperature of the eluate does not denature proteins. This device has the largest capacity of any electrophoretic instrument, with an ability to process as much as 150g protein per hour. It has had little application as yet to industrial separation problems but appears to be a promising technology. The CSS has recently acquired a BIOSTREAM and is currently comparing its performance to other methodologies. Its ability to purify antibodies is shown in Figs. 3 and 4. A plasma sample (3g), spiked with hemoglobin, was processed in 17 minutes. The results were analyzed by two-dimensional gel electrophoresis of selected fractions. The immunoglobulins were distributed over 10 fractions, showing minor overlap with albumin, the major plasma protein. While more detailed analysis is required, we estimate that 70% of the antibodies can be recovered with a purity greater than 95%. This performance is quite remarkable, as this was a preliminary experiment and no attempt has been made as yet to optimize this separation.

Recycling IEF. The CSS also has several preparative instruments of its own design. A unique recycling principle is the basis of two devices which are dedicated to isoelectric focusing. One of these, termed RIEF (35), is modular and in its most basic form includes a focusing chamber in which the separation occurs, a pump to recycle the solution, and a reservoir which serves to both dissipate heat and hold the bulk of the fluid being processed (Fig. 2). The heat exchange reservoir is a plexiglass box through which pass 12 glass tubes, 2 for electrolytes and 10 for the sample, connected to 12 cor-

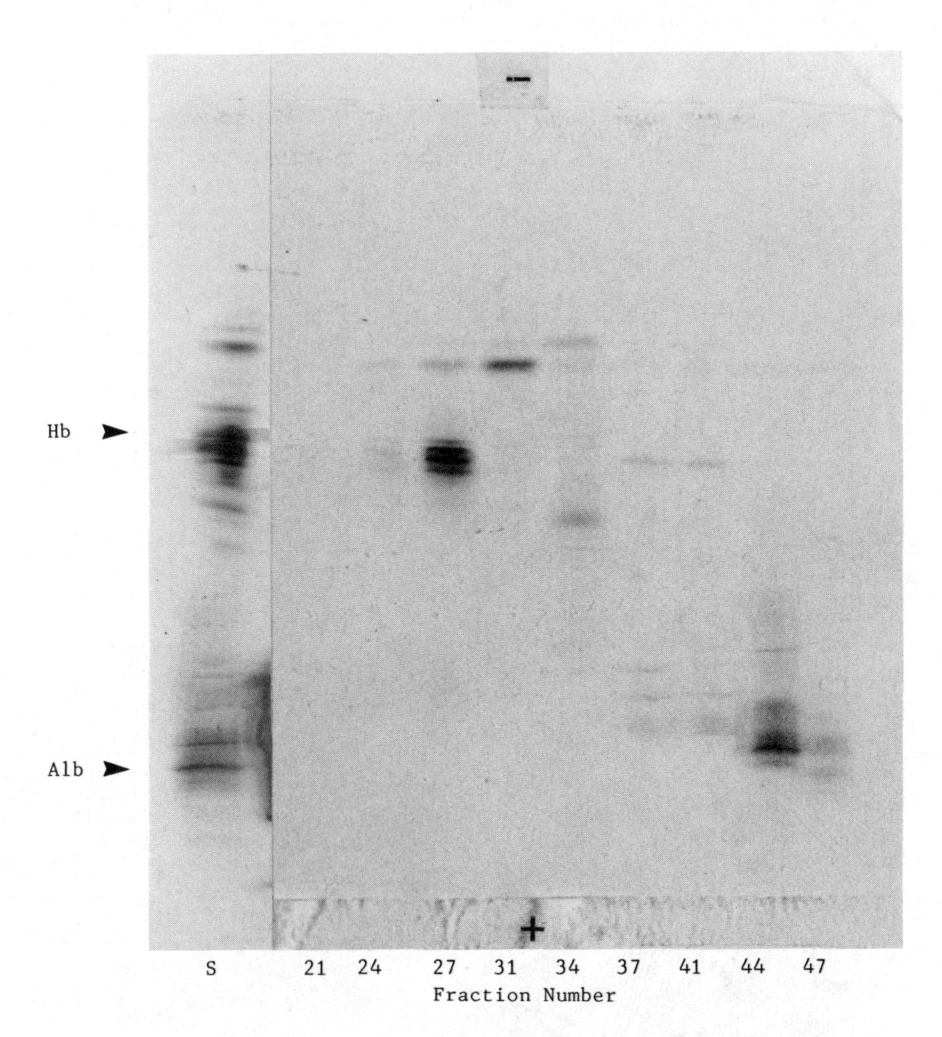


Fig. 1 Analysis by IEF in polyacrylamide gel (pH 3.5-10, 5% Ampholine, 3M urea) of the results of a fractionation of bovine brain homogenate with the Elphor VaP 21 operating in ZE mode. Sample: supernatant of 300g of whole brain homogenized in 300 ml water, electrodialyzed and centrifuged, input at 3 ml/hr in the center channel (46). Buffer: 10 mM sodium acetate, pH 5.0, 7 ml/min. Electrophoresis conditions: 218 V/cm, 85 mA, 6°C. The starting material is in the left most lane (S). The fractions analyzed are shown in number. This is a preliminary run and no optimization has been attempted. Hemoglobin and albumin from unremoved blood are indicated.

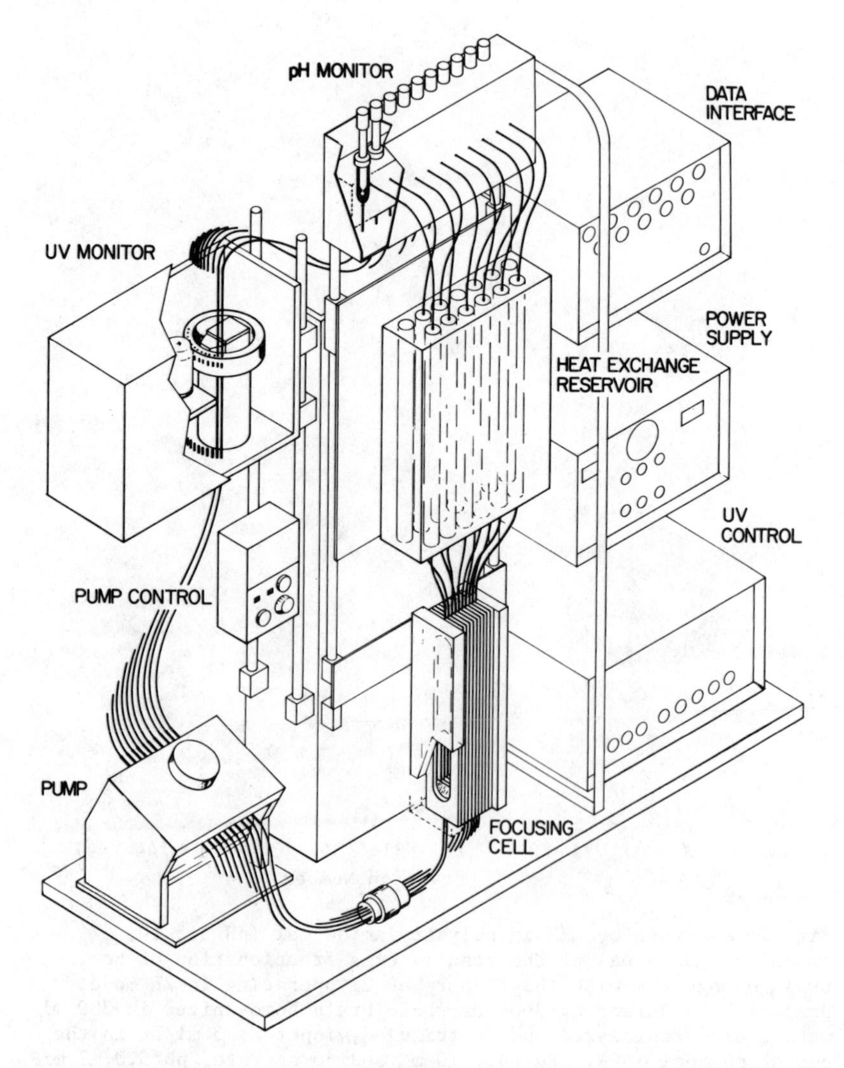


Fig. 2a. A schematic representation of the Recycling Isoelectric Focusing device (RIEF) (from Ref. 35).

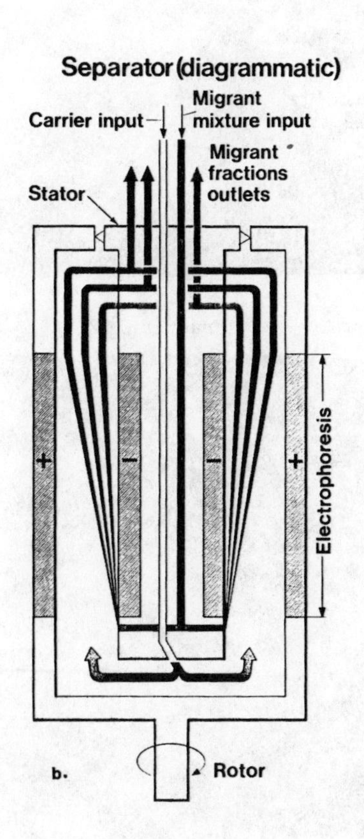


Fig. 2b. The BIOSTREAM separator (adapted from figure in Ref. 34).

Serum

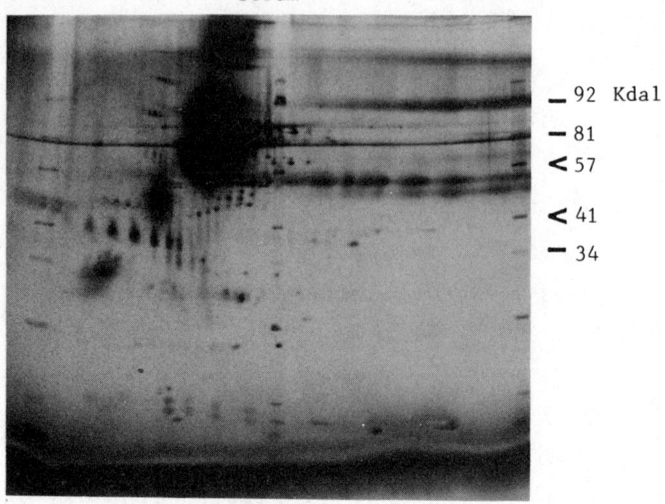


Fraction 12

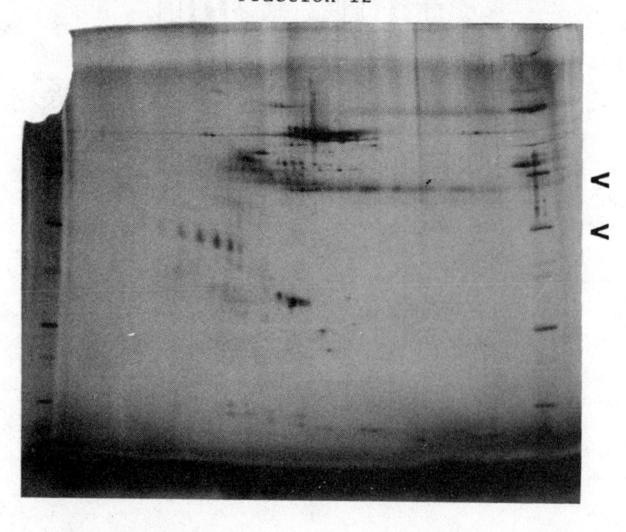

Fig. 3 2D gel analysis of the results of a fractionation of human serum with the BIOSTREAM Separator. Sample: serum augmented with hemoglobin was desalted to half the original conductivity, diluted 1:3 with running buffer and infused at 18 ml/min. Buffer: 10 mM phosphate, pH 7.1 infused at 750 ml/min. Electrophoresis conditions; 29V, 62A. All solutions were cooled to $0^{o}$C prior to infusion. Eluate temperature; $20^{o}$C. 2D gels were run as described (39) and stained with GELCODE. The molecular weights of several standards are indicated in kilodaltons. The pH increases from 3.5 to 7.5 from left to right in each gel. Relatively pure immunoglobulins (50 kd heavy chains, isoelectrically heterogeneous) were found in fractions 2-10, albumin (pI 4.8, 68 kd) in fractions 18-29, and the heterogeneous haptoglobin family, avg. pI = 4, avg. M.W. = 37 kd, in fractions 12-20.

Fraction 6

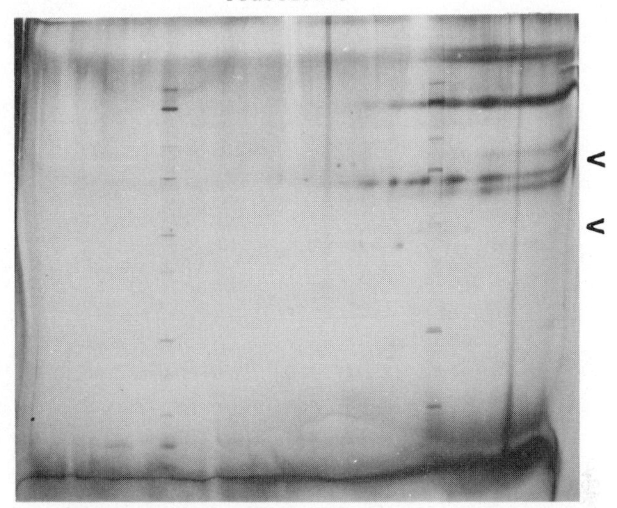

Fraction 26

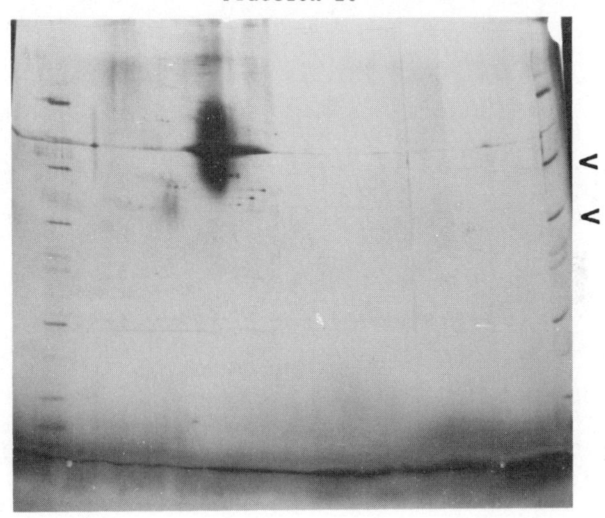

Figure 3. Continued.

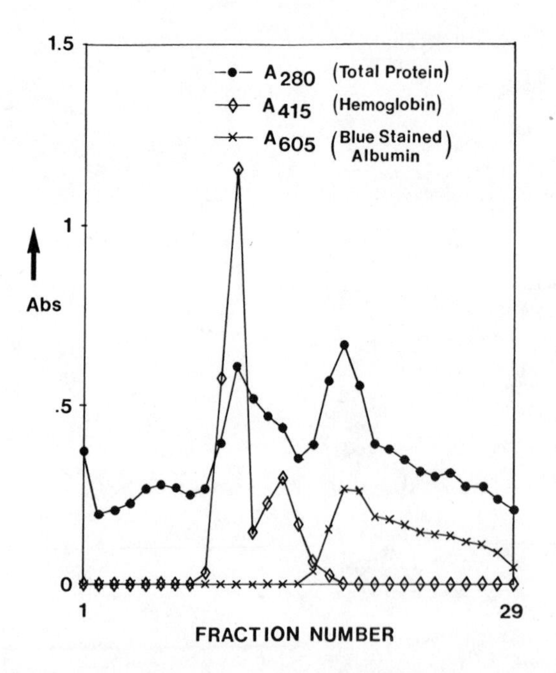


Fig. 4 The distribution of total protein, hemoglobin and albumin from the BIOSTREAM separation analyzed in Fig. 3. Immunoglobulins are distributed in fractions 2-10.

responding subcompartments in the focusing cell. These subcompartments, which are thin plexiglass plates (12 x 0.3 x 3 cm) with the centers removed, are separated by membranes (monofilament nylon screens). Coolant is continuously flowing through the heat exchange reservoir thus surrounding the tubes which contain the process fluid and the electrolytes. Sample solution flows from the bottom of the heat exchange reservoir, through silicone tubing and into the top of the focusing chamber. The solution then flows from the bottom of the focusing chamber to the pump and returns to the top of the heat exchange reservoir. Optional accessories include a UV monitor and flow through pH electrodes which allow computer controlled monitoring of these parameters during a separation. This automation provides a means to flatten the pH gradient and thus increase the resolution during a run by automated pumping of fluid from the extreme channels. The modular nature of this batch apparatus allows for easy scale up, simply by employing larger heat exchange reservoirs. This apparatus has been used to fractionate a variety of samples, including the purification of antibodies (36). Isolation of a single component from snake venom is shown in Fig. 5.

The CSS has extended the recycling principle to flowing thin films of fluid as in the Elphor VaP. The prototype shows a remarkable fluid stability with the chamber oriented vertically and the flow from bottom to top. When employed for IEF this device provides a continuous pH gradient within the focusing chamber rather than the step gradient obtained with membrane defined subcompartments. The absence of membranes permits the processing of particulate material as well as mixtures containing proteins which precipitate during focusing.

## Discussion

The high degree of purity required for some of the products of the biotechnology industry has led to increased activity in separation science. In particular, electrophoresis has received much attention due to its unsurpassed resolution for mixtures of proteins. This interest is exemplified by several recent developments including the EOS program of McDonnell Douglas Corporation and the recent funding by NASA of two university based centers of excellence in separation science.

One of these institutions, the Center for Separation Science, is a unique facility for electrophoretic study. In addition to its unparalleled collection of preparative electrophoretic equipment, the CSS has a state of the art two dimensional electrophoresis laboratory (Fig. 3), an active theoretical program including a computer model for all electrophoretic processes (37) and expertise and instrumentation in capillary electrophoresis (38). There is also an ongoing instrument development program. Its assembly of current preparative methodologies allows a thorough comparison of the capabilities of these instruments and permits the CSS to bring a variety of techniques to bear on separation problems. In general, these devices should be viewed as complimentary rather than competitive. The Elphor VaP is the most versatile of the instruments and due to its continuous mode of operation it has significant preparative potential. The BIOSTREAM, with it's large processing capacity, is a true industrial scale machine which is capable of excellent resolution at high processing rates. The RIEF operates in batch mode, but with an easily variable and potentially very large capacity. The

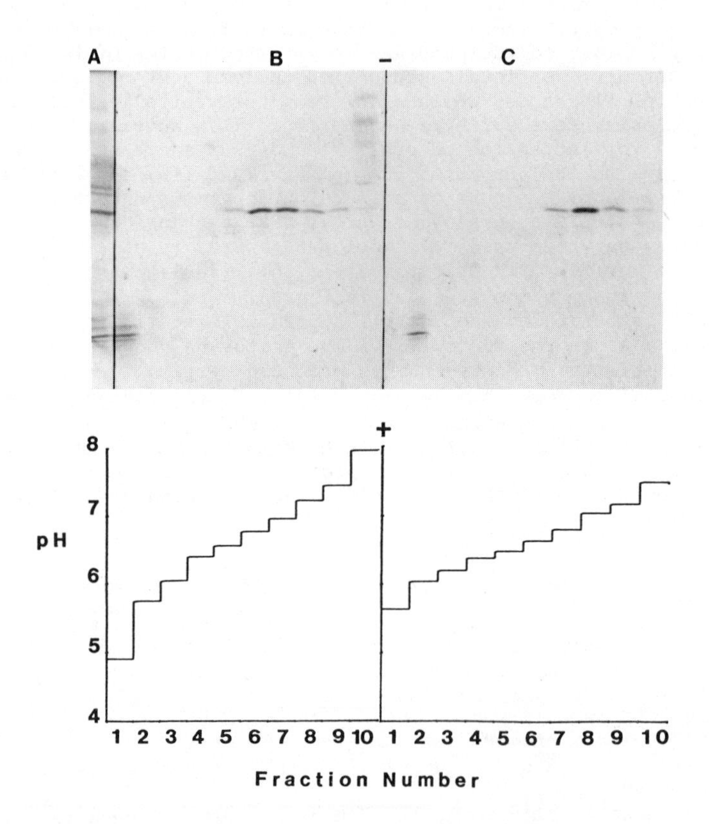


Fig. 5 Analysis by IEF in PAG (pH 3.5-10, 5% Ampholine) of two sequential RIEF separations (10 fractions each) of 1g of lyophilized snake venom. Panel A: whole venom. The experiments were optimized to isolate the major component indicated by the arrow. Panel B: crude venom fractionated in 240 ml of 1% pH 6-8 of the first run were pooled and refractionated with no further additions, 4 hr, same final voltage. 60% of the protein of interest was recovered in fraction 7.

online pH and UV monitors of the RIEF allow a separation to be feedback controlled. An additional attractive feature of the RIEF is that scale-up requires no sacrifice in resolution.

There is a clear need for the continuing development of preparative electrophoretic equipment. The two major problems which must be overcome are the dissipation of Joule heat and provision for fluid stability. These are engineering problems and their solution requires the establishment of collaborative efforts between electrophoreticists, fluid dynamics experts and chemical engineers.

## Acknowledgments

The authors would like to acknowledge the continuous support of Dr. Milan Bier and would like to thank Garland Twitty, Jeff Sloan, Terry Long, Arjan Ala and Lou Zawadski for skillfull technical assistance. This work was supported by NASA grants NSG-7333 and NAGW-693. The generous loan of the Elphor VaP 21 by Bender & Hobein of Munich, F.R.G. and of the BIOSTREAM Separator by CJB Developments Ltd. of Portsmouth, U.K. is also gratefully acknowledged.

## Literature Cited

1. Bjellqvist, B.; Ek, K.; Righetti, P. G.; Gianazza, E.; Gorg, A.; Westermeier, R.; Postel, W. J. J. Biochem. Biophys. Meth. 1982, 6, 317.
2. Anderson, N. L.; Taylor, J.; Scandora, A. E.; Coulter, B. P.; Anderson, N. G. Clin. Chem. 1981, 27, 1807.
3. Holloway, C. J.; Battersby, R. V. Meth. in Enzymol. 1984, 104, 281.
4. Bier, M.; Kopwillem, A. In "Electrofocusing and Isotachophoresis"; Radola, B. J.; Graesslin, D., Eds. Walter de Gruyter: Berlin, 1977; pp. 567-576.
5. Chrambach, A.; Nguyen, N. Y. In "Electrokinetic Separation Methods"; Righetti, P. G.; Van Oss, C. J.; Vanderhoff, J. W., Eds.; Elsevier: North Holland, 1979; pp. 337-368.
6. Lammel, B. Electrophoresis 1981, 2, 39.
7. Radola, B. J. Meth. in Enzymol. 1984, 104, 256.
8. Fawcett, J. S. Ann. N. Y. Acad. Sci. 1973, 209,112.
9. Basset, P.; Froissart, C.; Vincendon, G.; Massarelli, R. Electrophoresis 1980, 1, 168.
10. Righetti, P. G.; Gianazza, E.; Gelfi, C. In "Electrophoresis '84"; Neuhoff, V., Ed.; Verlag Chemie: Weinheim, 1984; pp. 29-48.
11. Janson, J. C. Ph. D. Thesis, Uppsala University, Uppsala, 1972.
12. Catsimpoolas, N.; Stamatopoulou, A.; Griffith, A. L. in "Electrophoresis '79"; Radola, B. J., Ed.; Walter de Gruyter: Berlin, 1980; pp. 503-515.
13. Rilbe, J.; Pettersson, S. In "Isoelectric Focusing"; Arbuthnott, J. P.; Beeley, J. A., Eds.; Butterworths: London, 1975, pp. 44-57.
14. Jonsson, M.; Stahlberg, J.; Fredriksson, S. Electrophoresis 1980, 1, 113.
15. Fawcett, J. S. In "Isoelectric Focusing"; Catsimpoolas, N., Ed.; Academic Press: New York, 1976, pp. 173-208.

16. Quast, R. In "Electrofocusing and Isotachophoresis"; Radola, B. J.; Graesslin, D., Eds.; Walter de Gruyter: Berlin, 1977, pp. 455-462.
17. Righetti, P. G. "Isoelectric Focusing: theory, methodology and applications"; Elsevier Biomedical: Amsterdam, 1983.
18. Bours, J. In "Isoelectric Focusing"; Catsimpoolas, N., Ed.; Academic Press: New York, 1976, pp. 209-228.
19. Valmet, E. Sci. Tools 1969, 16, 8.
20. Rilbe, H. Prot. Biol. Fluids 1970, 17, 369.
21. Jonsson, M.; Rilbe, H. Electrophoresis 1980, 1, 3.
22. Binion, S.; Rodkey, L. S. Anal. Biochem. 1981, 112, 362.
23. Binion, S.; Rodkey, L. S.; Egen, N.; Bier, M. Anal. Biochem. 1983, 128, 71.
24. Martin, A. J. P.; Hampson, F. J. J. Chromatogr. 1978, 159, 101.
25. Hampson, F. In "Electrophoretic Techniques"; Simpson, C. F.; Whittaker, M., Eds.; Academic Press: London, 1983, pp. 231-252.
26. Hjertén, S. J. Chromatogr. 1967, 9, 122.
27. Hjertén, S. In "Methods of Protein Separation"; Catsimpoolas, N., Ed.; Plenum: New York, 1976; Vol. II, pp. 219-231.
28. Lundahl, P.; Hjertén, S. Ann. N.Y. Acad. Sci. 1973, 209, 94.
29. Arlinger, L. In "Electrofocusing and Isotachophoresis"; Catsimpoolas, N., Ed; Plenum: New York, 1976; Vol. II, pp. 219-231.
30. Kobayashi, S.; Shiogai, Y.; Akiyama, J. In "Analytical Isotachophoresis"; Everaerts, F. M., Ed.; ANALYTICAL SYMPOSIA SERIES Vol. 6, Elsevier: Amsterdam, 1981, pp. 47-53.
31. Zhu, M.; Hjertén, S. In "Electrophoresis '84"; Neuhoff, V., Ed.; Verlag Chemie: Weinheim, 1984, pp. 110-113.
32. Hannig, K. Electrophoresis 1982, 3, 235.
33. Wagner, H.; Kessler, R. GIT Lab.-Med. 1984, 7, 30.
34. Thompson, A. R. In "Electrophoretic Techniques"; Simpson C. F.; Whittaker, M., Eds.; Academic Press: London, 1983, pp. 253-274.
35. Bier, M.; Egen, N. B.; Allgyer, T. T.; Twitty, G. E.; Mosher, R. A. In "Peptides: Structure and Biological Function"; Gross, E.; Meienhofer, J., Eds; Pierce Chemical Co.: Rockford, IL, 1979, , pp. 79-89.
36. Binion, S. B; Rodkey, L. S.; Egen, N. B.; Bier, M.; Electrophoresis 1982, 3, 284.
37. Bier, M; Palusinski, O. A.; Mosher, R. A.; Saville, D. A. Science 1983, 219, 1281.
38. Thormann, W.; Mosher, R. A.; Bier, M. J. Chromatogr. 1986, 351, 17.
39. Sammons, D. W.; Adams, L. D.; Nishizawa, E. E. Electrophoresis 1981, 2, 135.

RECEIVED October 16, 1986

# INDEXES

# Author Index



# Affiliation Index



# Subject Index

B



C

F

K



L



M

Z

*Production by Cara Aldridge Young*
*Indexing by Karen L. McCeney*
*Jacket design by Carla L. Clemens*

*Elements typeset by Hot Type Ltd., Washington, DC*
*Printed and bound by R. R. Donnelley Company, Harrisonburg, VA*

## Recent ACS Books

Personal Computers for Scientists: A Byte at a Time
By Glenn I. Ouchi
288 pp; clothbound; ISBN 0-8412-1001-2

Writing the Laboratory Notebook
By Howard M. Kanare
145 pp; clothbound; ISBN 0-8412-0906-5

The ACS Style Guide: A Manual for Authors and Editors
Edited by Janet S. Dodd
264 pp; clothbound; ISBN 0-8412-0917-0

Chemical Demonstrations: A Sourcebook for Teachers
By Lee R. Summerlin and James L. Ealy, Jr.
192 pp; spiral bound; ISBN 0-8412-0923-5

Phosphorus Chemistry in Everyday Living, Second Edition
By Arthur D. F. Toy and Edward N. Walsh
342 pp; clothbound; ISBN 0-8412-1002-0

Pharmacokinetics: Processes and Mathematics
By Peter G. Welling
ACS Monograph 185; 290 pp; ISBN 0-8412-0967-7

High-Energy Processes in Organometallic Chemistry
Edited by Kenneth S. Suslick
ACS Symposium Series 333; 336 pp; ISBN 0-8412-1018-7

Particle Size Distribution: Assessment and Characterization
Edited by Theodore Provder
ACS Symposium Series 332; 308 pp; ISBN 0-8412-1016-0

Radon and Its Decay Products: Occurrence, Properties, and Health Effects
Edited by Philip K. Hopke
ACS Symposium Series 331; 609 pp; ISBN 0-8412-1015-2

Nucleophilicity
Edited by J. Milton Harris and Samuel P. McManus
Advances in Chemistry Series 215; 494 pp; ISBN 0-8412-0952-9

Organ c Pollutants in Water
Edited by I. H. Suffet and Murugan Malaiyandi
Advances in Chemistry Series 214; 796 pp; ISBN 0-8412-0951-0

---

For further information and a free catalog of ACS books, contact:
American Chemical Society
Distribution Office, Department 225
1155 16th Street, NW, Washington, DC 20036
Telephone 800-424-6747